A CULTURAL HISTORY OF THE HUMAN BODY

VOLUME 5

A Cultural History of the Human Body
General Editors: Linda Kalof and William Bynum

Volume 1
A Cultural History of the Human Body in Antiquity
Edited by Daniel H. Garrison

Volume 2
A Cultural History of the Human Body in the Medieval Age
Edited by Linda Kalof

Volume 3
A Cultural History of the Human Body in the Renaissance
Edited by Linda Kalof and William Bynum

Volume 4
A Cultural History of the Human Body in the Age of Enlightenment
Edited by Carole Reeves

Volume 5
A Cultural History of the Human Body in the Age of Empire
Edited by Michael Sappol and Stephen P. Rice

Volume 6
A Cultural History of the Human Body in the Modern Age
Edited by Ivan Crozier

A CULTURAL HISTORY OF THE HUMAN BODY

IN THE AGE OF EMPIRE

Edited by Michael Sappol and Stephen P. Rice

BERG
Oxford • New York

English edition
First published in 2010 by
Berg

Editorial offices:
First Floor, Angel Court, 81 St Clements Street, Oxford OX4 1AW, UK
175 Fifth Avenue, New York, NY 10010, USA

Berg is the imprint of Oxford International Publishers Ltd.

Library of Congress Cataloging-in-Publication Data

A catalogue record for this book is available from the Library of Congress.

British Library Cataloguing-in-Publication Data

A catalogue record for this book is available from the British Library.

ISBN 978 1 84788 792 4 (volume 5)

978 1 84520 495 2 (set)

Typeset by Apex CoVantage, LLC, Madison, WI, USA.

Printed in the UK by the MPG Books Group

www.bergpublishers.com

CONTENTS

ILLUSTRATIONS

INTRODUCTION

CHAPTER 1

CHAPTER 2

CHAPTER 4

CHAPTER 5

CHAPTER 6

CHAPTER 7

CHAPTER 8

ACKNOWLEDGMENTS

We would like to thank the following institutions for their support in preparing this book: The National Library of Medicine, Ramapo College of New Jersey, the Wellcome Library, and the Clark Art Institute. We would also like to thank the volume's contributors for agreeing to be part of this project, and for their patience. Several individuals have been especially helpful along the way. Linda Kalof was always available for guidance and suggestions, Sita Reddy offered some great advice early on, Suzanne Chapman helped with the preparation of the final manuscript, and Taylor Migliorisi prepared the index. We are also grateful to Hannah Shakespeare, Emily Medcalf, and Tristan Palmer at Berg. Finally, we want to especially acknowledge Eva Åhrén and Jacqui Lunchick, whose generous warmth, patient nurturance, and sage advice helped us bring this book into the world.

SERIES PREFACE

A Cultural History of the Human Body is a six-volume series reviewing the changing cultural construction of the human body throughout history. Each volume follows the same basic structure and begins with an outline account of the human body in the period under consideration. Next, specialists examine major aspects of the human body under seven key headings: birth/death, health/disease, sex, medical knowledge/technology, popular beliefs, beauty/concepts of the ideal, marked bodies of gender/race/class, marked bodies of the bestial/divine, cultural representations and self and society. Thus, readers can choose a synchronic or a diachronic approach to the material—a single volume can be read to obtain a thorough knowledge of the body in a given period, or one of the seven themes can be followed through time by reading the relevant chapters of all six volumes, thus providing a thematic understanding of changes and developments over the long term. The six volumes divide the history of the body as follows:

Volume 1: A Cultural History of the Human Body in Antiquity (750 B.C.E.–1000 C.E.)
Volume 2: A Cultural History of the Human Body in the Medieval Age (500–1500)
Volume 3: A Cultural History of the Human Body in the Renaissance (1400–1650)
Volume 4: A Cultural History of the Human Body in the Age of Enlightenment (1650–1800)
Volume 5: A Cultural History of the Human Body in the Age of Empire (1800–1920)
Volume 6: A Cultural History of the Human Body in the Modern Age (1920–21st Century)

General Editors, Linda Kalof and William Bynum

Introduction

Empires in Bodies; Bodies in Empires

MICHAEL SAPPOL

THE ANATOMICAL PUZZLE

An illustrated theatrical flyer from 1886 tells something about the empires that were constructed out of bodies in "the age of empire," and the bodies that were forged within them. "Marinelli the Anatomical Puzzle," on a tour of the United States after a season or two on a European circuit, is shown striking a characteristically impossible pose (see Figure 0.1). Like an X-ray *avant la lettre*, an anatomical figure of his skeletal structure is overlaid by an outline of his naked body. His back is curved 180 degrees behind him. His head, placed behind his buttocks, seems poised to make an incredible journey up his ass, almost an Ouroboros. The small print explains that "this human marvel has appeared before all the leading continental physicians," including Rudolf Virchow, the celebrated German pathologist. The doctors have testified: The rearrangement of body parts, the "anatomical puzzle," is real. To further bolster its claims, the flyer boasts that its image was "taken from a photograph." All are part of the rhetoric of authentication.

In a small accompanying printed note, theater owner James L. Kernan invites the "Physicians and Surgeons of this City [Washington, D.C.]" to see the "world-famous Herr Marinelli, whose scientific and wonderful Performances have astounded the leading Professors of Anatomy throughout . . . Europe."[1] This "private exhibition" before Washington's medical establishment would bolster the credibility of Marinelli's claims and reassure a skeptical public that

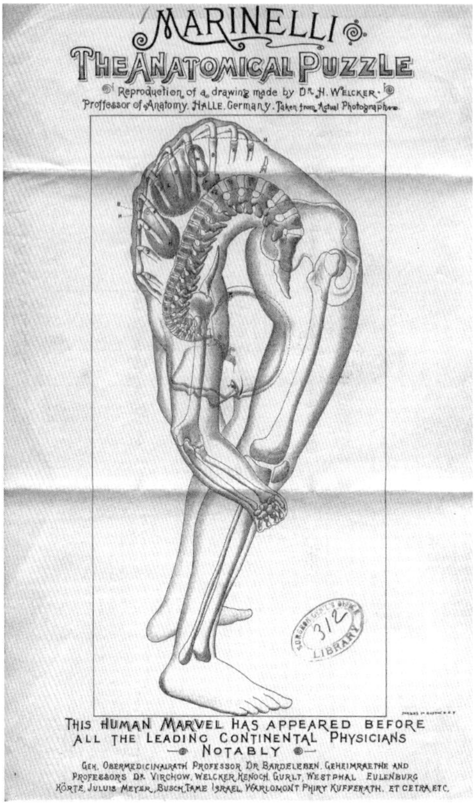

FIGURE 0.1: *Marinelli the Anatomical Puzzle* (Boston: Forbes Co., 1886). National Library of Medicine. Top caption: "Reproduction of a drawing . . . by Dr. H. Welcker, Proffessor [sic] of Anatomy. Halle, Germany. Taken from actual photograph." Bottom: "This human marvel has appeared before all the leading continental physicians, notably . . . Professor Dr. Bardleben, . . . Professors Dr. Virchow, Welcker, Kenoch, Gurlt, Westphal . . . etc." National Library of Medicine.

the Anatomical Puzzle was not a cheap hoax. After the medical preview, the show would open to the ticket-buying masses.

This volume is about the cultural history of the body in the age of empire, here defined as the period from 1800 to 1920. Although other ages were also ages of empire (including our own), this period was very much an era of state formation and expansion, geopolitical rivalry, and colonial and imperial wars and conquest in Europe, Asia, Africa, and the Americas. It was also an era in which empires were constructed in railroad, steamship, and telegraph monopolies; political parties; familial patronage networks; universities; prisons; academic disciplines like medicine, physics, art, and engineering; publishing houses; hospitals; mining; pedagogy; steel; tobacco; asylums; museums; bananas; patent medicines; fashionable dress; and electricity. Empires within empires.[2]

Since Michel Foucault, Alain Corbin, Thomas Laqueur, Mary Poovey, Roy Porter, Carolyn Walker Bynum, Catherine Gallagher, Judith Butler, and other scholars opened up "the body" as a field of inquiry several decades ago, the field has grown enormously in scope, covering an astounding number of body practices, historical moments, institutions, social settings, and geographical locations.[3] The scholarship in many disciplines—history, history of medicine, literary studies, art history, anthropology, sociology, architectural history, cultural geography, theology, feminist studies, and so on—using many different theoretical and pragmatic approaches—has proliferated. This volume therefore makes no claims to be a comprehensive survey. Our intention is, instead, to present a sampling of current historical work on selected aspects of the body in the period 1800–1920. Given the burgeoning diversity—even fragmentation—of scholarship, in this introduction I will read an idiosyncratic set of images (like Marinelli's theatrical flyer) to raise some of the larger issues at stake in current research on the body.

So, in what empires was the Anatomical Puzzle deployed?

There was a sprawling transnational empire of theaters, museums, tent shows, circuses, and so forth, where men and women flocked to see a diverse assortment of body performances: freak shows featuring people with congenital deformities, tattoos, and disfiguring diseases; displays of cultural exotics; and exotica. These featured amazing bodies of different sizes, shapes, and textures but also more normative bodies doing amazing things and amusing things: acrobatics, dance, gymnastics, boxing, slapstick, glass blowing, and so forth—and, at the most respectable end, dramatic and comedic stage plays with elaborate sets and star players.[4] Such shows were highly responsive to consumer demand: there was a diverse and growing public of various classes and social types who increasingly lived and congregated in burgeoning cities organized around the manufacture, sale, and transportation of goods and services—and who looked for diversions on which to spend their disposable cash.

Within the amorphous domain of popular entertainments there developed smaller structured empires. In 1886, James L. Kernan, the owner of the theater where Marinelli appeared, was just beginning to assemble a chain of burlesque and vaudeville theaters that would eventually stretch from Pittsburgh, Pennsylvania, and Buffalo, New York, to Baltimore, Maryland, and Washington, D.C. In succeeding decades he would join cause with a select group of theater owners in other North American cities and form a powerful "burlesque wheel," a monopolistic syndicate that tried to control the booking of performers in the eastern United States in order to dominate theatrical agencies and drive independent operators out of business.

In the process Kernan became a millionaire, a man of considerable social standing in Baltimore. Kernan owned an extravagant home, wore expensive clothes, and participated in social settings, where he displayed his wealth and performed his own set of calculated and uncalculated body performances before an elite audience of family, friends, business associates, and "society."[5] In retirement he became principal benefactor of Baltimore's James Lawrence Kernan Hospital and Industrial School of Maryland for Crippled Children—a medical, educational, and philanthropic empire of disabled and contorted bodies! Kernan's hospital–school, in turn, was but one of many new medical and educational institutions funded through the largesse of wealthy "great men" and "great families." In turn, these institutions were linked to voluntary charities, professional health and medical organizations, and the state (through funding, municipal courts, and social welfare agencies)—the empires and provinces of philanthropy and social welfare.

The Anatomical Puzzle also implicated another empire: the transnational, hierarchical domain of scientific medicine. This was the medicine of medical schools, hospitals, authoritative texts, credentialed professors, new discoveries, and interesting cases. In the nineteenth century, medical establishments proliferated in the major cities of Europe and North America and sent tendrils that reached out into the hinterlands and colonial periphery. Within these networks, medical knowledge was produced—and authority constructed—in dissecting rooms, lecture halls, clinics, laboratories, morgues, pathological collections, monographs, and journal articles. The prestige of the profession came in part from its ability to present, conceptualize, and explain the (living) clinical subject and (dead) anatomical subject—the human body and body parts to observe, cut into, and perform procedures upon; to display and circulate in glass jars and vitrines; to represent in models, photographs, drawings, engravings, and lithographs; and to describe in published and unpublished case histories. The guarantee of the doctor's expertise lay in his (and it was usually a he) mastery of clinical and anatomical subjects. Doctors conducted physiological research; contributed to a collective natural history of anatomical morphology, physiological function, and pathological dysfunction; and collected data, cases, and specimens for presentation. The spectacular case, the rare and freaky case, was particularly useful in dramatizing the itinerary of pathology in the body and advertising medicine's (and the individual doctor's) profound knowledge of it.[6]

In the 1880s, the highest medical authority emanated from Europe, especially Germany. There, Rudolf Virchow (1821–1902), the renowned author of *Cellular Pathology* (1858) and advocate of microscopy and standardized autopsy, was a venerated figure, the center of a large professional and political circle. Virchow was a major producer and collector of anatomical specimens.[7] His name glowed with an aura that was useful even to the likes of Marinelli, thousands of miles away in America. The very title of Marinelli's freaky act, along with the illustration he circulated to advertise it (and Virchow's testimonial), played on the long-established medical practice of anatomical collection and representation. Marinelli was his own anatomical specimen, his theatrical poster his own anatomical plate.

Beyond that, the Anatomical Puzzle fitted inside congeries of larger empires. All of this activity by Marinelli, Kernan, and Virchow was made possible by cheap mass-produced print technology; steamship, railroad, postal, and telegraph networks (regional, national, international); and the state, corporate, and financial monopolies and oligopolies that nurtured and harbored them. We know a lot about Virchow the German, less about Kernan the American, and still less about "Herr Marinelli," the German-titled, Italian-named performer from "the continent," who crossed borders and oceans to contort his body on European and American stages, and made a very good living doing so.[8] On contemporary maps the national boundaries of the United States, Italy, and Germany were clearly demarcated, but relatively new. In the 1880s, all three nation-states were consolidating their territories and governmental administrations, and the identity of each was far from secure. After a series of wars and insurrections, Italy had only recently been unified under Victor Emmanuel II of Piedmont. The force of Prussian arms had forged the German Empire out of a mosaic of statelets. The United States, a dynamically expanding nation on Europe's periphery, was divided into North and South, black and white, rich and poor, and multiple ethnicities and classes. It had grown to be an economic, industrial, and geopolitical giant despite nearly falling to pieces in its bloody Civil War.

So these were the empires in which the Anatomical Puzzle was forged, that structured his act and career. But what did the Anatomical Puzzle signify? What teeter-tottering cultural logic did it draw upon? What practices did it deploy?

Marinelli's act mobilized widely held and enduring structures of belief about mind and body, spirit and matter. The image in his flyer refers back to the centuries-old tradition of anatomical illustration inaugurated by Andreas Vesalius in the mid-sixteenth century. Vesalian anatomy employed dissection to open up dead bodies, and made profusely illustrated texts based on the observations of the anatomist, images that mapped an inner topography of rivers, canals, structures, and pathways. Such illustrations used illusionistic perspective to endow the interior of the human body with a highly articulated depth, as much geological as geographical, a stratigraphy of layers. And on these foundations, over several centuries, Vesalius's successors developed the notion of the body as a bounded (albeit porous) political economy, which contained within it sites of production, distribution, and control.[9]

In the nineteenth century the anatomical idea circulated far beyond the domain of elite medicine. Popular lecturers argued that an anatomical conception of self was part of the civilizing process: educated men and women must think of themselves as anatomical beings, with internal boundaries and place names, an inventory of parts, ruled by the laws of reason and health. Anatomical knowledge was a kind of moral discipline, a way that mind could govern body. Women, savages, children, and uneducated laborers, like beasts, all lacked an anatomical conception of self, and were therefore prey to ungoverned impulse and ill health. But they could be brought into the moral order through education: advocates of public school education in North America added anatomy and physiology to the curriculum; Christian missionaries in

Burma, Hawaii, and southern India translated anatomy textbooks into native languages and taught anatomy alongside Bible studies. In an age when many people were trying to remake themselves into "refined," "respectable," and "genteel" persons—a condition that was signified by the possession of homes with drawing rooms, libraries, objets d'art, prints, pianos, flower gardens, and picket fences—William A. Alcott (the author of many Sunday school tracts) wrote a popular illustrated anatomy book for families, teachers, and schoolchildren that reconceptualized the body as "the house I live in."[10] In its pages, the viscous mess of interior organs and fluids—and the appetites it engenders—was given order and containment through its skeletal architecture. Representation in words and images on the printed page domesticated corporeality, mind governed body.

The Anatomical Puzzle inverted this order. His head did not sit atop the sovereign self; his anatomy was not posed like a classical statue. His contortions bared and deformed the osteological structure that should have kept the body in order. His anatomical self was worn almost at skin level (much like stage bodybuilders or the hunger artists who billed themselves as a "living anatomy"). His interior and exterior converged. While Marinelli used Professor Virchow's testimonial to respectfully gesture toward high medical authority, he also travestied that authority. His body exceeded anatomical discipline. And this had a social dimension: the Anatomical Puzzle violated prescriptive ideals of masculine beauty, manliness, posture, even species. In the cultural logic of mind and body, the head was identified with regulation and control, humanity, the cultural and political elite, the text, the law, and reason. The rest of the body was identified with manual labor, the common people, the exotic, the beast, and impulsive desire. The image on Marinelli's flyer pointed to the flesh's perverse pleasure in contorting textual outlines and work discipline. The Anatomical Puzzle was unburdened by the requirement to live within his anatomical template, and he was thus joined to the exotic, the criminal, the effete, the working-class rowdy, and other wayward bodies—a lush and crowded assortment of social types. The body was a capacious signifier, a cornucopia of meanings, difficult to pin down, and impossible to fully domesticate, which was part of the fun of the fairground and stage-show attraction.

The case of Marinelli is instructive but opens up a host of problems. One is this: What relation does corporeal life have to the printed page of text and illustration? We can't bring Marinelli back to life to perform for us, or to submit to interview or forensic analysis. All we have are paper fragments: his illustrated poster and circular letter, brief mentions in American newspapers, a chatty French article with a half-tone photograph (which tells us that Marinelli ended his career in Paris, where he had a school for contortionists). Based on these scraps, and accounts of similar performers and shows, we can imagine how things went, but we don't really know who Marinelli was, what he did and knew, who came to see him and why, and how his audience responded and understood him.[11] We know that his poster connected his contortions to the textual world of medical discourse and anatomical representation. And we know that schools, legislatures, jails, cathedrals, and world empires were all founded on, or at least justified by, the assumption that civilization—the world of texts and learning and laws—was morally superior to undisciplined embodied life. In the years 1800 to 1920 there were few human beings who did not stand in some compelling relation to text and representation. We know this because texts and representations of the time tell us so.

Other documents say something different. We have words and pictures—novels, essays, newspaper articles, memoirs, court transcripts, paintings, photographs, engravings, and so forth—that describe or allude to an extravagant life of the body that resisted and exceeded textual regulation, even textual regulation backed by coercion and violence. And we have physical evidence to situate our bodies in space and spaces: human remains, buildings, farm sites, burial grounds, garbage dumps, and artifacts. So even if we need words to describe and organize the historical evidence, we also need to imaginatively reconstruct the embodied world beyond the text. There was a world of men and women and children, of sweat and toil, and of work at the plow, in the mine, workshop, plantation, and factory, in the kitchen and the laundry and the schoolhouse, on the railroad and the street corner, and aboard boats and ships. There was a world of speechifying and shouting and singing and gaming and dancing and flirting and fucking and fighting. A world of eating and drinking and spitting and smoking and chewing and pissing and shitting. A world of dirt and grime and stink and perfume. A world of birth and death, of health and disease and deformity. A world of love and play and violence in bedrooms, classrooms, barrooms, and barracks, on street corners and battlefields. The life of the body.

And, to swing another 180 degrees, if bodies stood in relation to texts, texts stood in relation to bodies. Text and representation only had force to the degree that they mobilized people to act in space and time. Marinelli's flyer and circular letter were designed to persuade people to buy tickets to his show; the profits from these (and similar) shows enabled James L. Kernan to found a hospital. Other contemporary texts were designed to persuade, coerce, and/or enable people to work in factories, exercise, wear modest and/or pretty clothes, pray to God, manufacture implements, perform surgery, sell newspapers, construct machines, kill enemies, grow crops, deliver babies, colonize foreign territories, emancipate slaves, bury their dead, and so forth.

THE BATTLE OF ULUNDI

If this is way too much to hold together, let's consider a smaller piece (which is actually quite large): military history—the history of war, armies and navies, organized and disorganized violence, military doctrines, and military cultures. This may seem a bit paradoxical: the field of body studies was partly constituted in *opposition* to the long-standing discursive habit of privileging the subject of wars, politics, nations, ideas, and great men. The Annalistes, Marxists, feminists, and Geertzian sociocultural historians argued that the grand tradition, because of its narrow focus and ideological blinders, had replicated the very power structure it sought to study and explain, and in the process ignored the much larger and richer history of everyday life at the level of embodied experience (which was important for its own sake and necessary for a better understanding of the history of wars, great men, politics, nations, and ideas). In response the body scholars produced an extraordinary corpus of writing on the production, discipline, interpretation, and performance of the body and body parts in other locales: the home, the prison, the city, the plantation, the pub, the stage, and so forth. But war and military culture play a vital role in the cultural history of the body, and cultural historians are obliged to consider it. So here's another image from the late nineteenth century that tells us something about the empires that were constructed out of bodies in the age of empire and the bodies that were forged (and mutilated and destroyed) within them.

Published in the *Illustrated London News* of August 23, 1879, Figure 0.2 shows a scene from the Battle of Ulundi, the decisive battle of the Anglo-Zulu War. In the foreground, we see officers on horseback, soldiers carrying a box of supplies, a fallen horse (left) and a fallen ox (right), a wounded man carried away by two native porters on a stretcher, three fallen and unattended native warriors (British allies or terrified captives?) next to a shield, a wounded man escorted by two comrades, many uniformed soldiers milling about, and urgent tasks being performed amid a chaos of work animals, barrels, and wagons. In the background, clouds of smoke rise up, and we can dimly discern ranks of rifle-bearing troops massed to fire upon onrushing Zulu warriors (even harder to discern) as the battle rages. The picture is busy and hard to take in; the editors have helpfully placed numbers within it, keyed to captions that describe certain aspects of the battle in greater detail. The point of view is entirely British, as seen from behind the front line (actually inside the square of the four connected front lines). We see the spectacle of an armed force engaged in combat; the image especially directs our attention to the British casualties, even though they were few. The battle—in which 17,000 British soldiers armed with large-caliber Martini-Henry rifles, Gatling guns, and artillery faced an army of 50,000 Zulu warriors armed with spears and rifles—lasted for about thirty minutes and was notable for its lopsided and decisive result. On the British side: ten killed and eighty-seven wounded. On the Zulu side: around 1,500 dead and 1,500 wounded, followed by the burning and destruction of the Zulu capital.

FIGURE 0.2: *The Zulu War. The Battle of Ulundi [July 4, 1879]. Inside the Square. Illustrated London News*, August 23, 1879. Wood engraving. Artist: Melton Prior. Wellcome Library, London. (V0015328); library ICV #15641

After the sketch was finished, it was taken from the battlefield, and delivered to London via an ocean-going ship. There, in the capital of the Empire, engravers broke down the image into square sections and engraved them into wood. The reassembled picture was then set onto the page along with the accompanying type for printing. About six weeks after the battle, it was finally published in the pages of the *Illustrated London News*, the most widely read and influential weekly graphic newspaper in all of Britain.

The engraving was merely one of a series on key episodes of the Anglo Zulu War—which itself was but one of a series of colonial wars conducted by British forces over the course of the nineteenth and early twentieth centuries.[12] What did readers make of it? And what does it tell *us* about the empires that were constructed out of bodies and the bodies that were forged within them? Several things. On first inspection, the picture suggests that the British Empire is men's business, that empire is forged in the crucible of armed violence between men. No women appear. But offstage the trope of assaulted female innocence played a key role in the larger war narrative. One of the British justifications for war—a moral imperative to conquest—was the Zulu slaughter of native women. According to key British officers, manly honor demanded that they protect the innocent women; so from start to finish the war was represented as a performance of masculinity. The larger narrative arc was that this was a "civilizing mission": a struggle between civilization and savagery, mind and body, reason and superstition, morality and brutality. But the triumph of civilization was no sure thing. With the scene of the contest so many thousands of miles away from the metropole, there was doubt about the outcome. In a previous battle, the Zulus had defeated the British. So the depiction of the rout at Ulundi reassured the British public: even though the British army was outnumbered three to one, technologically superior arms, bravery, discipline, and bold and intelligent leadership—civilization—would dominate.

The engraving is a variation on the genre of history painting, translated to the medium of newspaper illustration. It depicts history in the making; it tries to sweep the reader away in the grandeur of history—as history paintings typically do—and tries to foster some kind of imaginative identification with the hybrid cultural/political project of nation and empire building. The reader sees soldiers struggling to cope with the crisis, issuing commands, and obeying orders. Even if the figures are a bit wooden (an attempt to convey military bearing or just the artist's style?), we are expected to empathically imagine the embodied experience of the figures in the foreground: especially the wounded British soldiers, but also the dead horse and ox, and the fallen natives (unlike some other pictures of colonial battles that exoticize and even caricature the figure of the native). Here we see bodies in pain, confusion, and haste; men dressed in national uniform and native garb; and panicked animals. The image asks for an emotional response from the reader—who is not viewing a large colorful canvas in a public gallery or panorama, but is seated at home, or in some other interior space, viewing a rectangular, and very crowded, black and white engraving in the pages of the weekly journal. And this engraving is set alongside other scenes of politics, war, domestic life, fairs and balls, cute animals, natural wonders, ethnographic exotica, industrial progress, picturesque locations, and so forth. The newspaper is physically tiny compared with the scale of the events it depicts—its pictures are miniaturized and domesticated—but discursively it is vast, an expansive commons of views, stories, and description, a representational empire.

Read in the parlor or at the table over breakfast, the image thus contributed to what historian Benedict Anderson calls an "imagined community."[13] That is, it performed and helped to consolidate a shared idea of empire and nation, an imperial imaginary, which was linked to ideas about British masculinity and civilization. The domestic setting where the image was viewed places empire in the safe physical environment of the home, with its couches, carpets, drapes, and chairs; its differentiated rooms; and its differentiated family members—a space that has its own moral narratives, aesthetics, and ethos. The home may have been presided over by a family patriarch, but in the dominant bourgeois ideal of the era it was figured as the domain of woman and family. Cumulatively the cultural practices associated with domesticity—the sharp division between women's and men's spheres, empathy with the suffering of others, appreciation of art and art objects, pleasure in good manufacture, sentimental affect between family members, the protection of women and children—came increasingly to define what it meant to be civilized, to be human (or humane).

And these happy (even utopian) associations occluded the link between the violence of the depicted event—the killing, destruction, and subjection of the enemy other—and the appropriation of native lands and peoples for economic exploitation and geopolitical advantage.[14] The war's beneficiaries were largely the upper and middle classes of the metropole, but colonial wealth also trickled down to the broad mass of working people in the form of cheap staples and commodities, land, jobs and opportunities in the colonies, mass-produced goods and entertainments, even illustrated newspapers, which were also read outside the home, in pubs, libraries, and other nondomestic locations. So the occlusion was convenient; it fed a kind of moral complacency in British life, a willful pride in empire, which was performed bodily, in military parades, imperial rituals, and patriotic songs, and imaginatively, in representations of bodies, in Rudyard Kipling novels, engravings, celebratory verse, and so forth. At home, at school, and at the pub; in newspaper, novel, music hall, and church; and in town and city squares, uniform dress-ups, military postures, military metaphors, militarist sentiments, and cultural productions circulated.

Because this militarist culture was a shared identity—largely masculinist but one that many women supported and took pleasure in—it served to blur the vast gulf and razor-sharp tensions between working-class people (the soldiers, sailors, prostitutes, factory workers, domestics, and rank-and-file settlers who made capitalist accumulation and imperial adventures possible) and their social betters. Given this powerful class division—the two Englands—it was also possible for British readers to take the victory over the enemy at Ulundi as a victory over the disorderly, undisciplined, fearful, and desiring body at home.

The black inhabitants of South Africa were kept outside the circles that participated in the viewing of the engraving of the Battle of Ulundi: they experienced the consequences of the battle in a more visceral way. The war ultimately led to the development of a lucrative economy based on the extraction of gold, diamonds, and other commodities, mined by cheap native labor—and to the opening up of the interior of southern Africa to further exploitation and settlement.[15] This new political–economic regime, like the wars that established it, profoundly affected the native population: they were displaced and subjected to new forms of labor and juridical discipline that determined where they could live and what they had to do to survive. Farming and laboring, spaces of living, densities of population, familial relations, religious practices, modes of violence, customs of dress, and everyday speech all changed.[16] Bodies were punished, maimed, murdered, relocated, intermingled, and worked to exhaustion. And they were reconceptualized as racial bodies, Christian (or pagan) bodies, occupational bodies, and civilized or uncivilized bodies.

AN EVOLUTIONARY TREE

For readers of the *Illustrated London News*, the Battle of Ulundi was only one episode in the larger narrative of universal progress. By the 1870s, in Britain, Europe, and North America it was widely believed that a new era was dawning, an age of improvement, a time of onrushing breakthroughs in science, art, technology, philosophy and governance: modern times. Life was being visibly altered by industrial production and technological transformation: railroads, telegraphs, steamships, gas and electric light, and a succession of new inventions. The horizontal spread and vertical advance of civilization seemed almost inevitable, an essential property of history.

But, at the same time, there developed a counterbalancing notion of atavism, regression, and degeneration. Despite progress, life in the metropole was abundantly full of brutality, superstition, violence, immorality, opposition to innovation, madness, and so forth. The primordial was retained in the present. This could be projected outward, culturally, onto the so-called primitive races and regions of Africa, Asia, the Americas, Oceania, and the frigid poles: the naked savage, the aborigine. But it could also be figured socially, within the metropole, as the dirty manual laborer, the unlettered peasant, the prostitute, the stinking drunkard, the glutton, the lecher, the bratty child, the criminal, and a thousand other social types, all coded in some way as corresponding to the distinction between mind and body.

Such distinctions—the stuff of novels, newspaper stories, popular prejudice, and moral commentary—projected the body and desire onto the other. And yet we all have bodies and desires. Everyone experiences sweating, hunger, sexual pleasure, rage, thirst, lust, pain, and so forth. We all begin as unsocialized infants and children with intense bodily urges and emotional expressions, and we retain a sense of our original condition throughout our lives. Mind and body, self and other uneasily cohabit the same package of skin. And this troubling (and exciting) sense of internal division haunted the narrative of universal progress.

In nineteenth-century Europe, the vexed cultural logic of mind/body was already thousands of years old, but, over the course of the century, it was increasingly articulated within a framework that claimed to account for every aspect of human difference: evolution. A diverse body of thought—which aspired to be almost a master code—evolution talk circulated in many publications, many languages, many images, and many texts. The most innovative, influential, celebrated, and disparaged of these was Charles Darwin's theory of the evolution of diverse species from a common ancestor "by means of natural selection," which was articulated in his landmark *On the Origin of Species* (1859) and, with the addition of a theory of sexual selection, in *The Descent of Man and Selection in Relation to Sex* (1871).

But the concept of evolution long predated Darwin's publications, and had long been the subject of public controversy.[17] Not because of its epistemological status as a speculative theory, but because of its political, social, and theological implications. As theorized in the early decades of the nineteenth century by Jean-Baptiste Lamarck, Étienne Geoffroy Saint-Hilaire, and others, evolution was a radical doctrine. Evolutionary "transformism" or "transmutationism" disputed the fixity of species. Instead of God creating the world in roughly the form it is now (the scriptural account), species changed and evolved, and were still changing and evolving, through the direct influence of the environment on the body of the organism or on its behavior. Theologically this was troublesome because it seemed to posit the possibility of a continuing creation, ungoverned by God, and because it was entirely indifferent to Christian narratives and timetables of creation, fall, and redemption.

Politically this was troublesome because it seemed to undermine unequal social and institutional establishments. If the characteristics of species were subject to constant revision, leading progressively to the development of more complex and capable life-forms (and, in another register, if inventors could continually develop more complex and capable machines), then why couldn't divisions between human beings be revised, leading progressively to the development of more complex and capable human beings, and more complex and capable nations and laws? Transformist evolution, in other words, seemed to provide an intellectual and moral justification for reform and revolution. To make matters worse, it also complicated the categorical distinction between human and animal, between civilization and bestiality, and all the social corollaries of that distinction, which pervaded eighteenth- and nineteenth-century culture.

Darwin's work on evolutionary theory, carefully argued with a voluminous array of evidence, raised the stakes and altered the debate. The theory of natural selection tamed some of the wilder implications of transformism. It endorsed the notion of continuous change, but slowed the pace: change occurred in the *longue durée* of geological time, hundreds of thousands or millions of years, not lifetimes or centuries. It placed limits on the scope of transformation: change was incremental, accretional, and limited. A vertebrate could not turn back into an invertebrate. But in other ways, it was just as troubling, if not more so. Like transformism, it eroded the distinction between man and beast, and the moral privilege based on that distinction. Darwin argued that

> Man with all his noble qualities, with sympathy which feels for the most debased, with benevolence which extends not only to other men but to

> the humblest living creature, with his god-like intellect which has penetrated into the movements and constitution of the solar system—with all these exalted powers—Man still bears in his bodily frame the indelible stamp of his lowly origin.[18]

Darwin here begins by defining what distinguishes the human from other creatures: empathic identification, reason, instrumental power. But he ends by insisting that humans *are* animals. (Or, as it was often caricatured, "men come from monkeys" and so cannot claim any moral privilege.) Unlike transformist evolution, which in many versions retained some undefined metaphysical force that rigged evolution in the direction of progress, the theory of natural selection was indifferent to the moral or intellectual progress of humanity, or any other species. It could push us up or pull us down. Through natural selection over millions of years—survival of the fittest—the barnacle, originally a free-moving crustacean, lost its independence and adopted a passive tropistic lifestyle more akin to plants than animals.[19]

In the decades that followed the publication of *On the Origin of Species*, the concept of evolution gradually overcame determined opposition from organized religion and conservatives.[20] In the late nineteenth and early twentieth centuries, evolutionary doctrine came to dominate the newly emerging disciplines of the natural and social sciences.[21] But if Darwin himself rejected any evolutionary teleology and pointedly refused to equate evolution with progress, many evolutionists, even followers of Darwin, argued strongly for a teleological evolution with varying political implications, depending on the particularities of their arguments and the moment in which they were propagated.[22]

So here's another image (Fig 0.3), a lithographic plate from the first edition of Ernst Haeckel's *Anthropogenie* (1874), which tells us something about the empires that were constructed out of evolutionary bodies in the age of empire, and the evolutionary bodies that were forged within them. Haeckel (1834–1919) was an avowed disciple of Darwin, a prominent member of the first cadre of scientific amateurs and professionals who worked to establish evolution as a key concept within academic disciplines (biology, psychology, sociology), even literature and art, and to spread the idea of evolution to a wider public.[23] *Anthropogenie*, a profusely illustrated attempt to popularize the concept of human evolution, was enormously successful, went through several German editions, and was translated into many languages.

Figure 0.3 shows a chimpanzee, gorilla, orangutan, and "Neger" (German for "negro") posed in tree branches. The tree is an iconographic element that connotes nature and the primordial (the forest precedes human civilization and human evolution; trees outlive human beings and other species; Norse and other mythologies feature a "tree of life"; and a tree is, of course, prominently featured in the book of Genesis). Typically, in other illustrations and diagrams, evolutionary theorists represented evolution as a tree with branches, with humankind at the top.[24] Here, the human is uncharacteristically placed below, signifying that the Neger—who is not labeled with his species name, "human"—deserves to be morally and culturally classified with the great apes. Or perhaps signifying that human progress depends on climbing down from trees—which the jungle-dwelling Neger has not quite accomplished.

By placing the four figures as equivalents, and emphasizing the similarities of form (face, hands, feet, ears, and so forth), the plate makes the case for a familial relationship between human beings and forest-dwelling apes as descendants of a common ancestor. But the illustration not only depicts commonalities between apes and men but also marks physical differences: apes have fur and human beings do not.

And it marks a difference between negroes and whites that goes beyond skin color and hair texture. The naked negro lives in nature and belongs in a tree with apes. The (undepicted) Caucasian lives in civilization, wears clothes, belongs in a room inside a building, and reads the book with text and illustrations. In other words, the image responds to the Darwinian assault on human exceptionalism by relocating it. Haeckel called himself a "monist," a term that emphasized his opposition to the mind/body dichotomy. Like Darwin, he insisted on "the descent of Man from lower animals," a position that established a kind of universalist solidarity between humanity and other life-forms (Haeckel was famous for work on diatoms) and potentially overrode any morphological or cultural differences between human beings.[25] More profoundly, Haeckel's enormously influential dictum, "ontogeny recapitulates phylogeny" (i.e., the stages of embryonic development of an individual organism retrace the sequence of evolutionary development of the species), implies that evolutionary history and the succession from primitive to advanced lies within all of us, which potentially collapses the moral distinction between primitive and advanced organisms.

But here the plate cordons off the so-called primitive Neger from the civilized European and makes the Neger into a living fossil, an evolutionary dead end. Elsewhere, Haeckel argued that "the morphological differences between . . . sheep and goats . . . are much less important than those between a Hottentot and a man of the Teutonic race" and that, because "the lower races . . . are psychologically nearer to the mammals—apes and dogs—than to the civilized European, we must, therefore, assign a totally different value to their lives."[26] This was an argument for evolutionary teleology, with humanity the final stage of evolutionary progress, rather than natural selection. And for a moral calculus of race based on *civilization*, or progress in cultural evolution—with European male scientific authority as privileged arbiter (Haeckel, and by extension his colleagues, readers, and followers). The doctrine of *human* selection to breed an improved humanity, termed "eugenics"

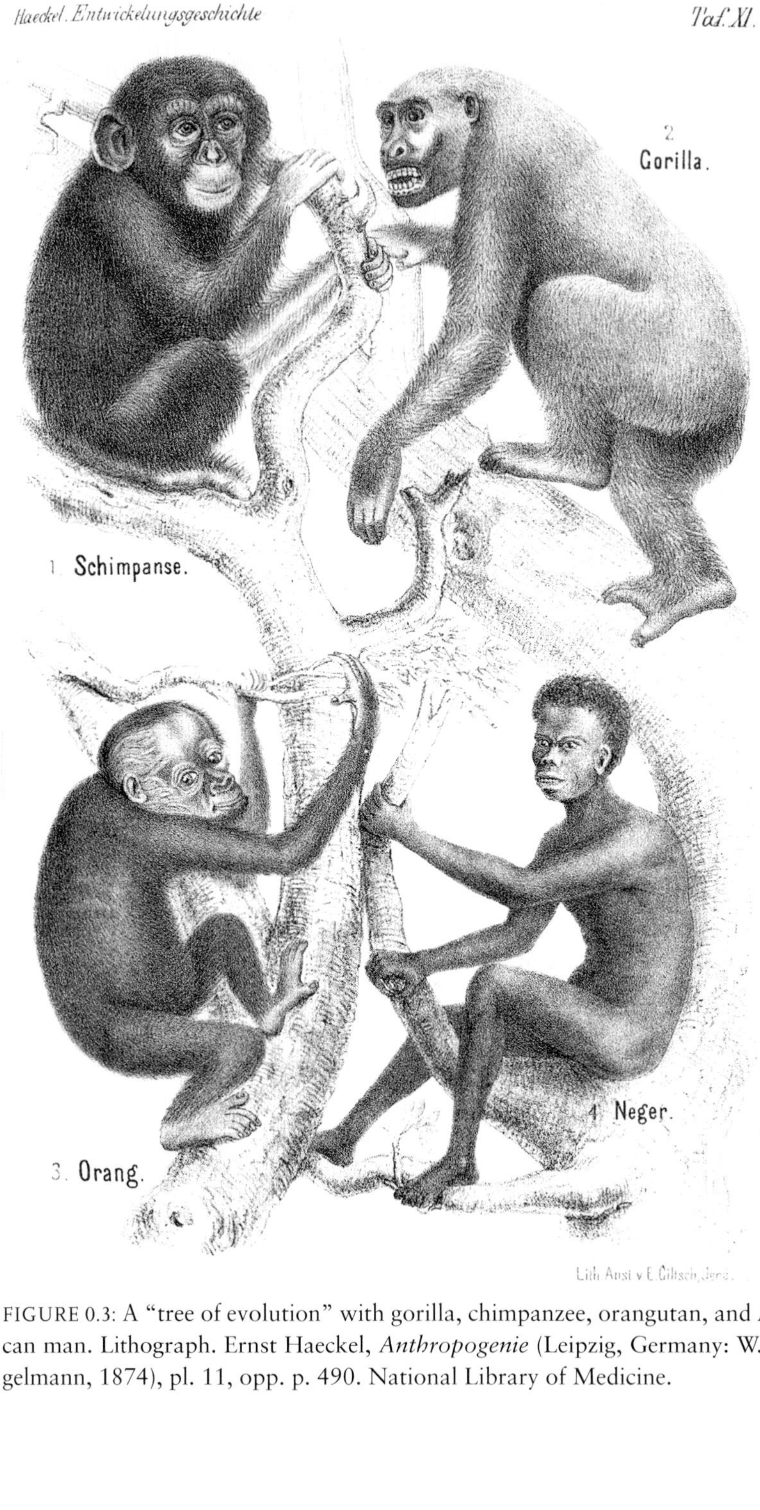

FIGURE 0.3: A "tree of evolution" with gorilla, chimpanzee, orangutan, and African man. Lithograph. Ernst Haeckel, *Anthropogenie* (Leipzig, Germany: W. Engelmann, 1874), pl. 11, opp. p. 490. National Library of Medicine.

by Darwin's cousin Francis Galton, gained a wide following in the late nineteenth and early twentieth centuries.[27]

So what does all this have to do with empire?

In the wake of *On the Origin of Species*, evolutionary theory became a vital doctrine for a new group of intellectuals who attempted to establish themselves as authorities over the body, people like T. H. Huxley and Haeckel, who made careers both within elite science and as popular authors and speakers. They helped to forge an empire of sorts, an informal domain of evolutionary writers and enthusiasts and their reading public. But the scope and reach of evolutionary discourse quickly became much larger than that: writers meditated on the implications and imperatives of evolutionary theory in articles and entertaining fictions that reached uncountable millions. If evolution reinforced and complicated the distinctions of mind/body and human/animal, and their corollary social distinctions, it also provided a logic for writers and artists as diverse as H. G. Wells, Sigmund Freud, Edward Bellamy, Herbert Spencer, Francis Galton, Jane Addams, Edgar Degas, W.E.B. DuBois, Theodore Dreiser, Cesare Lombroso, Andrew Carnegie, and Eduard Bernstein, and for cultural productions as ephemeral as popular songs, joke books, magazine articles, vaudeville skits, and political oratory, and as enduring as the zoo, the botanical garden, and the museum.[28] Evolution had a vast discursive empire: evolution talk became a powerful resource in the production of cultural authority.

And, via the doctrines of eugenics, evolution had a vast institutional empire: prisons, welfare agencies, and hospitals for the "insane" and "feeble-minded" administered by or run according to the precepts of the emergent professions of criminology, social welfare, and psychology. Inside these institutions, a eugenic calculus increasingly provided the basis for decisions on who would receive aid and who would starve, who was incarcerated and who would remain free, who was sterilized and who was permitted to reproduce.

Evolutionary discourse also had an emergent political empire. In late nineteenth- and early twentieth-century Europe, nationalist and anti-Semitic parties increasingly vied for power and mobilized eugenic arguments. Eventually, with the rise of Nazism, they implemented catastrophic racial/national eugenic policies: genocide.

But there was, obviously, also a link to literal empires. The social practice and cultural meaning of empire was built into evolutionary theory. The theory developed in an emerging colonial/global world system—remember the crucial role played by voyages of European travelers who made observations and collected specimens in distant locations, most importantly Darwin's voyage on the *Beagle*. Building on the Linnaean system, evolution represented all of life as falling into kingdoms and domains that could be construed as conceptually equivalent to political, geographical, and social divisions. These divisions were convenient models for those who wished to develop, naturalize, and deploy racial difference within colonial empires. Evolutionary distinctions could justify the exploitation of native peoples for cheap labor or their extermination (if they were deemed unfit to survive in the modern world). It also provided a justification for those who argued that, under generations of European tutelage, native peoples could be raised up from living fossils to a higher evolutionary level. So evolutionary discourse was mobilized to legitimate colonial conquests, social orders, juridical regimes, public education, and even religious missions. At a time when European powers were scrambling to divide Africa, Asia, and the Pacific, works like plate 11 of *Anthropogenie* were entirely congenial to colonialist and imperialist projects.

"THAT TROUBLES OUR MONKEY AGAIN . . ."

So evolution talk was powerful—but also problematic. If one followed its logic closely, contradictions appeared at every turn. Take, for example, the phenomenon of human hairlessness. Humans evolved from apelike creatures that were covered with fur; and evolutionary discourse, like older discourses, made hairlessness a signifier of human exceptionalism. But this argument goes entirely against the evolutionary hierarchies erected by racial science: East Asians, American Indians, and sub-Saharan Africans all tend to have *less* facial and body hair than Europeans.[29] (And at mid-century, it was the fashion for European men of high standing, or a certain age, to support a wild thicket of facial hair as a part of the performance of masculine authority.[30])

The hairlessness issue also complicated any evolutionary argument for male superiority over women: women tend to have less facial and body hair than men. Darwin, as we've mentioned, often argued against any necessary connection between evolution and progress, and he knew full well what a spanner he was throwing into the works of gender orthodoxy and fashion. In an 1860 letter to Charles Lyell, he twitted his critics with this remark: "*Our* ancestor was an animal which breathed water, had a swim-bladder, a great swimming tail, an imperfect skull & undoubtedly was an hermaphrodite! Here is a pleasant genealogy for mankind."[31]

And critics in turn twitted Darwin. A cartoon published in an 1872 issue of *Fun* magazine depicts a bearded Darwin as a furry monkey with a long curving tail, pedantically gesturing while he impertinently takes the pulse of a beautiful young woman, who is dressed up in a fashionable outfit and sports a very prominent bustle (see Figure 0.4). The cartoon refers to a passage in the 1871 *Descent of Man* where Darwin argued that the hermaphroditic "marine Ascidian" (a sea-squirt) was similar, in its larval stage, to humanity's distant invertebrate ancestor—and responds to Darwin's 1872 *The Expression of the Emotions in Man and Animals*. The caption reads: "Female descendant of marine Ascidian: 'Really, Mr. Darwin, say what you like about Man; but I wish you would leave my emotions alone!' "[32]

Darwin here is depicted as a hairy beast, a throwback—comic illustrators loved to depict proponents of evolution as apes and monkeys—but also as a type of the masculine. In Victorian society, "man" was the domain of the intellectual, cultural, and governmental institutions that govern the social and

FIGURE 0.4: "That troubles our monkey again." *Fun*, November, 16, 1872. Caption: "Female descendant of marine Ascidian: 'Really, Mr. Darwin, say what you like about Man; but I wish you would leave my emotions alone!' " Yale University Library.

political body but also of physical strength and sexual aggression. But Darwin is not represented as a governor of the body, but as body itself—a body of animal desire and aggressive impulse. And this impulse is linked firmly to Darwinian evolutionary theory, which unlike pre-Darwinian theory, rudely insisted that the fundamental principle of organic evolution had everything to do with the variation that arises out of sexual intercourse and reproduction (at least among creatures that reproduce sexually).

The "female descendant," in contrast, represents purity, civilization, delicacy, and refinement. She confronts Darwin as an exemplary feminine figure, morally and physically outside the evolutionary narrative. Everything about her is soft and gentle, even her admonition. She embodies a bourgeois ideal of Victorian womanhood that was ceaselessly iterated in sentimental poetry, etiquette books, fashion magazines, paintings, prints and engravings, and novels. Her porcelain complexion, her hair arrangement and hat, her politesse, and her fashionably silken and velvety clothes signal that she is virginal, almost immaterial—the surface of her body is entirely concealed except for face and hands. She is protected, sequestered from the rigors and demands of natural selection (and from sex talk). She does not struggle to survive; she is not "red in tooth and claw."[33] In *The Expression of the Emotions*, Darwin argued that the human capacity to feel emotion was an evolutionary trait of the species and had evolved, like all other human characteristics, from ancestral nonhuman species. "Man," Darwin insisted, is not an exceptional species; dogs, cats, and monkeys also feel emotion. The young woman in this image resents the suggestion: "Say what you like about Man"—that is the sphere about which Darwin, a man, can properly speak—but "leave my emotions alone!" She makes it personal: And the personal is her domain. Emotions, especially the fine emotions of love, faith, generosity, kindness, and aesthetic pleasure, are properly part of "woman's sphere." (From human exceptionalism we move to female exceptionalism.) Masculine intrusion into this domain is at best clumsy, at worst offensive. The distinction between emotion and reason is gendered, but here reason is identified with matter and bestiality, and emotion with spirit and humanity.

The only dissonant note (a loud one) is the silken bustle, which cloaks the lady's rear end but also magnifies it ludicrously. She is modest in the flashiest way imaginable, in a way that perhaps even suggests a categorical affinity to the enlarged buttocks of Saartjie Baartman—the famous "Hottentot Venus."[34] So, in claiming a privileged exemption from Darwin's theory of natural selection, the female descendant vindicates his theory of *sexual* selection. According to Darwin:

> Sexual selection depends on the success of certain individuals over others of the same sex, in relation to the propagation of the species. . . . The

> sexual struggle is of two kinds; in the one it is between the individuals of the same sex, generally the males, in order to drive away or kill their rivals . . . ; whilst in the other, the struggle is likewise between the individuals of the same sex, in order to excite or charm those of the opposite sex, generally the females, which . . . select the more agreeable partners.[35]

Thus, the fashionably dressed young lady presents a display that makes her sexually selectable (to young men), while the bald, middle-aged, heavy-browed monkey man presents a display that makes him sexually unselectable (to the fashionable young woman).

But gender is not the only social category in play here. The cartoon also suggests a class relationship between the two figures. Darwin, although not a pedigreed aristocrat, was a wealthy bourgeois gentleman of high status, a direct descendant of Josiah Wedgwood, the celebrated pottery manufacturer, who was also an intellectual, reformer, and patron of the sciences. Here Darwin is demoted to the status of a furry beast associated with jungle and zoo. And there is another association: monkeys were a familiar sight in Victorian London as antic street performers owned by Italian immigrant organ grinders. This is about as low as you can go: the organ grinder's monkey is the lackey of a social outcast.

This image, of course, had everything to do with empire and empires. First, as already mentioned, there was the increasingly globalized world of European dominion and trade, in which products and commodities from the periphery, and creatures like organ grinder monkeys from South America, Asia, and Africa, could appear in the heart of London. Travel and trade also proceeded in the opposite direction. Darwin the scientific gentleman traversed the world in a ship flying the Union Jack. His observations and findings depended on the growing tentacles of imperial commercial, political, and technical power.

Then there was a growing empire of mass-produced print publications—and within that, a growing domain of satirical cartoons printed on separate sheets and in weekly papers and monthly magazines that cropped up in Britain, France, Germany, the United States, and elsewhere in Europe. In Victorian Britain the two main satirical papers were *Punch*, a Tory weekly, and *Fun*, the Liberal equivalent. Satirical cartoons typically featured body distortions, body fluids, body functions, and body transformations—the grotesquerie of everything—as a retort to the pretensions of philosophy, politics, science, religion, and the social scene. In other words, the satirical cartoon, whatever the object of its derision, always played on the cultural opposition between spirit and matter, mind and body.

In this topsy-turvy logic, the figure (or position) of woman was highly unstable. On the one hand, polite usage insisted on the immateriality and purity of the morally exemplary woman, who in sentimental novels and poetry was so often depicted as the angel of the house, faithfully attending to father, husband, and family, which some historians have referred to as the "cult of true womanhood."[36] Woman was spirit; man was matter.[37] On the other hand, sentimental novels and other cultural productions also depicted women as a "child sex," impulsive and desiring, competitive with each other, prone to jealousy, easily distracted by displays of beauty, and tempted by displays of finery. The pure, morally virtuous woman was contrasted with the manipulator, the seductress, the shrew, and other female stereotypes. "Woman's sphere" was a nebulous category whose boundaries were endlessly contested and endlessly drawn and redrawn by social, cultural, and governmental policiers.

The sumptuous appearance of the young woman depicted in the cartoon—her beauty—references a growing female domain of fashion magazines, fabric and dressmakers, department stores, corset and hat shops, jewelers, balls, and parties. A domain that was then the cutting edge of a proliferating consumer culture. Or, from another perspective: a proliferating economy of factories, workshops, and shops, and a burgeoning social class of shopgirls, pieceworkers, seamstresses, domestic servants, laundresses, and so forth. Working-class women were not only producers of objects and services that signified femininity but also consumers with their own distinctive constructions of femininity. But all performances of the feminine were not equal: some people had the resources to buy fancy clothes and to hire domestic servants to assist them. Other people *were* domestic servants (a huge sector of employment in nineteenth-century Europe and North America). There was a proletarian "city of women."[38] Their access to the commodities, services, and events that signified femininity, which they were so crucial to the production of, was limited.

The cultural ideal of true womanhood valorized the hard work that women did in serving husbands and children (which was depicted in terms of sacrifice and redemption, an imitation of Christ), but also willfully ignored much of it (even the work of middle- and upper-class women). It was balanced by the anti-ideal of the fallen woman and the prostitute (in a period when many working-class women turned to prostitution as part-time work during some part of their life), the fishwife, the vulgar woman, the Irish servant, and so on—women who were not idealized and not protected.[39]

As is evident from the previous account, in the age of empire the masculine/feminine distinction—the *gender* system—however it was defined—structured, and complicated, every aspect of human endeavor. "Race, gender and class," Anne McClintock argues, "are not distinct realms of experience. . . . Rather, they come into existence *in and through* relation to each other."[40] Male/female difference was performed through body posture, clothes, hair arrangement, speech patterns, occupational and familial tasks, hobbies, aesthetics, and intellectual interests. And this difference was inscribed into nearly every social category and domain: fashion, social class, age, ethnicity, religion, racial identity, political beliefs, beliefs about the natural order, national identity, and so forth (more than just race/class/gender).[41] There were multiple and overlapping "languages" of gender. How *well* they were "spoken"—by women *and* men—was continually subject to critique. What did a person intend to signify with his or her gender performance? How successfully was it accomplished? In the course of a life, a person moved through many communities of opinion. No one was exempt from the gaze and assessment of others. And this was true for men and women, true in homosocial and heterosocial groups, and true up and down the social ladder.[42] Gender performance always signified itself and something else.

PHOTOGRAPHIC REALITY AND THE SHADOW INTERIOR

This social gaze was not panoptical. It came from many vantage points, employed many technologies, and was focused on many subjects. And, if socially organized, could be mobilized in many ways. In the late nineteenth and early twentieth centuries, the photographic essay and the motion picture played an increasingly important role in visually dramatizing the gap between Victorian domestic ideals and the embodied experience of the urban poor.

As a final visual document, see the photographs taken from *The Bitter Cry of the Children* (1906), a book by muckraking photojournalist John Spargo (1876–1966).[43] We start with a picture of a crowd of shabbily dressed children on a New York City sidewalk, mugging for the camera (see Figure 0.5). The picture emphasizes their "mass life," which takes place outside the home, on the street, with no adults in sight. The kids interact as a pack—there is no separation between their bodies—and practically mob the photographer. They exhibit some shyness, but also a kind of bodily pleasure in posing for the camera. These are children who are unprotected (at least so the photograph and accompanying text argue); accustomed to discomfort, dirt, and hardship; without much in the way of schooling; and by the look of things without much in the way of food. (The tall girl in front looks particularly frail.) And they often die young. According to Spargo, these are "lung block" children, prey to the ravages of tuberculosis, which, the caption tells us, has just taken one child.

A second picture shows a smaller group of children, "little tenement toilers," posed on the steps of a derelict building, which is a residence or a workplace or both, with many broken windowpanes, pieces of pavement, and garbage strewn around (see Figure 0.6). Again, no adult is visible. The cruel exploitation of the child laborers, and the utter desolation of their surroundings, stands in stark contrast to the setting where readers view the image: the protective cocoon of the bourgeois home, furnished with comfortable sofa, rugs, artistic prints, board games and toys, shelves of books, and objets d'art. And it stands in stark contrast to the domestic ideal of the sacralized child

A GROUP OF "LUNG BLOCK" CHILDREN

FIGURE 0.5: "A group of 'Lung Block' children. The white symbol of a child's death hangs on a door in the background." John Spargo, *The Bitter Cry of the Children* (New York: Macmillan, 1906), opp. p. 5. National Library of Medicine. (IHM A01356)

LITTLE TENEMENT TOILERS

With the exception of the infant in arms these are all working children. They were called away from the photographer to go on with their work!

FIGURE 0.6: "Little tenement toilers. With the exception of the infant in arms, these are all working children. They were called away from the photographer to go on with their work!" John Spargo, *The Bitter Cry of the Children* (New York: Macmillan, 1906), opp. p. 140. National Library of Medicine. (IHM A013572) who is carefully nurtured and disciplined, who delights parents with innocent remarks and nonsensical play, and who, in turn, is doted upon.[44]

A third picture, captioned "Night Shift in a Glass Factory," is one of many in the book that depicts child labor (see Figure 0.7). In a so-called civilized country, the wealthiest on Earth (the United States), the book argues, it is scandalous that children are forced to do hard physical work. Here we see four boys in a glass factory, in front of a hot oven. Instead of resting comfortably at home after a day at school and play (if indeed they go to school), they work the night shift. (Here too gender divisions prevail: This is man's—actually boy's—work. Other kinds of jobs are reserved for females.) The photograph asks the reader to focus on the contrast between the boys' youthfulness and the arduous adult work they perform. Above the oven, the boys have chalked words and drawings ("xmas," "hot air," "1905," and a crudely drawn face), childish graffiti that underscores the pathos of the scene.

A fourth picture, perhaps the book's harshest, shows the interior of a tiny, cheerless room in a tenement apartment (see Figure 0.8). It is utterly barren. In the foreground, a toddler sits atop a brass bed with frayed, dirty sheets. In the rear left, a mother anxiously tends to her piecework. On the right, an older daughter gets up from her piecework, perhaps startled by the photographer's flash. This is woman's work, shorn of its aura of feminine grace, reduced to killing drudgery. The domestic interior, like the destitute mother, utterly fails to protect.

These photographs should be placed in the context of a larger cultural shift, in which childhood and the figure of the child—especially the small child—was increasingly invested with a privileged moral significance: the child as moral innocent; childhood as a prelapsarian condition. From the mid-nineteenth century on, illustrated books for children became a popular genre, and games and toys began to be mass manufactured. Novels, plays, books, and articles focused on the condition of childhood and the lives of children. There were pedagogical works aimed at parents, teachers, and children. There were fictional entertainments about children at school, at play, and at home. There was academic research on the physiology and psychology of child development ("child-study"). And there were books, articles, plays, and later films that had a sociological dimension: exposés about children who roamed the streets in packs, sold vegetables, performed dances and tricks to get pennies from passersby, or who worked in factories, on farms, or as domestic servants. Almost all of these works focused on the subjective experience of childhood as a unique time of vulnerability; emotional intensity; plasticity; and intellectual, moral, and physical growth—and the parallel subjective experience of a redeemer adult who is especially moved by the plight of children (a surrogate for the reader). The last third of the nineteenth century saw the establishment of societies "for the prevention of cruelty to children," and the

NIGHT SHIFT IN A GLASS FACTORY

FIGURE 0.7: "Night shift in a glass factory." John Spargo, *The Bitter Cry of the Children* (New York: Macmillan, 1906), opp. p. 158. National Library of Medicine. (IHM A013568)

HOME "FINISHERS": A CONSUMPTIVE MOTHER AND HER TWO CHILDREN

Both of the children work and sleep with the mother.

FIGURE 0.8: "Home 'finishers': A consumptive mother and her two children. Both of the children work and sleep with the mother." John Spargo, *The Bitter Cry of the Children* (New York: Macmillan, 1906), opp. p. 172. National Library of Medicine. (IHM A030178) passage of legislation that contributed to the construction of a special moral, legal, and affective sphere around childhood and the figure of the child.[45]

In these four pictures, the photographic image serves as a kind of guarantor and a rebuke to the realm of the symbolic. The photos almost plead with the reader: "Look! Here are real children! Don't be deceived by the comforts and diversions of your civilized life! There is terrible suffering in the world, in your own country and city! The children cry out! Your aversion to seeing and hearing is criminal!" The pictures are cropped, and the child subjects carefully chosen and posed for maximum impact, but this artifice seems negligible compared with the evident realness of these scenes, the truthfulness of the photographic image and its sociological view.[46] Photography has an extravagant exteriority and faithfulness: it renders every extraneous detail, and every extraneous detail is revelatory—debris in a courtyard, a skinny arm, peeling paint, and the corpse-like complexion of a young girl. The realism of the images of victimized children acts powerfully upon the body of the reader: it induces a kind of empathic identification—feelings of loss, distress, sadness, and anger. The image-making practice here has an instrumental purpose, to educate the public about the "social facts" of the new industrial society, to mobilize support for progressive reforms that would protect and nurture the physical bodies, and spiritual life, of impoverished children: public schooling, bans and restrictions on child labor, welfare payments, provision of meals to the hungry, sanitary regulation of streets, residual buildings and workplaces, and health clinics.[47] A crusading reformer in the lineage that went from Henry Mayhew to Jacob Riis, Spargo wanted his photographs to prod his readers to action.

For both author and reader, these photographs powerfully signify in their mode of presentation and in the urban industrial subject: this is the condition of modernity, which is celebrated as an age of unprecedented technological progress, but also deplored as an age of unprecedented degradation. In Spargo's images, modernity is represented by the shabby tenement; the squalid, densely packed city; and the grimy factory where children and adults alike are exposed to dangerous machines, unhealthy air and temperatures, and industrial poisons. At the same time, modernity is signified by the mass-produced and mass-circulated book and the technology of photography itself, which in 1906 was receiving an ever wider circulation through the recent invention and ongoing improvement of half-tone printing, and, of course, by the thoroughly modern practice of the journalistic photographic exposé, and the movement for modernizing reforms that addressed modern conditions. Over the course of the nineteenth century, the desire (or anxiety) to be modern became an intensifying cultural imperative in personal dress, hygiene, state formation, slang, literary production, military and corporate organization, leisure activities, and the design and production of mass consumer goods. The idea of "the modern"—of the transformative impact of modernity on everyday life, industrial production, and the built environment—spread widely in the two decades before World War I,[48] and in World War I (1914–1918) itself, when one of modernity's great transformative effects was to maim and destroy bodies on a mass industrial scale.

Modernity was not just a condition, it was performed and experienced. Modernity certainly acted *on* the body: X-rays did that to patients, doctors, and researchers; artillery shells did that to soldiers; railway accidents did that to passengers and workers; and starvation, homelessness, overwork, and neglect did that to slum-dwelling children. But it was also acted *with* bodies and body parts (and representations and figurations of bodies and body parts) in jazz dancing, ready-made clothes, impressionism, amusement park rides, and brass bands. More mysteriously, it was enacted *within*, as a subjective state or condition, an inner structure of self-representation. Photographs and engravings of slum children, alongside the cumulative and accelerating barrage of experiences, induced a sense of exteriority that corresponded to, and perhaps even fostered, a growing sense of what Carolyn Steedman, Robert Miles, Charles Taylor, and other scholars have described as "interiority," "a world within": a heightened, complexly layered and divided sense of self.[49]

In the late nineteenth and early twentieth centuries, to be truly modern the self had to experience inner development from childhood to young adulthood to maturity, a mélange of sensual experiences, losses, and recuperations. Increasingly, the gaze turned inward and became psychological. But this interior self was always staged externally, was born on the stage and in books. The tragic bourgeois dramas of personal crisis, development, sexual repression and expression, and troubled family life modeled in the bildungsroman, and even more intensely in the plays and novels of Strindberg, Ibsen, Flaubert, and the Russians, were vitally connected to the dramas of subjectivity staged in the depths of the soul.

For Proust, this was the world we obsessively reenact in distant memories that continually force their way into consciousness. For Freud, it was the world we unknowingly reenact in neurotic behavior, under a regime of memory repression and control. In the works of Freud, the unconscious mind is almost the structural equivalent of the camera (or the phonographic recording device): it sees and hears (and records) everything, especially in the crucial stages of a child's development. According to Steedman, between 1900 and 1920, the "account of infantile sexuality and the process of repression," developed by Freud, "*theorized* childhood . . . , gave it another name as 'the unconscious.'" Steedman argues that childhood and the figure of the child—the dewy-eyed happy child, the exemplar of cuteness, but especially the neglected, traumatized child and the confused adolescent—played a crucial role in "a certain understanding of selfhood."[50] The child became a kind of homunculus that collapsed past and present and future. Haeckel's dictum, "ontogeny recapitulates phylogeny" became the foundational logic for a new conceptualization of self. "The child," Alexander Francis Chamberlain wrote in 1900, was "a study in the evolution of Man."[51]

The complex self came bundled with a complex history of unresolved conflicts going back to early infancy or even the womb. Childhood, like history, became the world we have lost—the world that continually haunts us. The self was transformed into our inner child, "the lost child within all of us."[52] This interiority was then elaborately figured and exteriorized in psychological and philosophical theory. The complex person became a person of complexes. Unconscious behavior was given a homeland—the unconscious mind; and a power supply—sexual drives and instincts; and an economy—the circulation and exchange of libidinal energy and repressed desires. Mapped against a corresponding life story and developmental sequence, the topography and economy of the psyche became a field of professional study and a social prescription: a modern person should act like a psychological person and should nurse and foster a sense of psychological complexity based on a deeply personal history that was a profound mystery. There was a correspondence: in Spargo's photographs, we identify with the lost children; in psychoanalysis, we *are* the lost children (and therefore partly the recuperable adults).

So there came to be a new empire within—a haunted house of subjectivity, opaque to photography, where the inward-looking self was figured as a child/adult who evolved and devolved in a dark, strange fragmentary analogue of the social world. This stream of consciousness (in William James's famous formulation) was replete with interior dialogues (spoken by internalized mothers, fathers, siblings, and teachers), maps, love affairs, wars, cities, jungles, factories, and colonies, in fact, all of the institutions, powers, social types, and relations that came to constitute human experience. In the interior self, all the ligatures and figures of empire were constituted in a drama, staged in an empty space where the child developed into an adult who was simultaneously a child. And, with the discovery of this new domain, there came exploration: psychoanalysis; phenomenology; the psychological novel, play, and film; surrealism; expressionism; existentialism; abstraction—constituent elements of the modernist aesthetics that extended far past 1920, throughout the twentieth century, and even into our own time. Like the image of Marinelli, which started off this essay, the self was an Ouroboros that connected the interior imaginary psychological self to the external material, historical world—and back again.

CULTURAL HISTORY'S TWO BODIES

Time . . . thickens, takes on flesh.

—Mikhail Bakhtin[53]

The foregoing discussion suggests some of the things that the body signified (and some of the things that some bodies signified) in the age of empire. What does it signify now? In this introduction, I've analyzed a few idiosyncratic images, and the matrix of cultural assumptions that they draw upon and transmit, and mentioned and cited some important works of scholarship. But, to be honest, we must acknowledge that body scholarship (and every other kind of scholarship) inevitably refers back to the conditions in which it was spawned—our time—which cultural critics have termed variously "postmodernity," "late capitalism," "the society of the spectacle," and "the simulacrum." The emergence of the body as a subject for historical research and analysis reflects our own preoccupation with our bodies, which is characteristic of life in consumer capitalism, with its continual exhortations to attend to the body, body parts, and bodies.

This emergence was fueled by the post–World War II expansion of university systems and student bodies in North America, Europe, and around the world. New scholars flooded the humanities professions, from increasingly varied backgrounds—women, working-class people, and people of color (African American, colonial, neocolonial, and postcolonial). They had one common experience: life in consumer/industrial capitalism. And they had one common obsession: the problem of identity. These scholars asked new questions that arose out of their own experiences and obsessions, and they sought out new areas of study that reflected their need and desire to differentiate themselves from (and supersede) their elders.

Beyond that, it's necessary to go beyond discourse about discourse. What happened to the body in the years 1800–1920, when everything changed again and again? In that long century there was a proliferation of bodies. By around 1800 a certain threshold had been reached. The population of the world had grown; cities of unprecedented size and density were springing up. To put a reductive point on it: there were more bodies and more diverse groups of bodies in ever closer proximity. In 1800 there were about a billion people on the planet—by 1920 the figure had almost doubled.[54] A new mass society emerged—linked by a technological infrastructure of postal systems; markets; roads; maritime, canal, and riverine shipping routes; railroads; telegraphs; and books and newspapers.

With the proliferation of bodies came a proliferation of body practices. Work was reorganized in ways congenial to the needs of industrial capitalism, the global economy, and the nation-state: this was the era of the factory, the plantation, the big city, colonization, the tenement, and financial exchange. Modes of consumption were reorganized: the apartment, private home, mansion, shop stall, marketplace, department store, and mail order catalog emerged or were reconfigured. There were new modes of cultural production: the stage play, the penny press, the minstrel show, the novel, and the motion picture. And there were new modes of state and empire formation: the legislature, the political newspaper, the ministries, armies, and police.

And this was the era of the growth of great nations (United States, Germany, France, Britain), great contiguous multiethnic empires (Russia, the Ottoman Empire, Austria-Hungary, Manchu China), great colonial empires (Britain, France), and something else (Spain, Italy, Portugal, Netherlands, Belgium). It was also an age in which military, police, bureaucratic, and corporate apparatuses grew in unprecedented size and complexity.

All devolved upon the embodied self. The body was programmatically and haphazardly militarized, aestheticized, theatricalized, politicized, colonized, exercised, exorcised, rationalized, textualized, eroticized, medicalized, racialized, evolutionized, and so forth. External apparatuses multiplied (the public school system, university education, the factory, the military, prisons, the dance hall, and so forth), simultaneously inciting and policing embodied desire. And these external regimes were taken up in the interior performances and ideational structures of the psyche. Public and private spaces proliferated outside and within the self. Social circles, political sovereignties and subdivisions, economic strata, academic disciplines, and a jumble of other cultural domains corresponded to and constituted a layered and jumbled self and an incoherent personhood—a personhood made up in part of a variety of doctrines, regimens, services, figurations—all designed to reduce embodied life to a system of coherence, but inevitably failing.

We tend to focus on supply, the production of body knowledge, body disciplines, and body punishments, as if an imperial will to power continually invaded the disorganized, barely sensate body, tyrannized the inert flesh, and subjected it to a succession of inscription devices like the tortures in Franz Kafka's penal colony—or made the body, breathing life into an assembly line of Frankensteins.

But there was also demand: mass markets and smaller markets arose to take advantage of anxious imperatives and cravings, the murky desire to define and script and legitimize performances of the embodied self. This describes the body in the age of empire. And it describes us, because one or two hundred years later, we have had more (a lot more) of the same. This, perhaps, is why the body in the nineteenth and early twentieth centuries seems so compelling and so confusing to us.

So, stepping back: much of scholarship on the body obsesses about identity—our identity. Were the Victorians like us? Unlike us? We define ourselves as modern (or postmodern) in opposition to the passé modernity of the Victorians (who defined themselves as modern in opposition to their nonmoderns: savagery, superstition, the medieval, the baroque, the agrarian). If so, then maybe modernity is a family tradition, and the Victorians were our parents or our grandparents or great grandparents. Or maybe we're the illegitimate issue of an entirely different lineage, barbarians at the gate, others unwillingly or willingly brought from outside into close and uncomfortable proximity with the Victorian world.

If so, that world contains *our* self and other: the body in the age of empire is our own displaced experience read back into the othered past. As Christopher Lasch once argued, we live in a "culture of narcissism" (bourgeois, consumer capitalist us).[55] We are compelled to seek our reflection in everything, but narcissism is lonely and boring, almost a nonexistence, so we seek out the other in everything to dramatize, model, or mark our boundaries. And for the scholar this becomes a genealogical quest, an archaeology of self-knowledge.[56]

Historically we know that the body can be made to serve as a usable other, and this is more than just a historical understanding. We have a feeling: the body is recalcitrant; it exceeds discipline and discourse, figuration and metaphor, and verbal, visual, and gestural rhetoric, even if we can only know or talk about what we know about our bodies ourselves in those idioms. We identify: our bodies ourselves. But the body tends to be a disloyal subject; it always has the potential to revolt, betray, and travesty. No system of training, whether a horse whisper or a totalitarian regime, can ever entirely make it behave. The body is always potentially in a state of insurrection, against the world and against the self.

With the publication of *The Birth of the Clinic* in 1963 (and its subsequent translation into English), Michel Foucault became the foundational figure in establishing the body as an academic enterprise.[57] And sometime in the late 1970s, Foucault's baby sprouted wings. The little angel became a *terme de rigueur*, *mot du jour*, flavor of the week. For a decade or two or three, new scholarship tumbled forth: more books and articles than anyone can keep track of. Then, like all fashionables, the body aged, was pushed aside (for the narrative turn, visual turn, memory, transnationality, post-everything-ism). But the body didn't die. Its élan vital was our cultural preoccupation with our bodies, which is not just a will of the academic wisp but the ingrained and self-reproducing pattern of our lives. Fueled by the unceasing torrent of commodified body representations, texts, and experiences—our everyday life—body scholarship continues to find an audience and still undergoes a proliferating cellular division; or, insatiably, annexes or colonizes every field of humanistic study in its path ("age of empire" anyone?); or has a diversified portfolio or an extended shelf life as an object of cultural exchange. The metaphors, like the domains of body scholarship, multiply, transmogrify (and refer back to our own experience): The body is protean, the body is a grab bag, the body is everything—or nothing.

A similar problem arises with that other protean everything/nothing term of discursive endearment: "culture," which, as everyone knows, has an etymology—"agriculture," "cultivation," the "cult."[58] In the nineteenth century, culture became synonymous with the domain of art culture: subject to historical development and refinement (and often enough decadence), a trajectory of improvement, loss, and recapitulation, as opposed to crude and childish pictures, folk tales, pornography, and low-brow sensationalist fiction. "When culture first emerged as a concept during the eighteenth century it was exquisitely informed by a sense of difference."[59] High culture might make you wet or hard but always claimed to be redemptive, transcendent, something higher than the body. And this gave the artist a certain amount of license (if the artist was legitimated by the critic, aesthetic colleague, and patron). Some people had (high) culture, others didn't. High culture corporealized ideas, depicted bodies, and transcended them.

Yet, almost in the same breath, culture acquired its other meaning, which said the opposite. Primitives/savages had culture: a culture of objects, rites, myths, and marriage and death customs, which took place within a system of belief that was outside history, timeless. Primitive peoples were frozen into a state of arrested development, were living relics of what humanity was like before it invented writing, metallurgy, and other technologies, before it adopted monotheism, before it had science, before it grew up. This notion of culture still persists in naive anthropology, but sophisticated practitioners (and their fellow travelers in cultural studies and cultural history) all abide by a historicist version: the people with culture (savages) have a history (or at least a putative historical development), and the people with history (the civilized) have cultural practices (not just high culture). Matter and antimatter annihilate each other, a big bang that resulted in the creation of a cultural studies universe and a new version of cultural universalism that relies on particulars—plural, local, partial—which is where we stand today.

The Victorians universalized to make a distinction: We have universalism, the other does not. Present-day scholars do much the same. Our contemporary scholarship is written in what purports to be a universalizing academic idiom, but it is actually a fractionated set of idioms: the historian only knows a bit of what the cultural studyist, sociologist, economist, philosopher, art historian, and anthropologist know, and vice versa. And it's increasingly hard to sort out the claims, the grand from the provisional.

So what can we say about our long nineteenth-century (1800–1920)? Was there any distinctive order of things? Body scholarship began with master narratives: class struggle, emancipation, a critique of reason, the spread of capitalism and the world system, and so forth. If Foucault was instrumental in fostering the idea of the body as a category of historical analysis with his insistence that practices constitute the subject, then Marxian social history, feminism, the Subaltern Studies Collective and postcolonial studies, and the history of medicine provided the troops.[60] Starting from a traditional Marxian "base, structure, superstructure" approach, in the 1960s, 1970s, and 1980s (and then proceeding further with a Gramscian analysis that collapsed the tripartite system), social historians argued that the dominant academic focus on ideas, politics, and elites (superstructure) had resulted in the neglect of economics, power relations based on violence, and the experience of "subaltern" groups: working people, women, children, and colonized and indigenous peoples. The mantra, coined by E.P. Thompson, was "history from the bottom up," and Thompson's *The Making of the English Working Class*, along with his essay "Time, Work-discipline and Industrial Capital," were immensely influential as models of how social history should be written and pointers in the direction of a cultural turn.[61]

In these works the study of the working class, and working-class bodies, began as a metonym of the social order, which was figured as a kind of mind/body duality. Traditionally, scholarship had privileged the history of ideas and politics and great events—the social historical critique went—at the cost of erasing or ignoring the history of everyday life as experienced within the body. Feminist and postcolonial scholars then got to work, insisting that the "personal is the political," and also imperial—and certainly historical—and that we need to study the history and political economy of gender, racial and national distinction, male privilege, and skin privilege. Body was coded, even constituted, in gender and race as much as in class (or even more so).[62]

And, under the tutelage of Foucault, the scholarship exploded. There was a certain geographical and chronological exoticism (lots of Asia and Africa; antiquity; the Middle Ages; early modernity; and long seventeenth, eighteenth, and nineteenth centuries) and an effort to create philosophical/theoretical/political foundations, originary narratives, or maybe just show off novel and obscure objects of cultural and historical study. In other words, an academic *nouvelle vague*—a long overdue diversification of subject matters, conceptual approaches, and staffing—which put the body at the top of the agenda. But what is the body? What bodies are worthy of study? Cadavers? Freaks? Wives? Workers? Indigenes? Royal families? Criminals? Slaves? Bankers? Prostitutes? Children? Work animals? Pets? What aspects of the body are worthy? Posture? Hygiene? Child rearing? Sexual intercourse? Hairdos? Wounds? And what parts? The bowels? The penis? The vagina? The nose? The eye? The hand? The skin? (Or problematically: the brain?) The scholarship proved many things and one thing: Whenever we confine ourselves to the body, an empire forms around it. Suddenly, here come texts, images, laws, laboratories, markets, museums, factories, and theaters. The body, it turns out, is a hall of mirrors: presentations, representations, re-representations, ad infinitum.

And so the body becomes a suitable subject for art history, iconography, and rhetoric—the body as representation and metaphor; and for the history of science, technology, sociology, institutions, and everyday customs—the body as a chain of mediations, as a social construction.[63] The body is the locus of discipline formation, state formation, and identity formation as well as architectural, religious, legal, and urban history, museum studies, and cultural geography (the spatial turn) as seen in topics such as sex in the city, stigmata in the shrine, fever in the sickbed, exploitation in the factory and on the plantation, incarceration in the cell, decomposition in the morgue, and refiguration in the museum. And all of these practices and experiences, along with totalizing doctrines (like civilization, evolution, science, and even the mind/body distinction itself) continually infect one another and are transmitted from one domain to the other. This "mixing" not only constitutes subjects with bodies, who have unique historically specific experiences, but it also produces "mixed up" people who are defined (and define themselves) in noncategorical categories of racial, class, gender mixtures, and who travel through, and inhabit, multiple geographical, hierarchical, and cultural zones. Does this seem confusing?

So, finally, as might seem obvious from the foregoing, here's a confession. Maybe I've lost faith in my pact with the body. The body has broken to pieces. What's left is a junk heap—out of which very talented scholars have made very pretty arrangements. Scholars originally promoted the body as an agenda for social, political, and cultural change (reform, revolution). That was ambitious, but, to be honest, body scholarship has tended to bear only indirectly on the great issues of our time. Poor body!—it's a category of displaced political analysis and struggle. We looked and looked at the body until we saw the universe: congeries of powers, relations, performances, narratives, figures, structures, and economies—an infinite regression of effigies, not coherent, and maybe not even subversive, but compelling. The field of scholarship is now littered with productive remains for readers and scholars to pick over. This volume contains some good bits. Enjoy.

And don't worry! Approaches and theories come and go, but my crystal ball tells me that the body won't, because it is nurtured by social and cultural obsessions that won't decline in the foreseeable future. If anything, they will proliferate. And so will body scholarship, which has already invaded every disciplinary domain in the humanities. Last word: The body won't decompose into the surrounding soil, because someone will always keep digging it up.

CHAPTER ONE

Birth and Death under the Sign of Thomas Malthus

THOMAS LAQUEUR AND LISA CODY

It may be true, as the caustic American wit Dorothy Parker supposedly remarked when she heard about the death of a friend, that all sorts of people are dying who never died before. It was not true—in general—about Europe and the United States in the nineteenth century. All sorts of people, especially children, were *not* dying who would have died before. And all sorts of people were not being born who would have been born before. In some measure this has to do with decline in mortality: lower infant mortality means longer intervals between births because women breast-feed longer. But by the end of the nineteenth century—and earlier in some places—something deep and mysterious in the human psyche had come into play. Couples, each making a decision for themselves, had fewer children and had them over a more limited span of a woman's life because of voluntary birth control. Sex became divorced from reproduction as never before; the great demographic transition was begun and largely completed. In short, the facts of life and death changed dramatically and probably irrevocably in the long century between the French Revolution and World War I.

That the facts of life and death could change at all was unimaginable at the beginning of the nineteenth century, for it seemed that the Malthusian trap—the perennial power of death to check population and the rambunctious sexual behavior that caused it—was a biological law, as impossible to change as Newton's laws of gravity. The Reverend Thomas Robert Malthus, author of *An Essay on the Principle of Population* (1798), who has come down to us as a hard-hearted flint of a man, distilled the essence of demographics to a neat arithmetic ratio: while population could grow "geometrically," resources could only expand "arithmetically." Whether this was really true or just a theoretical possibility, and whether population was in fact pushing on resources to the extent he believed, was debated at the time and has been debated by historians since. Malthus's trap was articulated just when the threat of mortality came to be less fearsome in Western Europe. But his rhetoric was very much of his age. Malthus did not primarily cast the battle between birth and resources, sex and death, in coldly rational equations, but in highly Romantic florid descriptions that constantly and poetically linked life and death.

To some degree the connection seemed to be grounded in indisputable facts: in 1800, it still seemed that when populations grew too speedily they indeed were checked by famine, disease, warfare, infanticide, or morally unacceptable human interventions like birth control. But more than these, demographic evidence forged the link between the sexual and the mortal body. There were deep cultural ties at the institutional, the cosmic, and many intermediate levels between birth and death, or, more precisely before the late eighteenth century, between baptism and burial. Midwives, for example, were generally licensed by the Church to play their role in bringing new life into the world, but they also often helped to prepare the dead to their final rest by washing and laying out the corpses of the deceased. The words themselves—baptism and burial—intertwine at the institutional and cosmic roots of their connection. Baptism, the Christian ritual of rebirth from the death of original sin, has obscure roots in the Old Testament Jewish connection between leaving behind wickedness and becoming a new creature (2 Kings 14) and probably in practices nearer to the time of Jesus as well. It is intimately tied with the coming of death into the world through the Original Sin of Adam and Eve and especially in the writings of St. Paul, through the death of Jesus, whose blood offered life to believers. The rituals of death, overseen by the Church, signal the end of one life and the beginning of another.

Judaism did not make the ritual connections between birth and death quite so explicit, but they were there; they were evident, for example, in the whole elaborate system of cleanliness and pollution that is grounded in the primal profound pollution of death that spreads in an attenuated penumbra to sexual fluids and acts—semen, menstrual blood, intercourse. Birth and death lived at home together in 1800, as they had before, and the romantics made much of their familial intimacy. Mary Shelley picnicked at the churchyard grave of her mother, the feminist Mary Wollstonecraft, who died of childbed fever days after giving birth to Mary in 1797.

A long century later, certainly by 1920, both birth and death had been largely removed from the rhythms of age-old practices and the cultural constellation in which they had been embedded. A 1902 British act regulating exactly what midwives could and could not do neatly encapsulated the twentieth-century shift: whereas midwives had once attended the dead and prepared them for burial, modern birth attendants were expressly forbidden from contact with the dead. There were reasonable public health principles at work here—commuting between cadavers and childbearing women spread microbes, as had been proven by the end of the nineteenth century—but this legislation also epitomized how death and birth were severed from each other, no longer ineluctably linked in religious rituals or in Malthusian math and rhetoric.

Global population would explode in the twentieth century, calling into question the reverend's cold-blooded equation: perhaps birth was not checked by the specter of death after all. But breaking that equation in empirical terms also relied on numerous technological, scientific, and cultural changes that would "medicalize" birth and death and shift intellectual authority over the beginning and end of life from priests and communities to experts. If such phenomena as the *global* population explosion and the extraordinarily low rates of maternal and newborn death rates in Western, industrialized cultures did not play out until the post–World War II period, the groundwork for these extraordinary developments were laid in the long nineteenth century from the era of the French Revolution to the end of World War I.

All countries in Europe experienced these changes to a greater or lesser extent and with considerable regional and class differences—this in itself became a subject of study as we shall see—but the overall quantitative trend is unmistakable. At the beginning of the century in Sweden, for example, 24.4 per 1,000 inhabitants died in any given year; by 1900 the number was only 15.5 per 1,000. During the same period mortality dropped in England from 27.1 per 1,000 to 16.1; in France from 30.1 per 1,000 to 19.6; and in Germany from 25.8 per 1,000 to 19.5. Life expectancy increased as well due largely to a drop in infant mortality. In Sweden, where records are especially reliable, a baby born in 1800 could expect to live only to his or her mid-thirties—36.5 years to be exact; by 1900 it was 54.00 years. Comparable figures in England are 36.9 years in 1800 to 48.2 years a century later—the average was dragged down by a rise in infant mortality in the new, rapidly growing cities of the Industrial Revolution and its penumbra. In France life expectancy at birth increased from 33.9 to 47.4 years. Even in backward Russia the terrifying prospect of dying, on average at age 24.4 in 1800—high even by old regime standards—became a little less terrifying at 32.4 years by 1900.[1]

To catch the full impact of the birth control and fertility revolution generally, we need to look a little beyond 1900 to 1910. Change is dramatic; in England the average number of children per woman—the total fertility rate—went

FIGURE 1.1: Childhood Mortality: Interior of a Nursery with Eight babies, Watched by Death Who Opens the Window from Outside and Reveals an Industrial Cityscape. Color lithograph by Alice Dick Dumas, ca. 1918. Wellcome Library (L0029945)

from 5.55 in 1800 to 2.84 in 1910; in Sweden from 4.27 to 3.31. In France, where fertility started to fall earlier than anywhere else in Europe, total fertility was only 3.38 in 1850 and 2.25 in 1910.[2]

But more than the facts changed. Birth and death and their relationship to each other came to be understood in radically new ways in the twentieth century. In certain realms—we will concentrate on political economy, demography, and epidemiology—they became newly joined in ways in which they had not been associated before. During the century that witnessed the birth of biology, the science of life—the word appeared nearly simultaneously in French with Jean-Baptiste Lamarck and in German with Gottfried Reinhold Treviranus and Lorenz Oken in 1802 (1813 in English)[3]—death and birth moved onto their own separate and distinct trajectories. They took on new and differentiated cultural work.

The year 1792 marks a critical rupture: baptisms and burials first became birth and death. In that year the new French Republic took vital statistics out of the control of the Church and made them a municipal responsibility, a matter of civil rather than religious record keeping. German states under French influence soon followed suit: 1792 in Rhineland, 1803 in Hesse-Nassau, 1808 in Westphalia and the Duchy of Warsaw, and 1809 in Hanover; in various parts of Italy under Napoleon the shift started in 1806. Civil registration of births, deaths, and marriages started in 1837 in Britain and in 1856 in some

Australian territories. Though some other European and colonial states did not take over the recording of births and deaths until considerably later in the century—not until 1867 in Canada, 1871 in Spain, and 1876 in Switzerland, for instance—people in the long nineteenth century tabulated birth and death in a way never before seen. Of course, religious bodies did not all of a sudden stop being the venue for rituals of passage; they continued to provide both the place and the ceremony to mark the social beginning and end of life. But words—baptism and burial—that had been about sin and the care of souls changed. Birth and death—biological terms that entered the language of the state in the interests of society and its improvement—came to describe our subject.[4]

This shift was felt well beyond the walls of registry offices. Even if places of burial continued to be mostly on sanctified ground—and, for the first time in almost two millennia, considerable and very public portions of it were not—they were regulated by state authorities and not by Church authorities. All over Europe the dead entered the administrative world of trash, water, and other waste. They entered the domain of new experts on what to do with matter. One of the very few realms in which the founder of modern Reform Judaism, Moses Mendelssohn, argued that civil law took precedence over Jewish law was in the matter of burial. It was traditional among Jews to put their dead into the ground within twenty-four hours of death, but it was in the interest of the Prussian state that bodies be kept in special "corpse houses" for three days to be sure that they were really and not only apparently dead. In other words, a novel anxiety about premature burial that came, within decades, to be addressed by secular legislation trumped several millennia of concern about the pollution of the living by the dead.

Of course death and to a lesser extent birth retained a hold on religious thinking and pastoral practice throughout the nineteenth century, but the genuinely novel development was the spectacular growth—technically, institutionally, and politically—of new disciplines that had as their subject matter birth and, even more prominently, death: medicine, public health and epidemiology, political economy, and demography. It has become almost a commonplace to regard these as the matrix for what the French philosopher Michel Foucault called "bio-power," the rise of new forms of knowledge in which individual human life was embedded and construed in such a way that individuals seemed to govern themselves in compliance with universal principles instead of being governed by sovereign authority imposed from outside. Well over a century before Foucault, William Hale, archdeacon of London in the 1840s and 1850s, captured some of the force of this claim when he argued that the civil registration of births and deaths construed as biological events—as opposed to the ecclesiastical recording of rituals (baptisms and burials)—was tantamount to giving the police, that is, the state, license to enter the most intimate moments and spaces of family life. By simply complying with seemingly neutral regulations, every man, woman, and child became imbricated in new, powerful fields of knowledge that were overseen by new cadres of trained professionals.

Of course life and death came to be connected in ways that went far beyond newly exigent fields of knowledge. The old cosmic connections were reconfigured in more secular language. William Wordsworth's 1798 poem "We Are Seven" exemplifies this shift. Here the poet's interlocutor insists on a community of the dead and the living—a family—that is not mediated by prayers for the dead but by an acute sense that the community of the here and now includes the dead imaginatively.

> "But they are dead; those two are dead!
> Their spirits are in heaven!"
> 'Twas throwing words away; for still
> The little Maid would have her will,
> And said, "Nay, we are seven!"[5]

But the text that signaled most clearly a reconfiguring of the relationship between life and death was Thomas Robert Malthus's *An Essay on the Principle of Population*. It is emblematic of how "age of life" was just as much—perhaps even more—an age of death. Unlike older views in which the size of the population was a sign of social health and fertility, from Malthus onward, it was a sign of danger and death as well. Death, and especially the death of children, was the reality principle on which late Enlightenment utopian hopes were dashed for Malthus and even for successive generations of Europeans, well into the twentieth century. We are arguing, perhaps paradoxically, that though life expectancy increased and it became increasingly safer to give birth and to be born in the nineteenth century, birth and death continued to preoccupy the Western mentalité, but in profoundly new ways.

The new sciences of life—and perhaps even more importantly of death—were the political weapons not only of the powerful but of the powerless as well, of workers and not just capitalists, of feminists and not just their opponents. Birth and death did different cultural work in the nineteenth century, but they did it in considerable measure—directly or indirectly—through the new disciplines that took them as their subject.

BIRTH AND NINETEENTH-CENTURY CULTURE

Few texts epitomize the collapse of Enlightenment ideals and the ascent of modernity as dramatically as Mary Shelley's *Frankenstein; or the New Prometheus* (1818). The allegorical tale of a scientist who wishes to create a new race of men, but manufactures a monster instead, was seen by Shelley's father

William Godwin (and others since) as symbolizing the violent, unnatural tendencies of industrialization. But Shelley's Prometheus was and has since been understood as also exploring the darkest fantasies and fears related to birth and creation. Before Dr. Frankenstein, few male characters were endowed with the power to make life, and in Shelley's horrifying tragedy, this new form of unnatural, technological birth was fundamentally intertwined with death as Dr. Frankenstein plundered the bones and organs of a charnel house to create his lonely creature, who in turn would fail to reproduce. Shelley's own sad history of pregnancy—her mother, Mary Wollstonecraft, died only ten days after giving birth to Shelley, and Shelley had her first late-term miscarriage in 1817—surely helped to inspire her morbid tale, but she converted these private female tragedies into a gothic parable for the nineteenth century. The gloomy novel, like Malthus's 1798 *Essay*, braided birth and death together, but more so than Malthus's foundational text in the history of political economy, *Frankenstein* would capture the distinctly ambivalent feelings many in the nineteenth century had about the costs of creation, birth, and parenthood, and the seemingly destructive aspects of what would later be viewed as the rise of male reproductive technologies and expertise.

On the surface, it might appear that Shelley and Malthus ran counter to dominant nineteenth-century cultural attitudes about birth, including the growing and glowing Anglo-American adoration of childhood and motherhood, categories that had not been eulogized so fervently and publicly in times past. Few would have expressed it so fervently before the long nineteenth century as the prolific family-values maven Mrs. Sarah Stickney Ellis who explained, "in the maternal bosom . . . is lodged an instinct stronger than any other which is associated with animal existence"[6] or little-known California diarist Georgiana Bruce Kirby, who reminded herself, "Since I was a girl of 18 I have been ever conscious of the most intense desire to become a mother."[7]

Yet, despite such exclamations, these women and others expressed rather more mixed attitudes about childbearing. Kirby admitted, "It is not that I am especially fond of little babies,"[8] and Mrs. Ellis did not even have children herself. More experienced mothers similarly dismissed raptures to birth and motherhood. Queen Victoria told her daughter Vicky: "What you say of the pride of giving life to an immortal soul is very fine, dear, but I own I cannot enter into that; I think much more of our being like a cow or a dog at such moments; when our poor nature becomes so very animal and unecstatic."[9] The result of birth was hardly compensation according to Queen Victoria: "An ugly baby is a very nasty object—and the prettiest is frightful when undressed—till about four months; in short as long as they have their long body and little limbs and that terrible frog-like action."[10] In other words, "a mother's love," whatever that might be according to Mrs. Ellis, was actually a more ambiguous and ambivalent state than some historians have claimed for the period.[11]

Perhaps Mrs. Ellis and Princess Vicky laid it on thick when pondering motherhood because they had comparatively little experience with the business of birth. Like other nineteenth-century elite women, Mrs. Ellis and the other reproductive romantics had few personal occasions to attend childbirths (other than their own, if that) or be involved in the actual management of pregnancy and childcare as had their great-grandmothers in the eighteenth century and earlier. Before the mid-eighteenth century, all women—even royalty—attended other women's births and dispensed copious advice about childbearing until the professional ascent of the male obstetrician throughout the eighteenth century in France, Britain, and North America challenged these customs. Though male practitioners never entirely took over the delivery of children in Europe (as they would in the United States by 1920),[12] they entirely reshaped the routine culture of childbirth for middle-class and elite women in the West, including discouraging women from attending any births but their own. As Dr. P. B. Saur admonished mothers in 1887, "one lady friend, besides the doctors and the nurse, is all that's needed . . . All chatterers, croakers, and putterers ought, at these times, to be carefully excluded from the room."[13] Giving birth no longer was a communal family affair but a medicalized one with a doctor at the helm, and after the 1840s, with the use of ether and chloroform, the laboring mother was barely present herself.

Because they were entering a traditionally feminine and domestic occupation, early men-midwives, as they were called in Britain and North America through the 1810s, had a hard sell not only to the public but also to their medical brethren. They consequently promoted themselves as amalgamating the feminine traits of tenderness and patience with the natural advantages of masculine intellect and ingenuity, and providing transformative tools such as forceps and anesthesia.[14] With the use of ether, chloroform, and other anesthesia in labor beginning in the 1840s, medical men associated themselves with erasing the physical pain of labor, a tremendous innovation that was at the time viewed as liberating by women (and potentially sinful by theologians, until Queen Victoria's physician publicized their use of the drug in 1853). By the 1910s, feminists clamored for "twilight sleep," a potent mix of morphine and scopolamine, an amnesiac, concocted by German and Austrian obstetricians. Unlike chloroform, which put a laboring mother to sleep, the scopolamine-based anesthesia kept women physically involved in their labors—indeed so violently and loudly that they were locked in padded delivery rooms—but would block all memories of the experience. Twenty-first-century feminists likely find it ironic that the leading advocates of this intensely disassociated form of childbirth in the 1910s were progressive advocates for a woman's right to control her body and medical care. Nonetheless, it is in part thanks to the work of the early *women's* movement that the ideal for giving birth after 1920 in the United States and Europe was entirely transformed: rather than be consciously present at their own children's birth, women were better off sleeping or forgetting. Puerperal amnesia became the twentieth-century gold standard in North American hospitals in particular, so much so that "natural labor" would be considered a radical and rare choice even as late as the 1960s for most U.S. mothers.[15]

Before 1800, at least 97 percent of deliveries were normal without complications or need for medical intervention; of the remaining 3 percent, most mothers would survive. There were a handful of desperate surgical techniques that saved mothers' lives (and once in a while a newborn) before the use of the controversial obstetrical forceps in the eighteenth century; it was not until the 1880s that medicine could save lives of both mothers and newborns through a Cesarean section. By then, a variety of surgical innovations gave mothers a four out of five chance of surviving a surgical delivery. By twenty-first-century Western standards, a 20 percent risk of death may seem appalling, but the possibility of triumphing over many reproductive complications altered the cultural sense that birth invariably led mothers to contemplate their mortality. The higher likelihood of surviving labor—and delivering a live child—thanks to nineteenth-century surgical advances helped to decouple birth from death in Westerners' imaginations. But by positing the solution to reproductive tragedy as the purview of the obstetrician, the process also "medicalized childbirth," a transformation epitomized by the piece of 1893 legislation in France that defined childbirth as an illness.[16]

As feminists have emphasized, labeling birth an illness best managed by men altered the traditional cultural beliefs surrounding human reproduction as a normal (if sometimes dangerous), feminine, communal event; but doctors' management of birth could have fatal consequences, too. Mothers faced the risk of postpartum infections, but in the early modern world (or the nineteenth-century rural world), where midwives typically spent hours or even days with each client at her delivery, the risk of cross-contamination from other sick or dead mothers was relatively low. Puerperal fever, the gruesome septicemic infection contracted most commonly from the *Streptococcus pyogenes* bacterium, results in abdominal swelling, intense pain, hallucinations, palpitations, and often death depending on the particular bacterial strain. Raving, drooling, dying mothers were first seen in epidemic numbers at the crowded seventeenth- and eighteenth-century maternity wards at the Hôtel Dieu in Paris, but such a ghoulish form of infection and death did not become widespread until the nineteenth century. Both in slums and in hospitals, puerperal fever became a regular epidemic for urban, poor women. At La Maternité in Paris, 180 of every 1,000 mothers who gave birth in the early 1860s died of infections. Ignaz Semmelweis, who charted maternal postpartum death rates in the 1840s at the Vienna General Hospital, recognized that the problem was with male practitioners spreading germs from the dead to the delivering: the obstetricians' wards saw a 16 percent maternal fatality rate compared to the 2 percent on the wards managed by female midwives.[17] His colleagues questioned his arguments and those of another hand-washing obstetrician, American Oliver Wendell Holmes, for another thirty years. Despite the triumph of germ theory in the late nineteenth century, maternity hospital rates of sepsis remained high—typically 10 percent or more—simply because hand washing and hygiene were not scrupulously maintained in hospitals or elsewhere. But even when giving birth at home, nineteenth-century women suffered from historically higher rates of infection than earlier generations, and the patterns could be surprising. According to Dr. Fordyce Baker in 1873, wealthy women in New York City contracted puerperal fever twice as often as poorer mothers, and home deliveries had four times the rate of infection compared with hospital deliveries.[18]

While the growing authority of the male obstetrician altered the nineteenth-century culture of birth, broader changes outside the clinic also altered the place of reproduction in peoples' daily lives, including, for instance, the architectural trend to favor private bedrooms separated from living spaces in nineteenth-century homes. Birth was split from family and community life with the social, economic, and spatial divisions of nineteenth-century urban industrial capitalism and the rise of privacy and the private home in western European and North American cities. This is paradoxical: in a bourgeois world that greatly valued the home and the sanctity of the family—at least in ideal—we find that the bodily and emotional experiences of birth were severed from family and community time. Nineteenth-century British women, for instance, who could financially afford to do, so kept in seclusion while pregnant, and even avoided naming their condition.[19] Compared with the British, the French were considerably more accepting of such reproductive mishaps as out-of-wedlock pregnancies, at least among the working class, but even the nineteenth-century French increasingly treated pregnancy as a delicate and very private condition. By the turn of the twentieth century onward, cultural critics viewed nineteenth-century silence about reproduction and childbirth as peculiar and perverted. Critics, including Sigmund Freud, such early twentieth-century "anti-Victorians" as Havelock Ellis, Lytton Strachey, and the novelist Samuel Butler, and others, considered nineteenth-century attitudes about birth and parenting as unnaturally repressed with potentially damaging psychological consequences.[20] Victorians may not have been as repressed as their Edwardian grandchildren claimed, but they did make reproduction a quite different category than had Europeans before 1800.

Obstetrics is perhaps the most obvious field that transported childbirth from the routine life of the home into the realm of science—the transformation that *Frankenstein* foreshadows. But other contemporary intellectuals and professionals transformed the category of birth and decoupled it from maternal experience. François Jacob and Michel Foucault argue that a fundamental epistemological shift occurred near the time of the French Revolution: "life" was discovered by the natural philosophers and social scientists. Carl Linnaeus and other taxonomists of the eighteenth century had neatly arranged living creatures as objects frozen in time, but zoologists, botanists, and comparative anatomists from the 1780s forward, including Georges-Louis Leclerc, the French naturalist known generally as the Comte de Buffon, Charles Darwin, and the temporarily forgotten monk and gardener of peas, Gregor Mendel, instead treated different creatures as moving, mutating, and evolving entities, in which reproduction was the switch-point for potential alterations. Less widely known at the time and since, there were also the scientists we would now call biochemists whose research illuminated how birth, heredity, and evolutionary change occurred at the most minute organic, chemical levels. In 1869, for example, Swiss physician Friedrich Miescher isolated DNA "from the pus-soaked bandages of German soldiers," postulating by the 1890s that what we now know as DNA contained a hereditary code "just as the words and concepts of all languages can find expression in 24–30 letters of the alphabet."[21]

The nineteenth-century growth of university science departments facilitating research in comparative anatomy, organic chemistry, biology, geology, medicine, embryology, and more led to discoveries that fundamentally altered how the educated viewed birth. For instance, Karl Ernst von Baer discovered an ovum in a canine bitch in 1827 and subsequently described the role of ovulation in conception. From the 1890s onward, European and American researchers found that reproduction and birth (as well as all other life functions) were orchestrated biochemically in the body through the endocrine system. Sexual lust, "mother love," and other deeply felt aspects of reproduction and birth were in fact dictated by a complex network of hormones. Reproduction—the very term marks a shift from the earlier word "generation"—was in fact viewed as no longer essentially the function of a thinking feeling human being; the truths of the birthing body were abstract, biochemical, and explicable only through the sciences. The discovery of X-rays—or Röntgen-rays as they were also known (named after physicist Wilhelm Röntgen who first realized that the light from a cathode-ray tube could penetrate and reveal the bones in his hand against the opposite wall in 1895)—enabled obstetricians to peer inside the pregnant body and bring the unborn to light with a previously unimaginable clarity and truth by the 1920s.

Malthus's arguments provided another paradigm shift in regards to reproduction. As pointed out earlier, until his blockbuster 1798 *Essay*, Europeans generally applauded a booming domestic population as a very good thing, the sign of a healthy nation, and the means to wealth and plenty. Malthus dramatically argued the opposite. He wrote in the midst of agricultural, economic, industrial, and international dislocation at the beginning of the nineteenth century, but in the coming decades his British acolytes focused particularly on the supposedly immiserating effects of state-supported welfare to the poor.

It was not just that Malthus inspired others to view birth negatively, but that political economists and bureaucrats transformed the category into an object, a statistic that could be tabulated, surveyed, studied, and perhaps even altered in its patterns. By 1830, when the liberal Whigs came to political power in England, they promised not only to expand the franchise but also to clean up old-regime, irrational institutions that made no political economic sense, including the Poor Law.

The "Old" Poor Law—distinguished from the "New" Poor Law with Malthusian thinking at its core—had allowed unmarried mothers to ask their parishes to pursue fathers and provide outdoor financial assistance, that is, aid to the poor given outside the confines of the workhouse. This changed thanks to the New Poor Law's authors who proposed all should be confined if they were to be assisted. Given the costs of supporting ever-growing numbers of fatherless children in the nineteenth century, local taxpayers despised bastard bearers and their offspring. The Whig point of view convinced much of the public, frustrated with the apparent growth of poverty and paupers, that the apparently free ride given mothers led to new mouths to feed, for which there would ultimately be no food. Reformers pushed through a hotly debated amendment that mothers could no longer name fathers or expect their parish to give them direct financial assistance. Instead, single mothers would either have to enter a workhouse or support their offspring on their own. The legislation passed—not without public outrage, as critics and nervous reformers wondered whether the new policy would lead to abortions, infanticide, suicide, and other misery. (Many contemporaries also viewed these single mothers as seduced victims.)

Not only should welfare recipients be prevented from reproducing themselves but so should lunatics and the ill. Before birth control would come out of the closet later in the nineteenth century with a heterogeneous and unexpected alliance of feminists, neo-Malthusians, eugenicists, medical men, and out-and-out racists who all aimed to lower the birth rate, bureaucrats and politicians of the 1830s hoped to do the same through welfare reform and architecture. The 1834 New Poor Law in England that revoked the ability of single mothers to obtain parish or paternal support for their bastards also helped to establish the specter of the Victorian workhouse with its sex- and age-segregated rooms, designed to prevent sexual relations (or any other form of pleasure). From 1834 onward working-class organizers invoked the notorious "bastardy clauses" as evidence that the propertied middle class had joined forces with elites to squash society's weakest members, and that the state wished to eliminate the reproduction of the imperfect and the impoverished.[22]

Although the bastardy clauses were motivated by liberals who did not want to pay for other peoples' premarital recklessness, quite another Malthusian movement gained steam in the nineteenth century: the campaign for birth control and family planning. Malthus himself thought artificial birth control was immoral; nineteenth-century couples may have publicly agreed with the reverend, but privately they took measures to limit their fertility. In France, for instance, despite a declining age of marriage across the period, "the birthrate declined from 32.9 per thousand in 1800 to 19 per thousand in 1910,"[23] and similar changes were seen in middle-class families across northern Europe and North America. Historians have attributed the decline to couples using various rhythm methods as well as condoms and caps, which, with the vulcanization of rubber in the 1840s, became more widely (if quietly) available and inexpensive. Abortion was also resorted to quite frequently, despite its universal illegality in the nineteenth century and condemnation from churches, doctors, and even some feminists. The technological causes behind a declining birthrate continue to inspire historical debate, but the economic and personal reasons for limiting family size are fairly straightforward.

The nineteenth-century feminist movement, unlike the earlier Romantic maternalist feminism of Mary Wollstonecraft and her peers, did not predicate its arguments on women's rights emanating from their childbearing capacities. That said, many supporters brought attention to the plight of mothers: George Eliot intended that *Adam Bede* (1859) inspire sympathy for single mothers. The aristocrat Caroline Norton, who lost custody of her toddlers in a divorce, campaigned successfully to gain modest custody rights for separated mothers. American Susan B. Anthony campaigned to outlaw abortion, because she viewed its main proponents to be adulterous men who did not wish to take responsibility for their acts of sexual seduction.

More novel and significant for the modern age than these incidents, however, was the feminists' position that maternity was unnecessary, and in some cases, an absolute negative. Advocates for female higher education and professionalization typically viewed childbearing and motherhood as permanently stunting women's intellectual and professional ambitions. Indeed, many of the most accomplished women of the long nineteenth century, from Jane Austen to Virginia Woolf, from Russian communist-feminist Alexandra Kollontai to radical Emma Goldman, did not themselves give birth. Others who did often delayed childbearing, as in the case of American physician and scientist Mary Putnam Jacobi, or handed their children off to servants or other relations. And yet others, including Margaret Sanger, the American birth control advocate, did give birth, but lost one of her children and recognized the toll that eighteen pregnancies had taken on her own mother. Her procontraception slogan that each woman should be "the absolute mistress of her body" (1914) became a cornerstone principle of post-Victorian feminism. Birth clearly was no longer conceivable as the inevitable lot of womankind. From feminists arguing that reproduction and motherhood limited women's opportunities to politicians and bureaucrats attempting to limit the birthrate of the poor, nineteenth-century

Europeans no longer viewed reproduction as an unquestionable good and no longer believed in the right of all to reproduce themselves.

Birth control, even in its rudimentary and not entirely effective forms of condoms and diaphragms and sponges, might be considered the most stunning technological and cultural breakthrough of the nineteenth century. Although the innovations usually attributed with giving birth to the modern age—from bills of rights to steam engines—harnessed and amplified power that already existed, birth control did something entirely new. It decisively divorced sexual pleasure from the burdens of reproduction and in so doing gave human beings a power over their bodies and economic lives that had never existed in human history before. The fact that women and men could physically enjoy themselves without the specter of birth disturbed cultural life irrevocably. Not only did all Christian theologians in 1900 condemn birth control for freeing sex from the entire basis of marriage and the family, but in 1929 Walter Lippmann described birth control as the greatest moral question of the day. Not only would birth control allow people to decide never to reproduce, but they might also extend their hedonistic worldview to all arenas of life and never fear the costs of pleasure of any sort. In Lippmann's age, influenced heavily by the psychoanalytic obsession with sex and violence in the wake of World War I, birth control generated disturbing questions about the manipulation of human life. Preventing life mechanically was a new moral problem to be sure, but it was simply the inverse of the one Mary Shelley raised when Dr. Frankenstein mechanically created life.

It may be said that the three inevitables were birth, death, and taxes, but nineteenth-century women and men, from Malthus forward, did their best to limit the extent and visibility of the first as much as possible. (There were of course pronatalist movements, to be sure, but the most powerful were in France, where family limitation had begun first and had the greatest impact on population.) Death, on the other hand, loomed large in the nineteenth-century imagination; it was far from hidden. But like birth, it underwent profound revolutionary changes in its demographic patterns and cultural meanings.

DEATH

Death, dead bodies, mourning, and memory are so thoroughly intertwined with the history of nineteenth-century personal, familial, public, and national life that it is misleading to tease out only one thread; it is the whole fabric that matters. Our large claim is that the dead body gained new meaning and power by appropriating a millennia-old history and by inserting that history in new ways of understanding. But we can only see how this works in particular instances. We can begin with what may be a universal fact: a dead human body is never *just* trash, and if it is regarded as such, that too has enormous significance. People make culture through what they do with, and how they regard, the dead. These facts mean that the dead body of a miner after an explosion meant something different in 1820, when the causal sequence of neglect and greed that produced the explosion that killed him was recognized, than it did a century earlier when such events were regarded as acts of God. More generally, epidemiology was predicated on the claim that death was "a social disease" with social and not only divine or idiosyncratic causes. The figure of a reactionary king watching respectfully as the body of a dead worker is borne through the streets of Berlin after a day of revolutionary fighting in 1848 has political power. New places for the dead—ordinary urban or suburban cemeteries and national sacred sites like the Pantheon in Paris or Vysehrad in Prague—created and made manifest new sorts of communities. New rituals—the elaborate bourgeois funeral of the nineteenth century—produced the pauper funeral that became a marker of class division. Mass literacy and universal education meant that death and the poetry of death was on almost every lip: Thomas Gray's "Elegy Written in a Country Church Yard" (1751) was the most reprinted poem of the eighteenth century right through to the early twentieth; Felicia Heman's "Casabianca," popularly known by its first line, "the boy stood on the burning deck" (1826), about a cabin boy who died with Admiral Nelson at the Battle of Trafalgar was the most used standard recitation for working-class schoolchildren in state schools. Death was in the fabric of nineteenth-century culture, and we can only gesture toward certain of its central features.

Epidemiology—the study of the incidence and distribution of diseases—is not just about dead bodies, nor is it strictly a nineteenth-century branch of knowledge. But as a matter of fact, from the earliest efforts to create a geography of illness and mishap in the late seventeenth century, death was its de facto subject, if only because it was the only marker of pathology that was systematically, if imperfectly, collected. And even if a diagnosis was vague and scanty—the classification of disease entities was far from uniform until the late nineteenth century—facts about a body were relatively clear and easy to ascertain: age, sex, employment when alive, former residence, current location, and condition. Malthus had articulated at the theoretically general level the powerful basic assumption of epidemiology: that there are patterns of death that have their origins in social facts. In the *Essay on Population* it was a miscalculation—in some measure caused by obfuscating social policy like the Poor Laws and agricultural subsidies—on the part of couples having intercourse about the possibility of feeding their offspring. Imprudent indulgence in sex resulted in dead children.

Developing out of Enlightenment projects to tabulate patterns of disease and death, epidemiology came into its own in the first decades of the nineteenth

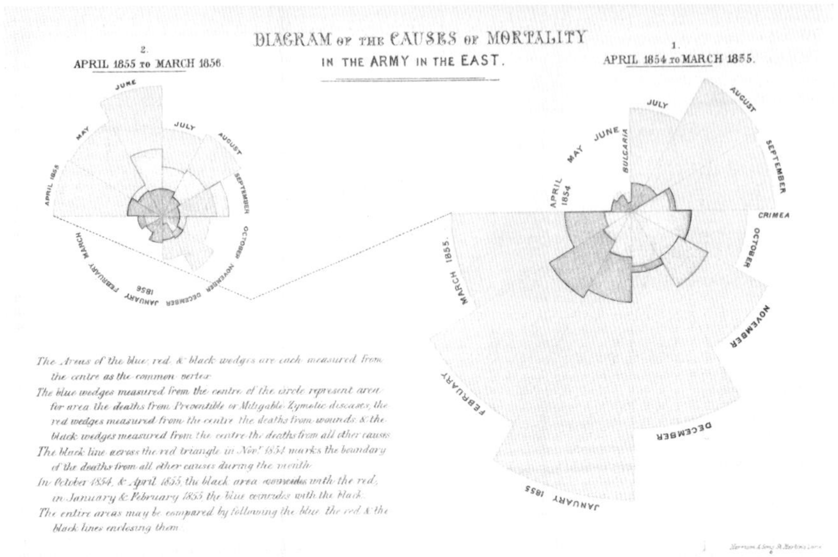

FIGURE 1.2: Two color charts showing the causes of mortality in the army in the East. From Florence Nightingale, *Notes on Matters Affecting the Health, Efficiency, and Hospital Administration of the British Army* (London: Harrison, 1858). Wellcome Library (L0041105)

century.[24] State registration of vital statistics, the efforts of scores of local statistical societies in gathering data, in addition to the findings of innumerable governmental investigations, provided the data for an ever more sophisticated analysis of the geography, sociology, and even pathophysiology of death in ever more contexts. Absolutely crucial to this was the development of statistics and demography as disciplines that created new analytic tools: life tables, general mortality rates, and the advent of such categories as "excess" mortality—the number of deaths beyond what was expected given life expectancy. The utility of each of these could be and was debated, but the intellectual stage was set for a new way of incorporating death into life.[25] Thus, for example, it was shown by Louis René Villarmé, a French doctor and what we would call a sociologist, that astounding variations in mortality between arrondisements in Paris—22.2 per 1,000 in the 1st versus 31.3 per 1,000 in the 12th—and between kinds of habitations—16.1 per 1,000 in buildings with the fewest untaxed rents versus 23.3 per 1,000 in buildings with the most untaxed rents—proved that degree of poverty translated directly into death.[26] Excess infant mortality over national averages in new urban centers became a marker of their uniquely insalubrious qualities and the spur to ameliorative efforts. The connection between impure water and death became clearer even if, as in Hamburg for example, an ideology of radical liberal individualism among the governing class delayed public responsibility for fixing the problem for decades.[27] Disease and, more specifically, mortality associated with specific occupations were studied with ever more precision, and the results of these hundreds of investigations were made known through the professional and lay press. How people died was a matter of public urgency.

The enormous growth of epidemiology and the allied discipline of public health that was meant to use its findings to effect changes in mortality made the dead body less evidence for individual negligence, unstoppable biological processes, or acts of God and ever more evidence for there being something wrong in the social order that could be collectively addressed. Mortality entered politics. Mine explosions, for example, could be explained as the result of specific actions or failures to act and not on unmasterable natural or mysterious supernatural forces: methane that could be detected had not been cleared by adequate ventilation; a fire that had gotten out of hand because mandated trap doors in a passage had not been installed. Indeed—and here is where new studies in pathophysiology come into play: miners, it could now be shown, might not have died of an explosion at all—these were relatively hard to control—but of heat and lack of oxygen that could have been mitigated even if there was an explosion. The dead body of a miner displayed not broken bones or crushed organs but a scorched esophagus. From this sort of evidence in tens of thousands of recorded examples grew political action to change the contours of death on an unprecedented scale.[28]

If, however, one were to identify the most visible, sustained public way in which the dead shaped the culture of the living it would be in the new spaces they came to inhabit beginning precisely in 1804. In that year a magnificent new cemetery opened its doors on the outskirts of Paris—Père Lachaise—and very soon became the world standard for this novel sort of place for the dead in an increasingly urban Europe and its current and former colonies. Unlike the churchyard it was not inextricably linked to a place—a sacred site—but could be anywhere. It was movable; the emperor of Brazil wanted one in Rio; the city fathers of San Francisco were similarly inclined. Cemeteries were parks—romantic, classical, or picturesque in style—designed often by renowned landscape architects. Unlike the churchyard, lumpy from centuries of burial, the cemetery was kempt and clean and large: scores and even hundreds of hectares in extent instead of one or two. Many became tourist attractions and places of technological innovation: the first macadam road in England, for example, was over the Egyptian Catacombs at Highgate in London.

The churchyard was communal property, and no individual had a permanent right to any one space. The graves in the new places of the dead, however, could be purchased freehold or rented on leases of various lengths; many cemeteries, in fact, were run by joint stock companies for a profit, even if some or all of their land was consecrated. Churchyards belonged to the parish; they were intensely local as well as communal. Cemeteries were the cities

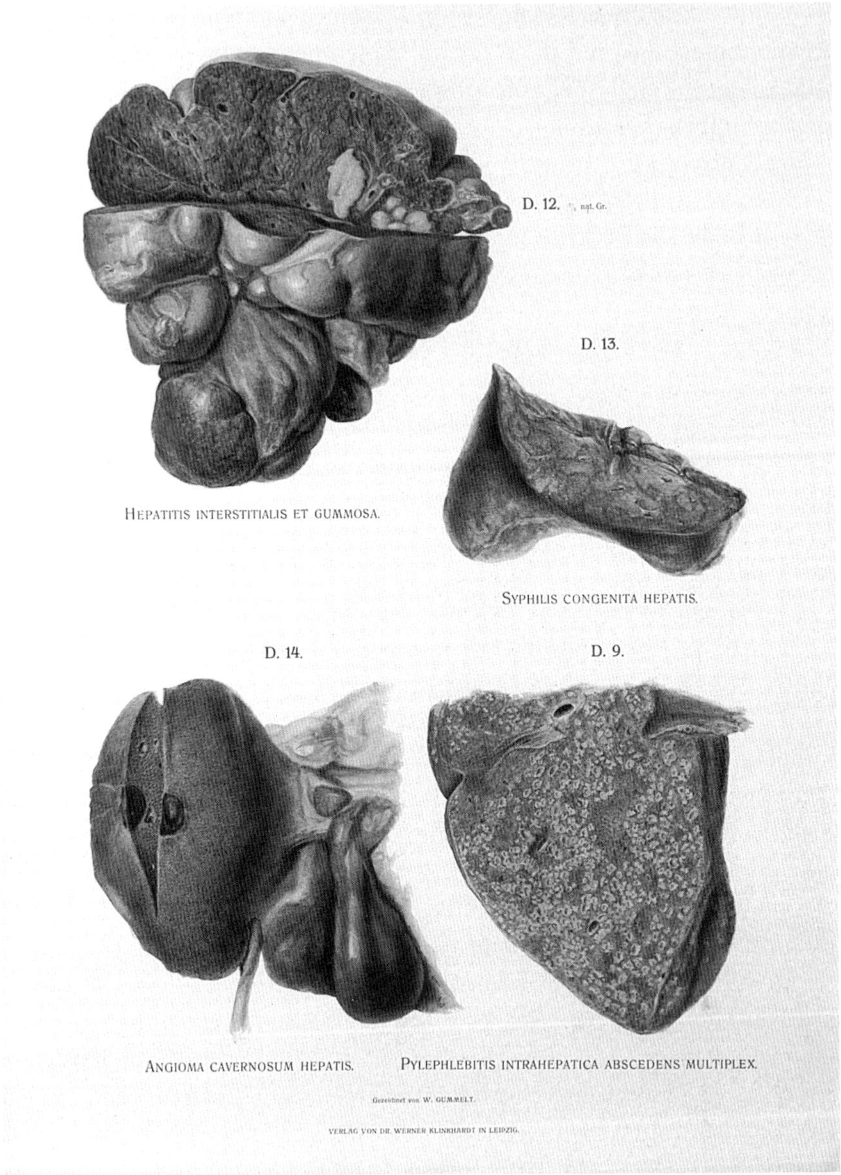

FIGURE 1.3: Chromolithograph illustrating liver disease in dissected specimens caused by hepatitis (top left), congenital syphilis (top right), angioma cavernosum hepatitis (bottom left), and pylephlebitis intrahepatica abscendens multiplex (bottom right). The plates were painted from hospital cases immediately after death. Chromolithograph by W. Gummelt, 1897. Wellcome Library (V0010282EL)

of a far more cosmopolitan dead; at least these were the only ones who were visible. In cemeteries grave markers and mausoleums faced paths, vistas, and broad avenues; in the churchyard graves faced east. Perhaps most impressively, the cemetery was chock-full of monuments of every description and historical style; churchyards were by comparison thinly populated by memorial art. If the churchyard was a place to imagine a congregation of the dead, in principle at least a community bound to a sacred ancient place and controlled by religious authorities, the cemetery was its antithesis: modern, open, secular in its management, and full of strangers.

By the mid-nineteenth century this shift in meaning was completely clear to contemporaries. In the August 1842 issue of *Ainsworth Magazine*, Laman Blanchard, a journalist, editor, and sometime secretary of the Zoological Society, wrote about his visit to Kensal Green and referred to the new "Asylums for the Dead." (That in itself is a telling metaphor in a world in which all sorts of people were being segregated in asylums. Cemeteries, in other words, were places where the dead were safely incarcerated.) Sorrows are soothed and anguish and terror softened by the well-kept garden. This is the sort of landscape before which one can calmly contemplate even one's own death. Crowded churchyards with their bones in the mingled heaps, the bones of one's forefathers gave way to "the pure and exquisite sentiment that should embalm the memory of the dead." Memory replaced bodies that were now decorously hidden. "Change [the capital C is made into an illustration of the gateway of a neoclassical cemetery gate through which a funeral procession is about to

FIGURE 1.4: Cemetery, Happy Valley, Hong Kong. Photograph by John Thomson, ca. 1868. Wellcome Library, London (V0036675)

pass]" writes Blanchard, "so busy in this eventful century with Life—is busier yet with Death."[29]

It was equally true that Death was busy with change. The dead lent their authority to social and religious pluralism—Jews, foreigners, and anyone else who could afford a place could be and were buried in cemeteries, at least in the nonconsecrated parts that every one of these new spaces of the dead provided. They made manifest new national communities: cemeteries in Prague and Budapest gathered together the great artists, intellectuals, and statesmen who collectively represented the nation. National cemeteries—Gettysburg is the earliest and most famous example, but the battlefield cemeteries of World War I soon followed on a much grander scale—took up the same classical, ancient model as had inspired civilian cemeteries to gather dead bodies together so as to create new solidarities. Individual plots met the demands of individual families for individualized memory. All sorts of political parties and voluntary organizations used the new, heterotopic places of the dead to make manifest and further their causes and cultures.[30] The dead worked for the living. Below the ground they became a powerful way of imaging and thus building new worlds above.

Burning dead bodies in the nineteenth century—cremation—did this and more. The national histories of cremation have important differences, but they all have certain features in common that attest to the cultural power of the dead. All begin with the fact that for almost two millennia Christian churches had had charge of dead bodies and that even in the most Protestant countries—places where purgatory was an anathema and saints' bodies in churches a vulgar superstition—the corpse retained something of its religious connotations. Churchyards, especially in the countryside, did not disappear, and even the new cemeteries—radical as they were—still maintained much of the old aesthetic: chapels on the grounds, earth on the coffin, and a body in the ground awaiting resurrection as it had always had. Burial, even in the new century, smacked of its Christian history.

Those who advocated cremation differed in many things, but they shared the hope that burning the dead body would help usher in modernity however they defined it. To some this meant lessening the influence of Christianity in public life. Already in the eighteenth century cremation was linked to a neoclassical aesthetic and a rejection of what was regarded as baroque darkness. In the context of the French Revolution this reversion to the ancients was given legislative form: cremation was explicitly made legal in 1797, and it became illegal in France again in 1804 as part of the Napoleonic rapprochement with the Church. Out of this moment grew the link between burning the dead and anticlericalism. The Church of Rome (and to a lesser extent too the Protestant national churches) understood it for what it was and for the first time developed theological objections that were little more than a cover for the real battle for the care of the dead. And advocates joined the battle. Francesco Crispi, the sometime prime minister of the new secular Italian republic and companion in arms of Garibaldi during the wars of unification, was a great patron of the international cremation movement. His friend Garibaldi's body shuffled around unburied as anticlericals fought to have it cremated as he had requested.[31] Freemasons, the great institutional foe of the Church, were also major proponents of burning the dead.[32]

Anticlericalism was not important everywhere—it played a minor role in England, for example—but cleanliness figured everywhere. Whatever else human remains might be they were dirt and needed to be treated as such. The full radicalism of this view, the ways in which it pushed the dead into entirely new spatial and mental directions, was clear at first only to cremation's bitterest enemies. What one does with the dead is a problem of hygiene, "the art of preserving health"[33] wrote one of the first spokesmen of cremation on an international stage. The simple fact with which Henry Thompson began his 1874 tract speaks worlds: all the rest is commentary. At death, he insisted the body is "instantly consigned to the action of physical and chemical agencies alone," however much we might want to talk about "the unknown and incomparable vital thing that is life."[34]

FIGURE 1.5: An anatomist's dissecting room with a man thought to be dead waking up and sitting up in his coffin. Drawing, 1830s. Wellcome Library, London (L0039194)

Colleagues filled in the many intermediate details. Once one thought of dead bodies as reagents in a laboratory reports—52,000 of them in London alone producing 2,572,580 cubic feet of gas that, if it wasn't absorbed by soil, went into the water supply—the terms of the discussion were clear.[35] The sooner society could rid itself of this odiferous garbage the better. In short, decay in the grave is but slow burning, as a sympathetic clergyman said. Or, alternatively, fire—and for that matter being devoured by wild beasts or being lost at sea—is just a faster way for bodies to be rendered "into their compound atoms," disbursed, and "re-contributed to the vast material world, to an ever living and changing Cosmos, from which they were originally borrowed."[36] What had once been the horrible fate of placeless dissolution would become a culturally resonant gloss on what happens when a dead body is ingested or cremated. But the new reality was that dead bodies were being claimed by sanitary engineers. The dead were like any other detritus.

William Eassie, who became secretary of the newly formed Cremation Society upon its founding in 1874, made his name first in the Crimea where he supervised the building of the Erenköy Hospital. Back home he was well known as the author of what became the standard handbook for building healthy houses, a technical work detailing proper drainage, ventilation, and "kindred subjects." But he thought of himself as engaged in a far grander, world historical project: "at length, the sanitary day is dawning." Humankind will at last be "restrained from committing suicide by setting at defiance all the laws of Health." The preservation of public health, he argued was crucial if we—we British—are "to hold our own in the rapid race for pre-eminence amongst the nations of the earth." "Science is to triumph"; "facts not fancies must be our guides."[37]

How this view translated into culture varied from country to country and region to region. In parts of Germany, for example, the sharp materialist views of hygienists with respect to the body made cremation appealing to the socialist and radical communist working classes. In England, where the labor movement was embedded in a dissenting religious culture, it was not. But however much the specific cultural work of the dead varied their transformation in the imagination of nineteenth-century men and women of all sorts was dramatic. The power of the dead body did not disappear; it was redirected.

Eugenics—the purported science of producing healthy offspring that was born of post-Darwinian hubris in the 1880s reached maturity in the early twentieth century and crashed in Nazi racial policy—brings together our themes of birth and death. Bertrand Russell, surveying the subject in 1929, admitted that there were difficulties with some of the claims of those who sought by positive means—education—or negative ones—sterilization or worse—to increase the number of healthy children in the population. But he remained clear on the principle: religion had been followed blindly for millennia, whereas science was new; if "there is to be tyranny, it is better that it should be scientific."[38] Each act of intercourse and each birth registered on the scale of biological life and death; the reproductive body seemed to hold enormous powers of good and ill. Early marriage, counseled a leading popular guide book published in 1916, "gives the world children totally unfit to struggle with its problems." According to the book, "Statistics show," that children of women marrying early come at shorter intervals and are twice as likely to die as those whose births are better spaced.[39] If the mother was eighteen the risk was even higher. Young parents bred idiocy and death. Pronatalist state policies produced the future and replaced those lost in war. The beginning and the end of life, as they were at the start of our period, are intimately linked but no longer, as they were, through the culturally refracted categories of baptism and burial. They were joined now through biology, social medicine, and the politics of the state and of civil society.

CHAPTER TWO

Medical Perspectives on Health and Disease

MICHAEL WORBOYS

In 1864, George Henry Lewes wrote in the *Cornhill Magazine* that "Few of us, after thirty, can boast of robust health."[1] This observation seems to support the common assumption that in the Victorian era sickness was rife and life short; yet, Lewes was reflecting on the chronic ill health suffered by his middle-class readers who undoubtedly enjoyed better health than most of the population. The oft-quoted observation that life expectancy in urban working-class districts in the 1830s was only around twenty years of age seems to suggest that few of the poor reached the age of thirty, but such a conclusion would be wrong. The figure of twenty years was an average, and this was low because of the statistical impact of the high rates of infant and child mortality. Indeed, the poor who made it to their late teens stood a good chance of a reasonable adult life, and many made it into their sixties and seventies, though usually with a burden of bodily ailments and chronic disease greater than their middle-class counterparts.

Reflections on health and disease in the nineteenth century that foreground chronic ill health contrast with the familiar story of high mortalities due to epidemics and infectious diseases produced by urbanization, industrialization, imperialism, and migration.[2] Historians have long argued that together these forces created susceptible bodies, malign environments, virulent germs, and conditions that facilitated the spread of infections, which took a terrible toll on poorly fed, weak bodies.[3] Although much more is known about industrial capitalist countries, we assume that the situation was as bad, if not worse, in the rest of the world because of the exploitation of peoples and land under imperialism, or because of low levels of technological development and material wealth. Yet, new research has shown that the position of communities across the world was complex and constantly changing, and that there was no typical experience.[4]

Patterns of health, disease, and death varied in every society by age, gender, class, ethnicity, and geography, and over time, but in industrialized countries over the nineteenth century, life expectancy rose across most social groups. Paradoxically, falling death rates (mortality) were accompanied by rising sickness rates (morbidity), even though being sick in the sense of taking to one's bed and having time off work was extremely difficult because of levels of poverty and the limited, if any, welfare provision.[5] Also, then, as now, the "sick role" was shaped by many factors. Nonetheless, people of all classes were not passive in the face of poor health and disease; they used self-help measures and the services of chemists, doctors, and alternative practitioners. Options for prevention and cure depended on resources and opportunities, but ran from nostrums, through whole-life regimens, to medically administered specific treatments.[6] The wealthy sought better health at hydros and spas, with changes of air, and by taking medicines prescribed by physicians; while the poor coped as best they could through better dietaries, self-medication, improved lifestyles, and dispensary services. Moreover, in an age of faith and religious observance, health was something that was prayed for as much as paid for.

MORTALITY AND MORBIDITY

What were the major disease problems of the nineteenth century? Many histories of the period tell us that industrialized countries were riven with major, countrywide epidemics that produced major mortality crises. This picture is very similar to that promoted by the nineteenth-century public health movement and its concern with insanitary conditions and filth diseases. However, the picture was partial then and remains so now.[7] Smallpox epidemics, while frequent, declined in intensity and duration over time, in part because of the spread of vaccination and perhaps because of changes in the virus.[8] Cholera was the shock disease of the century, a new plague spreading from Asia across the world in several pandemics, though again less frequently and with lower virulence on each sweep. The incidence of smallpox and cholera was usually local and sporadic, and although they were frightening diseases, most sufferers survived, albeit often scarred and debilitated. For most populations, most of the time, the main experience of infections was of the short-lived outbreaks of typhoid fever and typhus, food poisoning, and the cyclic toll of childhood infections—measles, scarlet fever, diphtheria, and whooping cough. Such en-

FIGURE 2.1: *A small girl recovering from an illness and being visited by neighbors.* Process print, 1907, after Emma Magnus, 1907. Wellcome Library, London (V0015099)

demic infections were culturally significant because they mainly affected infants and young children, who became more valued and sentimentalized over the century.

Although infections loom large in medical histories of the nineteenth century, at no time were they responsible for most deaths. A recent analysis of mortality in Britain for the period 1848–1872 showed that infections accounted for 33 percent of deaths, and by the decade 1901–1910, infections were responsible for only 20 percent.[9] Infections were certainly the major causes of death in children and young adults—up to 60 percent—but most deaths were in other age groups, and increasingly so as the century went on. These percentage figures count the largest single cause of death—pulmonary tuberculosis, which was responsible for one in seven deaths overall and one in four in adults—as an infectious disease. This was not the view of contemporaries, who regarded it as a chronic, inherited affliction, a view that only began to change toward the end of the century.[10] If we take pulmonary tuberculosis out of mortality figures for infections at mid-century, then catching diseases as contemporaries knew them accounted for around one in five deaths. The overall point is that Victorians would not recognize modern historians' accounts of their experience of disease and death as one dominated by epidemics and infections.

Widening our perspective to the many and varied diseases that Victorians suffered from and died of should help us better understand contemporary views on the determinants of health and disease. Historians are increasingly chronicling the experience and impact of endemic and chronic diseases, and their link to poverty, diet, housing, work conditions, fatigue, and injuries. In this context, it is significant that in Britain, bronchitis—a chronic, inflammatory, sometimes infectious, and ultimately degenerative condition—overtook tuberculosis as the single largest cause of death in the 1870s.[11] Although something of a catch-all diagnostic category, bronchitis was linked to many factors, including damp, dust, smoke, infection, physique, fitness, and degeneration of tissues, and it was experienced as progressively incapacitating. Bronchitis seems to have had just as low a profile in the nineteenth century as it does now among historians, but the death of one of its most famous victims—Karl Marx in 1883—reveals much of the mid-Victorian experience of chronic disease.

Bronchitis was the main cause of Marx's ill health for years and was made worse by the dampness and pollution of London where he lived in a tenement. Marx's poverty has been well-documented; hence, it is perhaps unsurprising that he suffered from many of the conditions that blighted the lives of the working-class poor. Yet, through friends such as Friedrich Engels, he had occasional access to money and to bourgeois lifestyles. These contradictions were evident in his final year, during which he traveled abroad for this health. Marx suffered principally from chronic bronchitis, though his condition was complicated by liver problems, pleurisy, pneumonia, and carbuncles, and it was not helped by his heavy smoking.[12] In the summer of 1882 his doctors sent him to Ventnor on the Isle of Wight, to Algiers, and then to Monte Carlo. His health improved slowly, and he was able to visit his daughter, who lived near Paris and was herself terminally ill with cancer of the bladder. Marx took the opportunity to call at the springs in nearby Enghien-les-Bains, before his doctors sent him to Vevey on Lake Geneva. He improved again but was made to spend the winter of 1882–1883 on the Isle of Wight on a strict regimen. He returned to London in January 1883 after the death of his daughter, still suffering from bronchitis. This was soon complicated by laryngitis and an ulcer in his lung; he rallied for a number of weeks before dying unexpectedly asleep in his armchair at the age of 65. In his *Condition of the Working-Class in England in 1844*, which was first published in English after Marx's death in 1887, Engels dwelled upon the chronic lung diseases that debilitated and killed the urban poor, so it is noteworthy that they also caused the death of one of their champions.

MEDICINE AND HEALTH

Our ability to chart the experience of health and disease in the nineteenth century is due in large measure to the data collected from the 1830s onward by middle-class social reformers.[13] In most countries the so-called statistical movement was headed by politicians, evangelicals, medical professionals, and civil servants, whose aim was to reform the way of life of the urban poor; hence, "public health" for some meant social and political health. Reformers were trying to understand the new social order and what they saw as the multiple determinants of individual and public health.[14] In most countries such groups did not just collect and analyze statistics on health, but they also campaigned for change and were activists in what became known as the public health movement. This was typically a coalition of professionals, experts, moralists, and do-gooders, who sought to improve and reform the lot of the urban poor and secure the wider social and political order.[15] In Britain, led by Edwin Chadwick, they focused on national and local epidemics, arguing that the best way to improve public health was through sanitary improvements, principally water supplies, sewers, drainage and paving, ventilation of homes and factories, and individual cleanliness. Their critics argued that what was needed were higher wages to allow families to afford better food (especially meat), housing, and rest from work and its disciplines, or that improvements could only be achieved if the poor and oppressed where afforded greater political rights.[16] Marx made these points in *Das Kapital* in 1867, though he was echoing the views of earlier writers, such as William Pulteney Alison, who attacked the narrow, sanitary, and politically conservative vision of public health promoted by Chadwick and other sanitarians.[17]

Over the nineteenth century, working conditions and wages improved for almost all groups in industrializing countries, largely as the result of political actions of welfare reformers and trade unions, but improvements came at variable rates for different groups. In the 1830s, epidemiologists considered rural areas to be "healthy districts" because of their lower mortality rates, but by the early 1900s there were fewer differences between urban and rural districts; indeed, with some diseases there was a distinct advantage of living in urban areas. With some groups progress on one front led to deterioration in another; for example, when African American slaves in the southern United States gained formal freedoms, many moved to the industrial cities of the North and suffered terribly there because they lived in the worst conditions and had little or no acquired immunity to prevalent infections.

The nineteenth century saw the medical profession develop a near monopoly in the labeling and management of disease.[18] This was problematic socially and culturally, for although greater authority and trust were accorded to doctors over the century, the gap between public and professional understandings of disease grew wider and so too did the potential for conflicts.[19] At the beginning of the nineteenth century, the public and doctors shared a broadly similar humoral model of the body, where health and disease were determined by internal balances and external interactions with the physical and social environment. Although doctors had their specialized terms, diseases were often defined by recognizable signs and symptoms that made sense to most people. By the end of the century, medicine had become more specialized, its understandings of disease were based on knowledge from morbid anatomy, physiology, microbiology, and other sciences. The profession and its understandings had become remote from that of laypeople as diseases were described in scientific terms and treatment regimens were more specialized. Doctors aimed to have naturalistic understandings, objective indicators, and knowledge of causes, whereas lay accounts showed a persistence of traditional, subjective understandings and a concern with the meanings of illness and prognosis.

Doctors had campaigned throughout the century for greater professional autonomy and self-regulation, while at the same time trying to secure occupa-

FIGURE 2.2: *An invalid with an undiagnosed illness, wrapped up and being pushed in a wheelchair.* Color lithograph by James Morison, ca. 1850. Wellcome Library, London (L0034503)

tional monopoly in prevention and treatment. Such moves were accompanied by the marginalization and stigmatization of alternative practitioners, along with their models of the body and its ills.[20] Both groups remained in the medical marketplace, but in most countries the orthodox profession enjoyed state licensing and, through the creation of other alliances, not least with the natural sciences, gained greater legitimacy and authority. In many spheres, not least the growing fields of health insurance and disease surveillance, doctors worked with and for the state. They decided who was sick in body and mind and who was not; who should be hospitalized, quarantined, incarcerated, and for how long; which treatments should be given; and who should be allowed to be an immigrant and who should be turned away. By the end of the nineteenth century, the social and cultural power of the medical profession had grown; indeed, historians usually date the beginning of the alleged medicalization of everyday life from that time.

The cultural history of health and disease in the long nineteenth century is a potentially huge topic, so I have chosen to focus on three diseases: tuberculosis, cholera, and insanity. As set out earlier, tuberculosis has to be approached in the first instance as a chronic, hereditary affliction. It was an iconic condition that had a similar position to cancer today; it was the dread disease associated with spontaneous onset and almost inevitable death, preceded by a gradual decline and suffering to the afflicted and their family. At mid-century the pallid face, gaunt physique, and heightened passions of the consumptive were associated with ideals of beauty, whereas later it was reconfigured as a disease of the poor and the dissolute. Cholera was the shock disease of the era, experienced in epidemic waves that spread round the world. Like HIV/AIDS at the end of the twentieth century, it was a new, exotic disease that produced graphic symptoms, dreadful suffering, and death. From 1820 to 1890, cholera was experienced as a disease of populations, communities, and environments, notably of the poor in urban settings, and was the emblematic foil of the public health movement. However, it was increasingly seen as a tropical disease that arose in the so-called primitive sanitary conditions of Bengal, India, and spread among filthy conditions across the world. Finally, the insidious and remorseless increase of insanity prompted questions over how and why, in an era of material, social, and intellectual progress, there should be deterioration in the minds of so many citizens. Indeed, toward the end of the century, the problem of insanity was constructed by doctors and others to be less a problem with individual minds and more one of the degeneration of the intellect and moral faculties of the public.[21] There was also the question of how to manage and treat mental disease: should this be through the mind, the body, changing behavior, or some combination of these? With each disease I begin with changing medical understandings and practices, moving out from the professional sphere to the ideas of lay groups and cultural meanings.

PULMONARY TUBERCULOSIS

Tuberculosis is an exemplary disease through which to explore changing ideas, practices, experiences, and cultural meanings of health and disease in the nineteenth century. Its prevalence, reputation, and long duration, where the patient declined toward death over four or five years, meant the disease and its sufferers were invested with many qualities and associations. Some of this was communicated in silences, as family and friends were reluctant to talk about such a terrible, terminal condition and because of the stigma associated with the hereditary and perhaps moral weaknesses that produced the condition. Among elites, romantic associations were linked to certain symptoms, for example, heightened sensibilities, attractiveness, and sexual desire.[22] Many famous writers and musicians died of consumption—Honoré de Balzac, the Brontë sisters, Elizabeth Barrett Browning, Robert Burns, Lord Byron, Ralph Waldo Emerson, Washington Irving, John Keats, Charles Kingsley, Sir Walter Scott, Robert Louis Stevenson, Henry David Thoreau, and Frédéric Chopin—which led some contemporaries to assume an association between the disease and creativity. Nonetheless, for most of the century and for most classes, it was a disease associated with poverty, suffering, and death.

Over the nineteenth century medical ideas about tuberculosis followed a reductionist trajectory from the whole body, to tissues, and then to a cellular construction. These shifts were manifest in changing terms. The century began with doctors mainly using the terms *phthisis* (Greek for wasting) and *consumption*, referring to how the body was seen at the bedside, weakened and seemingly consuming itself. The next term, *pulmonary tuberculosis*, came into favor in the 1830s and was a product of the hospital postmortem room. The term referred to the tissue changes seen in the lungs (pulmonary tissues): tuberculosis literally meant nodule production (from the Latin words *tubercle* and *osis*). In the 1880s the disease became *TB*; this was not a shortening of tuberculosis, but it came from *Tubercle bacillus*—the microorganism Robert Koch identified in 1882 as the specific and essential cause of tuberculosis. This finding was the result of research on animals in the laboratory, not from observing patients or postmortem findings. Koch had developed analytical techniques and experiments to show that the production of tubercles was due to the destruction of lung tissue by a specific microorganism. The sequence from *consumption* to *tuberculosis* to *TB*, associated respectively with the bedside, the hospital, and the laboratory, assumes a relabeling of the same underlying pathology; however, a closer look at their meanings for doctors and patients reveals a more complex picture.[23]

In the early nineteenth century, consumption or phthisis referred to a degenerative disease that was also quite variable in its development and duration. In northern Europe and North America the dominant medical and lay view was that it was hereditary, and sufferers were seen to exemplify types of persons—the consumptive, the scrofulous, or the tuberculous. In southern Europe the disease was seen to be contagious, and consumptives were people to be avoided. Famously, after being diagnosed with phthisis, Frédéric Chopin went to Majorca in 1838 to seek relief and a cure, but when his hosts discovered his condition they turned him out and burnt his mattress. In the medical culture of northern Europe and North America, consumption was assumed to arise from a particular diathesis or constitution, that is, certain inherited tendencies or proclivities that might remain latent but became manifest in certain circumstances, for example, when tissues were weakened by other diseases or by lifestyle changes. Indeed, a person's constitution was neither fixed nor wholly hereditary. A weak or strong constitution could be acquired through the influences of the environment, diet, behavior, and other illnesses. The incidence of consumption was highest among the poor, whose living conditions and lifestyle were thought to lower vitality and allow the tubercular constitution to express itself. The disease was also thought to develop in sensitive and artistic types, those whose creative and emotional energy drained their physical frame and powers, and allowed the physical body to wither from within. Some clinicians argued that consumption was not a disease at all, but "a mere mode of dying" and that all medicine could do was to "prolong a life naturally drawing to its close."[24]

Consumption came on insidiously over months and years; hence, doctors only tended to identify it at an advanced stage. Most treatments were ameliorative, aiming to relieve distressing symptoms with sedatives or to stimulate weakened bodies and devitalized tissues with tonics. If an inherited phthisical constitution could not be remade, then at least it might be compensated for.[25] Although doctors and patients often found it hard to talk about consumption, they had a shared understanding of its origins, character, and distressing progression. However, its meanings for the sufferer remained ambivalent between positive and negative, between the romantic aesthete and feckless poor.

In the second quarter of the century, work in hospitals and laboratories began to alter doctors' understandings of consumption and then, on a longer timescale, those of sufferers and the public. First, some doctors argued that tubercular lesions did not develop spontaneously, but were produced by irritation, or inflammation, due to physical, chemical, or biological agents. Second, a much smaller number of doctors, on the basis of experiments where they inoculated tubercular matter into healthy animals, suggested that the disease was catching. Contemporaries tended to regard this as "artificial tuberculosis," pointing out that it was difficult to see how, in ordinary conditions, tubercular matter might find its way from deep inside one body to another. Moreover, all doctors remained wedded to the idea that tubercles only formed in vulnerable tissues in susceptible people. The idea that tubercular disease

FIGURE 2.3: Parker's Tonic trade card. New York: Hiscox & Co., ca. 1880s. Wellcome Library, London (L0031706)

was "infective," in the sense of the spread of tubercles *within* the body—self-poisoning as some put it—was nothing new, as the disease was classified with cancer as producing new, internally disseminating tissue. There had been three critical shifts for doctors and patients: first, the focus on localized tissues rather than the whole body; second, that tubercle formation was due to inflammation rather than degeneration; and third, that there were now more options for treatment.[26]

So, rather than being offered sedatives and tonics, doctors now tried to remove irritants or counter their effects. The best forms of restorative, anti-inflammatory treatment seemed to be a good diet, cod-liver oil, exercise, a pure atmosphere, and bathing, which could be aided by climate, hygiene, and medicines.[27] The variability of the disease and the patient's background, temperament, heredity, surroundings, and income meant there was no formula; everything depended on the judgment of the doctor. A separate line of therapy was the use of specific anti-inflammatory treatments, which involved inhalations with the vapors of carbolic acid or creosote. The role of climate in treatment became a highly complex matter, because success depended on matching the patient, the personality, and their illness to a place and regimen. Someone with active disease might be recommended a long sojourn in a sedative, Mediterranean climate, both to lower the system and to avoid irritating chills.[28] On the other hand, patients in the early or quiescent stages were recommended the bracing air of the Alps, either to stimulate the whole system or to elicit specific improvements in lung expansion and pulmonary circulation.[29]

The notion that pulmonary tuberculosis was catching and produced by an infective agent that inflamed tissues attracted more and more support within medicine after 1870, principally from laboratory work on the disease in cattle. In the late 1870s, German workers reported finding a possible causative micro-organism in tubercular lesions.[30] From this time there was a growing contradiction where, as one doctor put it, "pathology undoubtedly points to a virus as the cause of consumption, and clinical facts point to the state of the system as at the bottom of it."[31] The view from pathology received important new support in 1882 with the claim that the *Tubercle bacillus* was the specific cause; however, what this meant for prevention and treatment divided the profession. At one pole, doctors claimed that the demonstration of contagion was "revolutionary" and that there was no precedent for "so sudden and complete casting aside of tradition."[32] At the other pole, doctors maintained that the bacillus had been expected and that it could be easily adapted to prevailing ideas in pathology and clinical experience, principally with the seed and soil metaphor, where the infective agent only developed in a vulnerable body. Most doctors could not accept that tuberculosis was contagious on the model of smallpox or measles, it was not specific in the same way, rather its causes and development were variable things in which the bacillus was likely to be one among many factors.[33] Doctors worried that talk of contagion would produce phthisisophobia among the public, leading to the shunning and isolation of sufferers, and creation of a new group of social outsiders.[34]

The major novelty in medical discussions after the spring of 1882 was not contagion and prevention, but the fact that TB, as it was increasingly called, was a germ disease and might be treatable medically. One goal was to try to destroy the germ in the lungs, and doctors tried antiseptics introduced by inhalation, injection, and irrigation.[35] These treatments seemed to offer patients some relief, though they seemed not to arrest the disease. In 1891, the tuberculin cure was announced, another product of Koch's laboratory in Berlin.[36] This event was culturally important and can be seen as the world's second medical breakthrough—the first being Pasteur's rabies vaccine in 1885. In both instances, medical progress made front page news, and it was stressed that the laboratory rather than the clinic was its source. Medical men and patients flocked to Berlin from Europe and beyond, and initially there was great enthusiasm. Euphoria soon turned to despair. Evidence accumulated that although tuberculin improved a few patients, it had no effect on most, and in a significant minority made the disease much worse.[37] Despite the "shattered hopes and expectations," to Koch personally, and the bacteriological understanding of disease more generally, medical researchers emerged with their standing only slightly damaged. The reputation of laboratory medicine was promptly salvaged with the introduction of a different so-called miracle cure—diphtheria antitoxin—and by vaccines for cholera and plague.

Although the laboratory had failed to produce a cure, the new bacteriological understandings informed other treatment plans and reshaped the cultural meaning of the disease. The new aim in treatment was to strengthen the body to resist germs by hardening the lungs, and the most common regimen was a combination of hygienic measures, which followed the accumulated experience that people with TB did best with a change of air, rest, a good diet, and clean living.[38] Clinicians became much more optimistic that TB was curable or at least arrestable. They had always recognized "spontaneous cures," seemingly inexplicable recoveries where patients defied the odds. Now, the combination of postmortem examinations and bacteriological testing had revealed that most adults in urban or industrial communities had been infected with the bacillus at some time in their life, but the disease developed only in a minority.[39] The general point was that with most people, most of the time, the balance of forces was with the power of the human soil to resist the growth of tubercular seeds.

For the wealthy, the treatment of choice became a change of air in the mountains, now sought because it was germ free and rich in disinfecting ozone. However, in Germany in the 1880s the treatment came down from the mountains and down market, as doctors offered a more accessible regimen. They downplayed the value of altitude as such, and instead promoted the virtues of open-air life in pine forests and the systematic application of hygienic rules. The main features of what became styled as the sanatorium treatment were living day and night in the open air, overfeeding, rest with some controlled exercise, and strict discipline. Of course, for the wider public, the opposite of sanatorium life defined TB-producing conditions: indoor factory working, public houses, poor diet, drinking, overwork, and dissolute behavior. On one reading, sanatoria represented a return to nature and holistic therapies, and reaction against reductionist, laboratory medicine; indeed, this was undoubtedly part of their attraction. Yet their meaning was ambiguous; for example, in 1898 the treatment was described as "the open-air life of the savage, combined with the hygienic comforts of our own age, and systematized with military precision."[40]

In the 1890s, those promoting the development of the sanatorium treatment were absorbed into wider movements to control the disease, in what was the first large-scale campaign to mobilize medical, voluntary, and state agencies against a single disease. Indeed, it was only at this late stage that TB became a public health disease and governments took an interest. National associations against TB were founded in most countries, inspired by the claim that the disease was "curable and preventable." The aims of these organizations were to make the sanatorium treatment more widely available, to educate the public on the principle and practice of prevention, and to tackle tuberculosis in cattle. Initially, most organizations were voluntary, but many became associated with, or were taken over by the state, either in social insurance schemes or as state welfare.

At the end of the century, some doctors wondered whether early-century consumption and late-century TB were in fact the same disease, a view that many modern cultural historians of disease would empathize with. Rather than assume a continuous history of a single disease, is it more accurate to accept that there were three distinct conditions? Consumption, tuberculosis, and TB were experienced differently by doctors, sufferers, and the public, and had many different meanings. In addition, they had been framed by doctors within distinct medical cosmologies, which meant not only that they had different etiologies and pathologies but also different assumptions about the very nature of disease and the human body. Against this relativist reading, is the evidence of doctors and patients being pluralist in their knowledge and practices, of the similar signs and symptoms of the different conceptions, and of patients in the late nineteenth century feeling they had tubercular constitution, and had weak lungs and had been infected.

CHOLERA

Cholera has been termed the shock disease of the century. In modern terms it would be called an emerging pandemic. Although tuberculosis was a chronic, endemic, insidious, and largely private affliction, cholera spread from country to country, causing violent symptoms and rapid death to those who succumbed, and led to emergency actions by local and central governments. The first epidemic spread eastward from India and across Asia in the 1820s, reaching Europe and North America in the early 1830s.[41] Cholera wrought its greatest toll among the poor, although all classes could be vulnerable as in November 1831 in Germany when it was responsible for the death of Georg Hegel, who was a major influence on Marx, and the military theorist Carl von Clausewitz. There were further pandemics in every decade in the nineteenth century, and the one in the late 1840s was the most severe in terms of lives lost.[42] Yet, the specter of the disease was present in every decade as newspapers reported cholera somewhere in the world.

Cholera was one of a group of so-called zymotic diseases—the others included smallpox, scarlet fever, typhus fever, and measles. The term derived from fermentation and was the root of the modern notion of an "enzyme"—to induce fermentation. The medical view was that some agent caused the body to "ferment"; hence, with cholera the body would heat up (giving fever and swellings), which led to sweating and diarrhea, and then quickly to the characteristic symptoms of blue lips, gray skin, and wasting. Other zymotic diseases had different symptoms and pathologies because each was assumed to be produced by a different complex chemical "ferment."

FIGURE 2.4: *A young Venetian woman, aged 23, depicted before and after contracting cholera.* Colored stipple engraving. Wellcome Library, London (V0010485)

Although cholera caused suffering and disease for individuals and families, it was approached as a disease of populations and environments. Responses were largely the preserve of the state public health agencies, many of which had been rudimentary but were transformed in the 1830s in part as a result of cholera. The disease was also a stimulus to the reform of urban environments and, as noted earlier, the creation of public health movements. In some countries these movements were national; in others, they were based in individual towns and cities. The doctors, engineers, and bureaucrats that were the frontline public health workers were known as sanitarians and practiced sanitary science, a subject that was seen to be both very ancient and very modern.[43] Hippocrates' airs, waters, and places was usually cited as its foundation; indeed, nineteenth-century approaches still looked to air (meteorology, ventilation, and pollution), water (plus beverages and food), and places (soil and habitations). Sanitarians aimed to remove "bad water, bad air, defective drainage, overcrowding, dirty and irregular habits"—the conditions that produced and propagated zymotic diseases.[44] They looked to statistics and epidemiology to help them understand the patterns of disease and their location in geographical space, social structure, and historical time, and in turn to guide preventive actions.[45]

Historians often write that the public health movement did the right things for the wrong reasons. The right things were removing the sources of disease by building sewers, draining streets, removing waste, and halting the transmission of disease by providing clean water and improving housing and ventilation. The wrong reasons were basing their work on the miasmatic theory of disease, which is often said to be vague and prescientific. Miasmatic theory assumed that diseases could arise spontaneously in malign conditions and spread through the air in an unpredictable manner. The right answer–wrong reasons argument is flawed historiographically and empirically. First, historians should not be judging past ideas right or wrong, but instead should be assessing how they functioned theoretically and practically for contemporaries. Second, miasmatic theory was not vague and unscientific; it is a caricature to suggest that it saw elusive poisons wafting around immaterially. Miasmatic theory was sophisticated and up-to-date science and, rather than showing uncertainties, was based on the well-grounded assumption that epidemic and zymotic diseases had multifactorial origins and spread in diverse ways. Moreover, understanding this complexity allowed public health reformers to target both specific points of passage of disease poisons and the general conditions that produced disease-causing ferments in the first place and allowed them to spread. All zymotic diseases, be they miasmatic, infectious, contagious, epidemic, occupational, or nutritional, were assumed to have external exciting causes, as such they were in principle preventable. Doctors often contrasted such diseases with the constitutional afflictions, like tuberculosis, that were nonpreventable because their origins were internal and spontaneous.

Cholera was the model miasmatic disease, its poison spreading on air currents that settled in particular areas and induced decaying matter to elaborate poisonous "cholera-stuff."[46] In the 1820s and 1830s, cholera was firmly associated with conditions of filth and overcrowding, not least because it has proved impossible to contain with quarantines and isolation. There was an ongoing debate in medicine and public health agencies over the degree to which zymotic diseases were contagious, and most adopted policies based on anticontagionist principles; namely, a rejection of quarantines and other methods of isolation, which they saw as futile because airborne and spontaneously generating diseases would not be contained by such measures. Nonetheless, anticontagionists accepted that many diseases were highly contagious, smallpox for example, and that all zymotics could be contagious in certain conditions. The goal of sanitarians was to create the conditions in which zymotic diseases could not arise or spread, in other words, to adopt inclusive measures that made peoples and places less productive of, or receptive to, epidemics. The dominant view in medicine was that cholera was contingently contagious, in other words, catching in certain conditions, which was congruent with wider experiences that even during the most severe epidemics most of the population remained uninfected and only a third of sufferers died.

By the time cholera threatened Western countries again in the late 1840s, preventive measures had switched firmly away from quarantines to programs of reducing "receptivity" of local environments through nuisance removal, cleansing, and other improvements.[47] In Britain, this approach was championed by Edwin Chadwick, and its narrow environmental focus made public health an increasingly technical area that was divorced from issues of poverty and social justice.[48] However, in Mediterranean countries anticontagionism did not attract as much support, and governments there continued to implement quarantines. In most countries the death toll was higher than in the 1830s, and the relative failure to control the epidemic led to the exploration of alternative ideas on the nature and control of the disease.

From the mid-1850s there was a movement of opinion toward cholera being spread, at least in part, via water contaminated with human wastes.[49] Some doctors began to advise the need for "special precautions of cleanliness and disinfection . . . with regard to infective discharges from the bodies of the sick," and that all water supplies should be examined so that "no foul water be drunk."[50] Such changes were influenced by explanations of cholera transmission by the fecal–oral route and the importance of drinking water in spreading infection.[51] It would be wrong to think of an opposition between miasmatic theory on the one hand, and emerging germ theories on the other. For example, the report of Britain's chief medical officer on the 1865–1866 cholera epidemic, which accepted the notion of a cholera germ and the water carriage theory, still concluded that our "knowledge remains unchanged; and unchanged remain also our practical means of applying it. . . . Excrement-

FIGURE 2.5: *Patients suffering from cholera in the Jura during the 1854 epidemic, with Dr. Gachet attending them.* Pencil drawing by Amand-Désiré Gautier, 1859. Wellcome Library, London (L0012076)

sodden earth, excrement-reeking air, excrement tainted water, these are for us the causes of cholera."[52]

On the international front, the evidence that the cholera had spread from east to west through the Mediterranean led the first International Sanitary Conference in Constantinople in 1866 to recommend the use of new quarantine measures.[53] These were now redefined and would no longer halt the movement of all people and goods for a fixed time; rather, people and goods would be held long enough for inspection and perhaps disinfection, with only confirmed or highly suspect cases being confined. Crucially, the understanding of cholera was changing from being a disease of communities and localities to one carried by individuals and goods. When cholera spread again in the 1870s and 1880s, outbreaks were localized, and there was not a global pandemic. Moreover, the threat was less to Western industrialized countries and more to colonial territories and to trade, the latter because of the switch to quarantines. Yet, among the doctors working in public health in India most continued to be anticontagionists and resisted quarantines.[54]

When the disease reached the Mediterranean coast of Egypt in 1883, European scientists seized the opportunity to pursue laboratory researches to find the cholera germ. None other than Robert Koch led the Germany group, while Pasteur's assistant Emile Roux (1853–1933) headed the team from Paris.[55] Koch, fresh from his success with the *Tubercle bacillus*, reported in September 1883 that a small rod-shaped bacillus was the essential cause of cholera. The French group was unconvinced, especially as neither group managed to reproduce the disease by the inoculation of germs in animals. Research was soon halted by the decline of the epidemic, though undeterred, and with the permission of the Indian government, Koch went on to Calcutta to continue his work and found his bacilli again, now described as comma-shaped. He was again unable to produce the disease experimentally in laboratory animals, so the claim about its specific germ causation remained controversial. Nonetheless, he returned to Germany triumphant, lauded by the state and the Berlin medical establishment.[56] Subsequently, his findings were given many meanings and were used to support all control options, from strict quarantines to nonspecific sanitary improvements.[57]

Apart from the major outbreak in Hamburg in 1892, cholera moved increasingly to become a problem for imperial governments in their tropical colonies, especially the British in India. The international pressures for quarantines and their resistance by Britain, the largest imperial power, ensured that the disease remained a matter of debate. In India, localism and anticontagionism was still favored and bacteriological understandings eschewed; hence, it is not without irony that Waldemar Haffkine, working from the Pasteur Institute in Paris, made his first anti-cholera vaccine in 1893 with cholera bacilli supplied by Anglo-Indian doctors.[58] Vaccination to protect populations against smallpox had been promoted in India for many decades, and there was an infrastructure to support the policy.[59] Pasteur's appropriation of the term "vaccination," from just the inoculation of cowpox, to all attempts to produce artificial immunity by the inoculation of modified germs, was controversial. His methods were laboratory based, seemed to be grounded in theory rather than experience, and had not been clinically tested. Pasteur's laboratory used live, attenuated (weakened) germs as vaccines, and these were seen to be risky—what if the germ regained virulence? Haffkine demonstrated his vaccine in England in 1893 and was given the opportunity to undertake trials in India in 1894.[60] The initial results were mixed. He returned to India for further trials in 1896 at the same time the plague arrived. Using now-established Pasteurian techniques, he quickly produced an antiplague vaccine, though this was based on killed organisms rather than the live modified ones favored in Paris.[61] The plague epidemic, and especially Haffkine's vaccine, caused such shock waves, that the Indian government established a Plague Commission to investigate the disease in India and report on the best means to control it. Plague and then malaria replaced cholera as the major focus of imperialist concerns about health and disease, but by 1900 there seemed to be a clear divide between temperate and tropical diseases, though the latter were defined largely around health threats to European colonists and only to indigenes when their health threatened imperial ambitions.

MADNESS

One of the best-known deaths from cholera in the nineteenth century was that of the Russian composer Pyotr Ilich Tchaikovsky. He died in November 1893 in St. Petersburg, which was then in the grip of an epidemic. Although cholera was the official cause of death, an alternative version of events in which the composer took his own life has been canvassed. The least plausible is that Tchaikovsky obeyed the sentence of a "court of honour" of the St. Petersburg School of Jurisprudence, which had censured him for his homosexuality. This is unlikely because the composer's sexuality was well known, and although homosexuality was against the law, there were few prosecutions and general tolerance, especially in artistic circles. The second explanation is that the composer, known to have suffered mental breakdowns and to have tried to commit suicide previously in 1877, poisoned himself after the poor reception of his *Sixth Symphony*. Those favoring suicide at the time of his death, and since, have pointed to the somber final movement of the symphony, especially its fading ending, as an indicator of the composer's mental state. The balance of opinion among Tchaikovsky's biographers is now with death due to cholera; however, the assumption of suicidal intentions was illustrative of the new interest at the end of the century in borderland between sanity and insanity. Indeed, the sense that mental illness was becoming more insidious, and that its milder forms were affecting more people, was very much in tune with fin de siècle social and cultural pessimism. The nineteenth century saw an expansion of mental illness in two senses: a huge growth in the numbers certified as mad and incarcerated in asylums and, as madness became "mental illness," the inclusion of a range of neurotic and hysterical conditions.

While bodily health improved over the century, there was a clear deterioration in mental health. In 1800, across Europe and North America, there were a few madhouses and each counted their inmates in tens. By 1900, there were hundreds of large asylums that had thousands of inmates. Rates of insanity in the population rose fifteen-fold from 2 per 10,000 to 30 per 10,000.[62] In many ways, insanity was the great new health problem of the nineteenth century—it appeared to medical and social commentators that urban, industrial society, with its new disciplines and freedoms, was literally driving people mad. Moreover, when the new psychiatric professionals shifted their gaze from incarcerated inmates to the wider population in the late nineteenth century, they found that certified insanity was the tip of an iceberg of mental illness and instability.[63] Whether there was an actual increase in psychotic and neurotic diseases over the century, or whether their seeming rise was due to relabeling, changing attitudes toward difficult people, or psychiatric imperialism, or other factors is the subject of ongoing debate among historians.

Definitions of insanity and the management of the insane became the preserve of doctors over the century, in particular those who managed asylums and were known as mad doctors, then alienists, and later psychiatrists. However, there continued to be a significant lay component in decisions on the incarceration and discharge of asylum inmates, because institutions were part of state welfare systems and to counter claims of possible wrongful confinement for ulterior motives.

Doctors assumed that the mind and body interacted in complex ways and, as Rosenberg observed for the nineteenth century, "all medicine was 'psychosomatic'; there was no need for a special term."[64] With many asylum inmates the bodily causes of their mental condition were only too obvious, for example, those who were brain damaged, drunkards, and in the final stage of syphilis. There were also those whose physical symptoms showed a body marked by a seemingly uncontrolled mind, for example, those with repetitive movements and who self-injured. Of course, doctors and patients saw quite specific mental components to many illnesses, from hallucinations and convulsions produced by fevers to the belief that extreme emotions, such as hydrophobia after being bitten by a rabid dog, could produce the physical disease.[65] In many cases, mental illness was inscribed on the body of the mad and seen in their physiognomy, physical stature, and deportment.

For much of the nineteenth century madness was a problem defined by an institution—the asylum—that haunted the imagination and represented the

FIGURE 2.6: *Seven vignettes of people suffering from different types of mental illness.* Lithograph by W. Spread and J. Reed, 1858. Wellcome Library, London (L0006935)

antithesis of the dominant bourgeois ideals, such as, self-help, restraint, and respectability.[66] The insane emerged as a category in the attempts of reformers after 1800 to replace disgraced madhouses, like Bethlem (commonly known as "Bedlam") in London, which was widely caricatured and condemned, with new, state-regulated asylums modeled on the Quaker-run York Retreat.[67] The hope was that the new asylums would provide sites where disordered minds could be restored or, at the very least, inmates would be given sanctuary and would not be ill-treated or exploited. The reformers promoting asylums, many of the same evangelicals and utilitarians who campaigned against slavery and the exploitation of children, were often antimedical, arguing that madness was a problem of the mind not the body.[68] Reformers took a much stricter Cartesian view of the separation of mind and body than most of the doctors who had run eighteenth-century madhouses and had often treated the mind through the body. Reformers argued that doctors were only trained to deal with the body, and that sympathetic laypeople would be better at disciplining deranged minds, restoring rational thought, and ensuring that inmates lived orderly lives in a humane environment.

Asylums were founded with great optimism that the mind was malleable and that insanity could be cured by "moral treatment," a term that seems to have been coined to be deliberately oppositional to the physical treatments and immoral operation of madhouses.[69] Public asylums were created for those who were mad and poor, or poor because they were mad, or people whose family or community could not manage their behavior. In Britain they were known as pauper lunatic asylums, and potential inmates needed to be designated in receipt of public aid before they could be a lunatic. Indeed, asylum inmates were a very mixed bunch and must not be thought of as analogous to those suffering from mental illness today. There were people with disabilities, such as deaf-mutes; those with conditions like epilepsy and chorea, the old and infirm suffering from dementia, and some, especially women, who had shown their irrationality by breaking moral codes, for example, unmarried mothers. The mad of the wealthier classes were boarded in private asylums, where standards of care were generally higher than in public institutions, or managed at home, often kept out of sight like Bertha Mason, the first Mrs. Rochester, in Charlotte Brontë's novel *Jane Eyre*.[70] The sources of her condition were typical of the combination of inherited and acquired with which contemporaries accounted madness; Mr. Rochester observed that, "she came of a mad family;—idiots and maniacs through three generations! Her mother, the Creole, was both a mad woman and a drunkard!"[71]

Reformers' hopes that laypeople rather than doctors would run asylums were not met.[72] First, mad doctors had a vested interest in this area of medicine and fought to keep their living in what was a highly competitive medical marketplace. Second, the diverse and seemingly ever-growing group that constituted asylum inmates had many bodily illnesses, not least infections and injuries; hence, every institution needed the day-to-day support of a medical doctor. Third, no other occupational group emerged to replace mad doctors, so as much by default as anything else, the medical profession maintained its role with madness. Moreover, asylums became places where doctors could develop expertise, in part medicalizing moral therapy. The creation of very similar institutions nationally and internationally allowed alienists, as they termed themselves, to share and codify ideas and practices. Last, the new alienists assimilated the pathology of mental illnesses to broader medical trends, suggesting that such conditions had a physical basis in structural or functional abnormalities of the brain or nerves. Indeed, they maintained insanity was due to specific diseases, rather being simply abnormal emotional, behavioral, and mental states.

Alienists dealt with the mind–body problem in two ways. First, they agreed with the reformers that madness could be cured and that engaging with the minds of the mad in an orderly asylum was important, and that medical men were best placed to do this work because of their experience with the insane. Second, they continued to treat the body as a way to affect the mind, albeit using less heroic measures in line with the general moderation of treatments from the early nineteenth century, looking to new ideas on mind–body interactions.

In the 1820s and 1830s, alienists were enthusiasts for the new science of phrenology, which was popular across all social classes for a generation.[73] The core premise of phrenology was that the brain is the organ of the mind. To many contemporaries this was a radical, materialist doctrine privileging brain over mind; yet in medicine and its popular forms this theory was interpreted as dualist and safe; indeed, those most respectable Victorians, Queen Victoria and Prince Albert were aficionados. The second premise of phrenology was that the brain was a congerie of organs or localized faculties, each producing a different feature of intellect, affectivity, and character. Moreover, the size of each faculty, both absolutely and relative to others, equaled its strength and influence in thought and feelings. A third premise, that the size of faculties could be read from the topography of the skull, was central to phrenology's following in popular culture but relatively unimportant to alienists. What attracted alienists was the way phrenological principles could inform their view of insanity, which in the 1830s was that most lunatics suffered from mania, and especially monomania—irrational beliefs or excitement or obsession about one thing, which was also exhibited in repetitive, singular behaviors. These conditions were seen as entities and offered alienists the nosologies—disease classifications—that characterized advanced ideas in the rest of medicine. Alienists mapped different manias and mental symptoms onto different parts of the brain, and rather than adopting a pessimistic view that anatomy was destiny, they argued optimistically that the development of other faculties could divert and moderate monomaniacal tendencies. As well as providing guidance for treatment, phrenology was used to legitimate alienists' claims to expertise by aligning them with a modern, albeit controversial science of mind.

So rapidly did the number of inmates in asylums grow that very few alienists had the time or opportunity to provide sustained treatment regimens for individuals. Instead, moral treatment became moral management, and alienists complained that they functioned as administrators rather than doctors. With the demise of phrenology, alienists' dominant explanation of madness switched from brain to nerves and their functioning, particularly the view that the thoughts and behavior of the insane were governed by reflexes and were impulsive, whereas sane people had greater, if not total, self-control. Such a model of madness was readily linked to evolutionary ideas, to those of Herbert Spencer more than Charles Darwin, and the idea that human evolution had been due to races controlling primitive thoughts and impulsive behavior through the development of higher mental, emotional, and moral powers of control. Hence, insanity was considered a return to an earlier stage of mental evolution, and some conditions were associated with so-called "inferior" races, like the Mongols with mongolism.

Alienists, like other nineteenth-century doctors, and indeed the public, viewed the body as having a finite amount of energy at its disposal. Hence, a person who shouted all day, or had a permanent tic, was probably mad in

FIGURE 2.7: *Mentally ill patients dancing at a ball at Somerset County Asylum.* Process print after a lithograph by Katherine Drake. Wellcome Library, London (L0000508)

part because he or she had no energy for other thoughts and actions. Indeed, certain types of madness were said to be produced by lack of sufficient energy to animate a person's moral sense, a phenomenon seen in the numbers of sad or bad people in asylums. This view of the relationships between nerves, energy, body, and mind was not only applied to the insane, but it was also increasingly used to understand and manage all manner of ailments, and to understand various illnesses and conditions in the general population. It blurred the boundary such that by the end of the century, the category of neurotic illness had been expanded from physical impairment of the nervous system to include a range of emotional and affective conditions. Women were seen to have less energy than men; hence, they were perhaps more prone to neurotic illness.

This development is best illustrated by neurasthenia—nerve weakness or exhaustion—which was commonly found in women after mid-century and showed itself in a myriad of symptoms. It seemed to be most prevalent among Americans; indeed, the psychologist William James styled it "Americanitis."[74] The diagnosis was invented by George M. Beard in 1862 and was the central theme of his 1881 book *American Nervousness*.[75] Beard contended that each person has a particular nervous force, the power and form of which is determined largely by heredity and which acts through reflex centers, wherein lesions and abnormalities produce functional differences and the wide variety of symptoms seen in the condition. Beard saw neurasthenia as a disease of modern civilization, especially urbanism, industrialism, and the competitive culture of capitalism, though he hoped future technologies would relieve the stresses of modern life. Given his views on the bodily origins of neurasthenia, it is perhaps unsurprising that Beard favored physical treatments. Some colleagues tried electrotherapy to stimulate and enhance the sufferer's mental force. The most famous, perhaps notorious, treatment was the "rest cure," invented by Silas Weir Mitchell.[76] This required the total inactivity of body and mind, and was especially recommended for women. It became best known as the subject of Charlotte Gilman Perkins story "The Yellow Wallpaper," published in 1890, which portrayed the treatment as producing rather than curing madness, and symbolizing masculine power over women.[77]

Ideas on the loss of higher functions and neurasthenia were elements in what became the dominant explanation of insanity in the last quarter of the nineteenth century—degeneration.[78] The term entered medicine in 1859 through the work of the French physician Bénédict Augustin Morel, who linked the insanity of individuals to the deterioration of the national and racial stock. He wrote, "The degenerate human being . . . becomes not only incapable of forming part of the chain of transmission of progress in human society, he is the greatest obstacle to this progress through his contact with the healthy proportion of the population."[79] The key idea was that inherited and acquired mental defects would accumulate over generations, producing almost subraces of people with insane, imbecilic, or antisocial characteristics. Such views have been termed psychiatric Darwinism, though this is only a useful term if we remember that in the last quarter of the nineteenth century Darwinism stood for all forms of evolution, not least, Lamarckian and Spencerian ideas of the inheritance of acquired characteristics.[80]

A key feature of psychiatric Darwinism was its emphasis on the hereditary character of insanity, which fostered pessimism at two levels. First, it suggested that with an individual insane person there was little alienists could do to treat and cure a condition rooted in his or her constitution and where nature was more powerful than nurture. Second, it posited that the mental health of society and the strength of its culture were threatened by the accumulation of people moving from mild neurotic conditions to serious and chronic insanity.

In the 1880s and 1890s, medical discourses on degeneration influenced cultural critics such as Max Nordau, whose influential 1892 book, entitled *Degeneration*, bemoaned the effects of urbanization and industrialization on mind and body. The notion was also important in the new Gothic fiction, in such classics as Robert Louis Stevenson's *The Strange Case of Dr. Jekyll and Mr. Hyde* (1886) and Bram Stoker's *Dracula* (1897), in which mad obsessive creatures haunt modern lives. Nordau, who was a physician by training, drew on Morel and used medical metaphors to characterize what he saw as the unhealthy features of modern, progressive culture, especially the tolerance of sexual deviancy and the flight from positivism.[81] In Gothic fiction, atavistic characters haunted narratives as throwbacks, creatures that had lost civility and perhaps their mind, and who threatened the future progress of humankind. Indeed, such concerns were linked to the feeling that the industrial civilization manufactured madness.

CONCLUSION

In his 1864 article, George Henry Lewes also wrote that "Who so speaks on Health is sure of a large audience," pointing out that understanding how to avoid disease and achieve better health was an important aspect of the life of all ages, genders, classes, and ethnic groups, and of Victorian culture more widely.[82] By the end of the century there was evidence of improved physical health for most social classes in industrialized countries, though there was a more mixed picture internationally. In Europe and North America, life expectancy was rising, principally due to fewer deaths from diseases in childhood and early adulthood and more people surviving serious illnesses. The evidence that morbidity was also rising does not contradict this picture of improvement; rather, it indicates that improvements were in part due to better access to doctors, better diagnoses, better care, and the beginning of social welfare. The importance of chronic illness in the nineteenth century must be recognized; it was seen in the large numbers of consumptives; the disabilities produced by difficult births, injuries, and poor nutrition; and the debility seen in such conditions as bronchitis. Nonetheless, increased life expectancy was widely cited as a symbol of progress in the Victorian era, alongside such qualitative improvements as the conquest of pain through anesthesia, the better control of wound infection, and the reform of nursing.[83] By the end of the century, diseases were things that could be conquered and perhaps eradicated. Diphtheria antitoxin, which miraculously saved children close to death, promised a new era in effective therapies, and there were hopes that cholera and plague vaccine would repeat the benefits of vaccination against smallpox. One new hope was that malaria—a zymotic disease that was the scourge of imperial ambitions—might be eradicated because of a new understanding of how it was spread by mosquitoes.[84] However, progress seemed to be leaving many behind; this was seen most clearly in the rise of insanity. Asylums continued to be overrun, and there were those in psychiatry and elsewhere who believed madness was spreading insidiously down the generations and across society, and new mental conditions, such as hysteria and phobias, were affecting an ever larger proportion of the population. Such observations fed the wider notion of degeneration that was associated with the development of mass societies and with fears about the future of the white races that were refracted in new policies to improve motherhood, childbirth, and the health of children. Those reflecting on health and disease at the end of a century of spectacular material progress looked forward to further improvements, though not always with great confidence.

CHAPTER THREE

Othering Sexual Perversity

England, Empire, Race, and Sexual Science

RICHARD C. SHA

Historians have long pointed to a decisive break in the history of sexuality, the moment when the homosexual became a person.[1] Sexology, the scientific study of sexuality, broke down sexuality into various character types—the invert, the homosexual, the pervert. But before the late nineteenth century, sexology did not exist. Rather, categories were based not on personality or individuality but on immoral acts: the sodomite, the pederast. In the starkest version of this argument, before sexology, one had sex, whereby sexual identity was defined in anatomical and pathological terms, and after sexology, one could have sexuality, the connection of sexual identity to "impulses, tastes, aptitudes, satisfactions, and psychic traits."[2] The corollary to this claim is that "there were no perverts before the existence of a concept" of perversion as a psychiatric disease category.[3] This essay refutes that argument by showing that if perverse or nonreproductive sexuality could be projected onto the other, as it was especially since the development of trade routes in the eighteenth century and imperialist ambitions exposed Europeans to countless others, then one needs neither psychiatrists nor sexologists to connect sex with tastes and inclinations. Founded relationally, the self is constituted in reference to an other that defines the subject. The seemingly essential perversity of the other, for instance, mandates that the self conceive of perversity outside of identity, sometimes in terms of acts. Yet this denial is nonetheless a crucial part of identity formation. To make matters worse, the other's perversity can screen the perversity of the self.

By understanding the vexed role of the other in the history of sexuality, one gains an appreciation for the costs of identity, even as sexology enables sex to subsume identity. As a container for identity, then, sexuality is never quite so porous as when it functions as an identifying essence.[4]

I will show how sodomy, pederasty, and lesbianism were foisted onto Italy, Greece, Turkey, and Revolutionary France at the same time that sex was limited to acts and that anatomy explained sex.[5] So long as the other could be a pervert, thus insulating the European from a perverted identity, then the argument that the pervert did not exist until the rise of psychiatry in the nineteenth century seems like wishful thinking.[6] And although perversity was considered a vice—an ethical or theological issue—while perversion became a medical problem, the province of psychiatrists, the fact that sexual perversity was made other suggests that perversity has something to do with identity (perversion) insofar as immunity from perversity is a claim about the self.

The category of perversion was used to consolidate, legitimate, and naturalize the British Empire, and it could do so to the extent that perverse sexuality helped to constitute both native identity and a new imperial national identity for native-born Britons. Since the persuasiveness of European narratives of native sexuality was contingent upon the colonizer's need to maintain his or her own respectability and to have this respectability recognized by the other, the sexuality of the one had implications for the other. The colonizer's often lower-class social origins or status as a displaced person or debtor further heightened the stakes in making the natives sexually perverse. Imperialism was simultaneously a means for the empire to dispose of its unwanted[7] and a means for the unwanted to dispose of their social identity as the dregs through Britishness. Like disease, then, perverse sexuality is a dialectical structure that allows the colonizer to fear his or her collapse, project it onto a negatively characterized object, locating it in the diseased body of the other.[8] Thus framed, perverse sexuality is a placeholder for the kinds of identity that are intelligible as identity only through denial.

Finally, in keeping with my skepticism about a definitive break between sex and sexuality, perversity and perversion, I conclude with sexology. Psychiatrists had difficulty convincing others of their expertise over sexuality because they could not locate "mental disorder as an organic disease of the brain and the nervous system,"[9] and this difficulty further suggests that sex cannot be separated from sexuality. As sexology made sexuality into the central truth of identity, it simultaneously collapsed self and other so that humanity became defined as essentially sexual, and even perverted, beings. I therefore show how the late nineteenth- and early twentieth-century writers who invented the field of sexology, like Iwan Bloch, Havelock Ellis, and Richard Krafft-Ebing, and so on, needed to frame reticence and modesty as active denials of the truth of sex. Their overcoming of such denials spoke to their professional power, even as their marginality in medicine and society mandated their performance of authority. Moreover, whereas historians of sexuality credit sexology with welding sexual acts to identity, sexologists developed concepts like pseudo-homosexuality or acquired homosexuality, concepts that began to undermine the relationship between sex and identity. Such undermining acknowledges that there are considerable costs to identity. For instance, psychiatrists considered that when construed as a congenital part of the self, homosexuality could not be cured. And, if congenital, how can homosexuality be thought of as radical, as a way of liberating people from normalizing regimens? Sexologists' fundamental disagreements about whether or not homosexuality and inversion[10] were pathological or normal further meant that perversion itself could be normalized. Such doubts imply that any alleged relation between sex and identity—pathological or normal—has more to do with who has the authority to tell the truth about sex than the actual relation.

Hence, sexual identity is much more incoherent than historians have wanted to acknowledge because it also traffics in race, national identity, pathology, gender, class, and species.[11] Such trafficking reminds us of how identity looks fixed and natural even though its components are highly mobile and responsive to historical pressures. Such trafficking further allows sexuality to masquerade for such other essences, such as race and class, thus extending its leverage—making it more powerful—while keeping incoherence at bay by hiding it under another frame of analysis. What we have too often mistaken for fixed essences, immune to the pressures of history, are in actuality undergoing essentializing.[12] Health and disease bind sexuality to the self or other, for instance, so that the Victorian lower classes become guilty of promiscuous combination. Sexuality, then, adapts to the changing needs of identity, often by expanding its metonymic reach to encompass identity or contracting its purview under race, class, or gender, and this process has been hidden by the fact that sexuality has become the central truth of identity.[13]

MAKING PERVERSITY OTHER

The ascription of sexual perversity onto the colonial other was always inconsistent, dependent upon contingent historical circumstances, but it was always motivated by a need to have a discursive figure against which to define the normative masculine (and less consistently feminine) self, and by the need to have a discursive figure against which to define the normative national self. My story begins with the unstable social identity of the British middle classes in the early nineteenth century. Sandwiched precariously between the aristocracy and the lower classes, the middle classes turned to the sexual perversions to consolidate their position and status, to give their social identity an essence.[14] Whereas the bourgeoisie equated aristocracy with sterility, impotence, and adultery—itself an excess sexuality that was thought to make one sterile—they represented the lower classes, because of their crabbed living spaces and shared beds, with being licentious and promiscuous. By connecting excessive sexual desire to extinction, Thomas Malthus in his *Essay on the Principle of Population* drove home the economic consequences of marrying before one could support one's family, and he effectively blamed the lower classes for overpopulation.[15] In Victorianism, rich and poor became two nations. Not only were lower-class gatherings sexualized in terms of promiscuous combinations, but the middle class also feared pollution by their inferiors. Embodying a healthy, fecund sexuality, the middle classes scorned the indulgences of the upper and lower classes. Thus, male middle-classness could come to stand for the embodiment of all virtues and the major barrier to the spread of venereal disease.[16] The Enlightenment gave birth to sexual science, and initially focused on the differing sexual customs of races only then to solidify the incommensurate biological differences between men and women.[17] By making English middle-class domesticity antithetical to sexual passion, licentiousness could be projected onto peoples living in the torrid zones. However, because women were thought to have bodily torrid zones, women could contaminate men no matter where they lived.[18] The ideal of English female passionlessness nonetheless intensified both the eroticism and innocuousness of flirtation.[19]

In a larger view, Europeans mapped sexual perversion onto such geographically distant countries as Turkey, Persia, India, Africa, Italy, and Greece. Father Jean Antoine Dubois reported that pederasty and male prostitution flourished in Hindu and Muslim areas of the Indian subcontinent.[20] According to a Dutch admiral, "the sin of Sodom is not only in universal practice among [the Moghul Bengali], but extends to a bestial communication with brutes, and in particular with sheep."[21] After the Indian Mutiny of 1857–1859, the identification of the Asian other with extravagant desire intensified because the mutiny was figured in terms of rape and sexual mutilation.[22] The British were thus forced to replace the fantasy of an essentially effeminate Indian, one especially vulnerable to sodomy, with another fantasy: the Indian rapist. This complete about face in stereotype occurred because vulnerable English women were an especially effective siren call for British men to rally around Empire. The discovery of Herculaneum in the eighteenth century solidified connections between hot climates and perverse sexuality. Obscene artifacts from Ancient Rome, especially phalli for the purposes of religious worship, led to the creation of secret museums, where only gentlemen were permitted entry.[23] These artifacts cemented the notion of Italy's perversity. If the Victorian cult of Greece made it possible for Victorian men to indulge in unphobic delectation over the male body,[24] such homoeroticism, thanks to Johann Winckelmann and Walter Pater, could now become aesthetic and thereby part of one's identity. Oriental societies were imagined as places where women were treated badly in harems and in polygamous marriages, and tyranny itself was orientalized and eroticized, as voluptuous indulgence fostered absolute rule. African women were also seen as "wanton perversions of sexuality."[25] Mungo Park lamented in his best-selling *Travels in the Interior Districts of Africa* (1799) that Moorish women regarded voluptuousness as their "chief accomplishment" (151).[26] Systematic comparisons, empirical detail, and theoretical arguments combined to make widely accepted the savage, decadent, and uncivilized sexuality of the other.[27] Ann Stoler sums this up: "adherence to middle-class European sexual morality was one implicit requisite for full-fledged citizenship in the European nation state."[28]

If European sexual morality was necessary for European citizenship, such morality did not prevent Europeans from tarring each other with the brush of sexual perversity. From a decidedly English perspective, a den of sexual perversion lay just across the English Channel. Within Europe, perversion could help consolidate national differences simultaneously as empires were built. The French Revolution reverberated well into the nineteenth century, threatening to unleash lesbianism, sodomy as the *vice aristocratique*, and other forms of sexual libertinism onto the shores of Britain. Pornographic pamphlets about Queen Marie Antoinette depicted her as a tribade, and because these successfully undercut court culture, British radical publishers like William Dugdale adapted revolutionary techniques between the 1820s and 1830s for use against King George IV.[29] Medical literature described sexual practices in foreign languages, especially Latin and French, and this further made perversity other, even as those languages conspired to prevent the poor and uneducated—economic others—from being seduced by the very sexual perversion they allegedly embodied. Perversity was further made other by the transformation of poor bodies into clinical material while alive, to be poked and prodded in hospitals, and into objects of anatomical study after death.[30]

That sexual perversity could be made other in multiple explains why it has taken so long for historians to find it at home and why those living in the nineteenth century could remain blind to even their own perversions. Because perverse acts were framed as moral acts that had no necessary connection to identity, persons could then deny their own perversity. We can see this logic informing the Victorian physician William Acton's insistence that male clients were the victims of prostitutes: as victims, men neither had their male identity nor their morality impugned. Perversity, then, has always been useful for its ability to detach acts from identity, often by emplotting such acts as sins to be redeemed. Simultaneously, perversity was racialized and foisted onto the lower classes, thereby ensconcing the English middle class in sexual health and purity. If their innate purity came under too much threat, campaigns against onanism, purity campaigns, and the Society for the Suppression of Vice, founded in 1802, would help preserve that purity.

After Darwin, racialization of sexual perversity acquired the sanction of evolution. Victorian evolutionary theory related all forms of desire to each other, thus subverting difference and otherness.[31] Paradoxically, it related these desires in terms of a developmental hierarchy of primitive and developed races, and this provided otherness with additional leverage in part because struggle is now the very condition of progress.[32] On the one hand, as organisms evolved they showed greater sexual differentiation. Sexual ambiguity thereby became necessarily primitive.[33] On the other hand, once ontogeny recapitulated phylogeny, then primitive savage sexuality lay dormant within the self. To the extent that the primitive other is always threatening to break out of the self, Darwin leads to Freud. Indeed, Freud argued that the "conception of the sexual perversions in terms of developmental arrests approximat[ed] the fixed adult stages of our animal ancestors."[34]

Once othering is understood as the crucible of identity, perversion becomes visible long before the rise of psychiatry. Lord Byron and his Cambridge circle turned to an elaborate code including botany, Horace, boxing, and criminality to communicate his bisexuality.[35] Steven Angelides has argued that the speciation of bisexuality has been long resisted because it "disrupts the very classificatory alliance of sex/gender and sexuality."[36] As recently as 1998, Randolph Trumbach argued that sapphism did not exist as an erotic identity in England until the 1770s.[37] One can trace sapphism as an erotic identity back much further, if only one thinks about how anatomical explanation works by metonymy and how female friendship was used to actively deny lesbianism.[38] From the seventeenth century onward, an elongated phallic clitoris begins to explain the tribade, a woman who rubs her genitals against another woman. Yet anatomy itself does not quite account for the fact that lesbian desire is heterosexualized; the transformation of the clitoris into a virtual penis implies that even anatomy begins to spill over into the psyche.[39] More to the point, if anatomy must rely on bodily metonymy to capture pleasure and desire insofar as body part equals tribade, then the gaps between pleasure and body part are the places we can trace erotic identities before sexology. Although female friendship outwardly idealized female passionlessness, this ideal made all eroticism between women surplus, especially when one of the ladies was known to dress in male clothing. The fact that sexual perversion and class/race/gender/nation/species have often operated as metaphoric substitutes grants sexuality power across all forms of hierarchy. The downside of substitution is that the relation of sexuality to hierarchy can remain invisible.

This racialization of perverse sexuality is obscured by the linking of sexual differences to differences of climate: hotter climates fostered sexual licentiousness, while England's temperate climate meant the English were modest. If climate explains the racialization of sexuality, then how can otherness speak to identity? The key lies in the fact that perverse sexuality adheres to the racialized other even when the other migrates to a more temperate climate. And whereas the other could not lose the marks of perversity, the European, by contrast, was always under threat: European blood offered no immunity to perversity. This "asymmetrical grammar" of race implies that blood works flexibly; its ability to confer identity has an on–off switch.[40] Moreover, as Ann Stoler observes, "sexual promiscuity or restraint were not abstract characteristics attached to any persons who exhibited these behaviors, but as often post hoc interpretations contingent on the racialized class and gender categories to which individuals were already assigned."[41] Finally, because the term "tropics" shifted from a climatic term to a medical one, resulting in the idea of the tropics as a place of sickness needing to be cured, even environmental terms took on the weight of identity.[42]

No one was really immune from sexual perversion.[43] Masturbatory insanity evolved into neurasthenia, which derives from that pathological catch-all term "nervous diseases" in the late eighteenth and nineteenth centuries in England, and the upshot of all this disease was universal vulnerability to perversion. Localizing disease in the nervous system gave physicians all the benefits of localization—the aura of a cure, therapy—without needing proof. Sexual perversion thereby gained the allure of embodied identity, even as the nervous body functioned as a kind of black box, specifying a cause without having to offer anything more than an assertion of cause. The fact that medical jurisprudence taught physicians how to read the male anus for signs of forced sodomy, listing "inflammation, excoriation, dilation of the sphincters, ulceration, a livid appearance, and thickening" as signs of "unnatural rape" further gave the lie to such immunity.[44] Immunity would seem to preempt the need for diagnosis. William Acton, the Victorian specialist in genital–urinary disorders who famously argued that women are not "troubled by sexual feeling of any kind" yet blamed female prostitutes for ensnaring men, rebutted the common English perception that venereal disease was far more common in Paris than in London by noting that London Guards had four times the cases of venereal disease than Parisian troops.[45]

Thus, European bourgeois sexuality and racialized sexuality overlapped.[46] Indeed, the standard classical education in English public schools made it clear that Ancient Greeks and Romans found little wrong with supposedly perverted forms of same-sex acts. Along with the public school custom of two males sleeping naked in the same bed, who knows what forms of combustion Horace, Petronius, Catullus, Martial, Plato, and Sappho could inspire or legitimate.

The July 8, 1810, police raid on the White Swan in Vere Street, a known hangout for sodomites, would show just how intertwined homosexual and heterosexual eroticism were. This coterie of men mimed heterosexual marriage, performing ceremonies complete with bride maids and bride men. Yet if homosexual relations travestied heterosexual norms, this group nonetheless disturbed the cultural linking of effeminacy with sodomy. The author of the *Phoenix of Sodom* notes that "it is generally received opinion, and a very natural one, that the prevalency of this passion has for its object effeminate delicate beings only: . . . Fanny Murry, Lucy Cooper are now personified by an athletic bargeman, an Herculean coalheaver . . . the latter of these monsters has two sons, both very handsome young men, whom he boasts are full as depraved as himself."[47] If manly men could prefer sex with other manly men, then desire could not be intrinsically heterosexual, and effeminacy could not explain a man's desire for another man. The Earl of Haddington likewise acknowledged that women could prefer to be with women. Although he insisted that both men and women were passionate, no one could predict the object of eroticism. "And I am told some women can / Do very near as much as man; And others have found out a sport, / To please them, of another sort. But these secrets are past my skill," he wrote.[48] By figuring lesbianism in terms of hearsay and secrecy, he made lesbianism a kind of ghostly apparition whose tenuous materiality made the very possibility of lesbianism deniable.[49]

INFLICTING COLONIALISM, NATIVE SEX

Sander Gilman identifies why sexuality had such a powerful role in Empire when he claims that "the Other's sexuality, labeled as perverse because it was seen as retrogressive, was soon identified with all modes of sexuality other than those prescribed for adult Europeans."[50] Sexuality and degeneration became intertwined in nineteenth-century thought, so much so that the Hottentot Venus, along with her large buttocks and so-called primitive genitalia were offered as evidence of primitiveness that help to justify European colonization and conquest of Africa.[51] Upon her arrival in London in 1810, Saartje Baartman became the star of one of its most successful shows, and viewers could poke and prod her buttocks for an extra fee. Although her protruding buttocks were thought to compensate for racial limitations,[52] her elongated labia were perhaps explained by masturbation. After her death, she was dissected in France by none other than the famous French comparative anatomist, Georges Cuvier, who donated her genitals to the French people. He lavished attention on her genitalia, breasts, buttocks, and pelvis, devoting nine pages of his sixteen-page autopsy to her sexuality.[53] Her enlarged clitoris could point to a primitive sexuality because less dimorphism was equated with savagery, and scientists implicitly argued that Africa needed to be settled by the racially, morally, and culturally superior white races. This othering of sexual passion from the female European onto the African native was in part a projection of profound psychological conflicts within European women that pitted the role of spiritual mother against the procreator.[54]

The denial or othering of perversity thus becomes a key strategy for maintaining empire.[55] The colonizer could now see himself as a masculine, even fatherly figure, taking possession of the wayward female native.[56] Such poster-boy masculinity was intensified by the colonist's effeminate status as the detritus of empire or of debtor. When perverse sexuality did not stick to the colonial body, the colonizer began to look colonized. Moreover, sexual perversion could fuel racism that was itself merely a heightened nationalism, an exalting of one's own nation over all others. Simultaneously, nationalism redirects men's passions to higher control.[57] Sublimation, in turn, made it all the more imperative to foist perverse sexuality onto the native. Trainee missionaries knew they had to marry before emigrating to secure them from the sexual licentiousness associated with the East and with slavery and Africa.[58] Victorian belief in the overcoming of the sexual instinct helped to sanitize the colonizer even as promoting public health and preventing disease helped to legitimate British imperialism.[59] Sexuality becomes such a useful tool because of its ability to screen one's own desires: the explorer Sir Richard Burton's translation of *The Kama Sutra* (1883) into English, for example,[60] emphasized the fact that Hindoos considered biting and scratching, not to mention fellatio, to be erotic acts. If this helped to make Hindoos into perverts, did English voyeurism make the English perverts too?

As Burton's example already indicates, British colonial rule in India was buttressed by the sexual barbarisms of the natives. Preventing Indian widows from immolating themselves in suttee and eradicating female infanticide helped to justify British civilizing of India.[61] Initially, because the English knew of the Indian practice of purdah, female seclusion, Indian women were thought to be passive.[62] Yet as the Begums of Oudh, elderly princesses who defied British rule, demonstrated, Indian females could feign passivity to manipulate the British into providing certain forms of protection for them.[63] Nearby, the British effeminized the Bengali male, claiming that such effeminacy originated in sexual overindulgence and by the premature consummation of marriage. Such effeminacy mandated virile British rule. However, many a manly colonial official were stymied by the fact that homosexual practices existed even among the most outwardly virile natives.[64] The virile yet homosexual Bengali undermined a European linkage of homosexual practices with an effeminate male personality, one that preserved the essential heterosexuality of desire. This violation made Bengali men irretrievably perverted. By the end of the nineteenth century, "tropical sex, like tropical disease was . . . unchecked by Western mores and malignant. Both—sex and disease—required the full power of medical, military, and judicial force to control the potential for contamination."[65]

Of course, much needed to be denied. Homoeroticism fostered group loyalty, however sublimated to concepts of duty and virtue, and fervent male friendships played key roles in the governing of empire.[66] Former public schoolboys went off to join regiments in India and lived in similar male societies abroad that they had at school.[67] As these examples suggest, male–male eroticism was threatening precisely because it so resembled the homosocial[68] colonial administrative world. In 1885 in England, the Labouchère Amendment sought to sever the resemblance between homosociality and homosexuality by making acts of "gross indecency" between men, even those in private, criminal acts.[69] Outside of England, boy brothels in Tianjin, China; Naples; and Karachi catered to foreigners along with 40,000 Shanghai prostitutes. Colonial narratives teem with accounts of the other's preoccupation with sexual excesses; nonetheless, it seems that no sooner than the European went south so did his morals and sexual proclivities.[70] Blame it on geography.

In Victorianism, perverse sexuality was embodied in the harlot, not to mention the decadent fop, and the unusual sexual proclivities of the upper classes were splashed throughout the pages of *The Times*.[71] Gentlemen moved into the slums of London beginning in the 1880s, and settlement houses of young bachelor Oxford and Cambridge graduates permitted same-sex desire under the guise of fraternal benevolence and religious instruction.[72] The prostitute's body, whose features were erased by syphilis, emblematized the prostitute's invisibility, her "strangely fleshless fleshiness."[73] As part of this erasure, Victorian prostitutes were depicted as being black primitives, throwbacks to man's origins in the chimpanzee.[74] This need to make prostitutes primitive, however, was belied by the fact that, as William Acton noted, no rigid demarcation existed between respectable society and the depraved underground as many English prostitutes lived in externally respectable establishments and not brothels.[75] Along with English prostitution, the feminist idea that marriage was a form of sexual slavery began to undermine distinctions between self and other because Western marriage was thereby likened to Oriental despotism. As sexology eroded feminist beliefs in women's sexual difference, purity, and continence, the notion of English purity was further strained.[76] The fact that British pornographers George Cannon, William Dugdale, and Henry Hayler were among the world's leading exporters of nude erotic photographs made English claims to purity a joke.[77] The London Police confiscated 130,248 obscene photographs from Hayler.[78]

Because the English had long perceived the Catholic Church to be a den of iniquity, with Catholicism being England's ultimate other, English conversions to Catholicism in the Victorian era demonstrated the extent to which the other could become the self. Britons prided themselves on their anti-Catholicness, and they equated the pope with the antichrist. Whereas Oscar Wilde was drawn to the ritual and eroticism of Roman Catholicism in part because homoerotic Christian icons intensified desire and priests approached the dandy, Radclyffe Hall, herself a lesbian and a convert to Rome, asks, "what of that curious craving for religion which so often went hand in hand with inversion."[79] The vexed relation between self and other was intensified by the concept of Aryanism. The very idea of the Aryan people was deeply embedded in the Vedic tradition, and this meant that the European self was not only indebted to the other for its self-conceptualization as Aryan, but it, as a result, urgently needed to make sexual perversity stick to the East. If the Eastern origins of Aryanism could elevate India into British brethren, that bond could be broken when Indian rebellion led to an emphasis on India's spiritual poverty, its lack of Christianity.[80] Once racialized as Aryans, Indians after the Mutiny were transformed into "niggers."[81] Once reracialized, Indians could become known for their perverted effeminate sex.

Despite the native's alleged perversity, sexual relations between colonizer and colonized could be rationalized, as it largely was in the British Empire, until 1909. Whereas the French thought sex would provide an easy means of gallicizing West Africa, British sexual contact with colonial Indians was legitimated on the grounds that this could provide essential knowledge of India.[82] Indeed, the Hindustani word *bibi,* meaning high-class woman, came to mean native mistress, and she was also referred to as a "sleeping dictionary."[83] In India, concubinage became the norm until the purity movements of the 1880s—which emerged from the movement to repeal the Contagious Diseases Acts of the 1860s, acts intended to ensure a healthy prostitute population for the British military—because British soldiers had to have their sexual needs met, and because intermarriage might help strengthen the army. Sexual continence of British soldiers was thought impossible; therefore, between 1850 and 1888 regulated prostitution was available in seventy-five cantonments where the Indian army was stationed.[84] That the British incorporated and regulated prostitution in India meant that their purity was questionable. Once in the East, however, the role of presiding over the exotic other made it possible to bracket sexual practices outside of one's European identity.

But sexual contact with native populations and miscegenation became a problem when British rule insisted upon well-trained administrators, and when the purity movement in England successfully persuaded the English that such contact was essentially corrupting.[85] Whereas they were initially embraced because they helped populate the Empire abroad, mixed races became a threat to white prestige and now embodied European degeneration and moral decay.[86] Of course, racial mixing was a challenge to the very idea of racial identity formation. The creation of the category of memsahibs—European ladies in Asia—provided an important buffer between civilized and native sex.[87] By 1864, English army admissions to hospitals for gonorrhea and syphilis reached 290.7 per 1,000 of total troop strength.[88] Because medical doctors defined women as the carriers of this disease, and men were depicted as its unwary victims, the colonizer's infections of natives along with his pathology were screened from view.[89] Once again the dialectical structure of perverse sexuality helped to sanitize the self while projecting disease onto the other.

SEXOLOGY, RACE, AND OTHERING THE SELF

First associated with a group of German scientists examining sexual disease, sexology developed a descriptive system to classify sexual types, like the homosexual, as well as forms of sexual desire, like fetishism, masochism, and sadism. Many were homosexual and sought to repeal legislation condemning homosexuality. Because the discourse of "perverse sexuality" situated the colonizer/colonized within the self, sexologists therefore needed to distinguish themselves from their objects of study by either taking a scientific role of urologist or endocrinologist or by labeling the patient deviant, perverted, or degenerate.[90] Race could further immunize the sexologist from perversion. Moreover, manifest intolerance to racial others often helped to deflect attention away from sympathy toward sexual others. When Havelock Ellis, for example, the author of the first serious study of homosexuality published in England, insinuated miscegenation and primitive sexuality into his study of same-sex relationships, he used the combined weight of race and primitive sexuality to absolve himself from contamination.[91] He predicted that "the question of sex—with the racial questions that rest on it—stands before the coming generations as the chief problem for solution."[92] Among "various lower races" and "warlike peoples" "there is a widespread natural instinct impelling men towards homosexual relationships."[93] Lumped into this category were the Ancient Greeks, the Chinese, American Indians, and Eskimos.[94] Closer to home, "a considerable lack of repugnance to homosexual practices may be found among the lower classes" and here he specified soldiers in Hyde Park.[95] For all his efforts to distance himself from his subject, *The Lancet* still denounced his work as "odious."[96] No doubt the fact that 33 of 36 of his case studies were British had something to do with its dismissal.[97] His sense that homosexuality was a statistical abnormality and not a disease further mandated denunciation.[98] Yet he supported the decriminalization of homosexuality on the grounds that it bore "for the most part its penalty in the structure of its own organism."[99] Both homosexuals and the "lower races" thus would put an end to their impurities by putting an end to themselves.

If Ellis demonstrates that the sympathetic study of homosexuality and racism could go hand in hand, the American Elizabeth Willard, the woman who coined the term "sexology," linked blackness with sexual savagery, even though she deplored slavery. She argued that "there is a much greater difference between Julius Caesar and the lowest Hottentot, than there is between the lowest Bosjeman and the highest chimpanzee."[100] Willard suggests that evolution made it possible to gesture toward equality and yet to essentialize blackness in terms of primitiveness and sexual savagery. If racial inequality were pushed back to a distant past, so as not to obscure the equalizing of the present, that very primitiveness implied that any equality would have to be earned. Because she considered sex without reproduction to be antithetical to health and bodily strength, pathology could further justify inequality. When medicine and race no longer supported racism, Willard turned to Christianity to justify inequality. She concludes by singing the praises of missionaries throughout the world who instruct "the ignorant, and hard-hearted."[101]

Likewise, sex researcher Magnus Hirschfeld argued for decriminalizing homosexuality, while simultaneously defending eugenics, the science of improving stock. By insisting that homosexuals could procreate if cured by psychoanalysis, and that any attempts to cure them would lead to further degeneration, Hirschfeld warned that the role of psychology was to make homosexuals happy by accepting the nature of their condition. Nonetheless, while homosexuals were not degenerates themselves, they were "substitute[s] for degeneration" and thus their children bear "the stamp of intellectual inferiority." Warning that "the marriage of a homosexual man or woman is always a very risky venture," Hirschfeld argued that society need not punish homosexuals because the very "nature of homosexuals serves as a preventive means against degeneration."[102] In any case, he suggests that even homophobia could help liberate homosexuals from psychiatric cures.

Yet too much otherness could prove a liability. If otherness enhanced the illusion of objectivity, it could also estrange the scientist from his subject, especially when the subject's psychological identity became the crux of his sexuality. If the other became too other, how could one know his or her psychology? Hirschfeld thus emphasizes his *connaitre* over his *savoir* in his first study of sexuality: the former implies knowledge of people, and the latter suggests clinical knowledge.[103] Sexologists needed to demonstrate some authority over sexual perversion and to persuade that the perverse had some truth to tell about the normal.[104] As the opposite to normal, perversion had no claims upon the population at large. Especially because sexologists operated at the margins of medicine and society, they needed to make the abnormal speak to the normal or to promise cures for abnormality.[105] While Ellis argued that "the stream of Nature still flows into the bent channel of sexual inversion," Mrs. Elizabeth Willard linked the laws of sexual order with those of the physical universe.[106] Together, they make perversion a natural part of human behavior. Another way to manifest authority was to frame modesty as repression, a repression that only a sexologist could recognize because repression could look like absence. Repression collapsed the distinction between self and other because there was now no position outside of sexuality; moreover, because modesty was considered to be a form of repression, the most normal-looking person now had the greatest potential for perversion. Whereas repression enabled a sexologist to extend his or her authority endlessly, it had the unfortunate result of making it impossible to immunize the sexologist from the contamination of perversion. Such extension, however, stretched the sexologist's professional training thin.

The sexologist therefore acquires mastery over human sexuality by reading modesty as sexual repression, and by claiming to inform his readers while titillating them. Thus, the sexologist's opening gambit was often the need to overcome the reticence of the public and provide information. In *Sexual Inversion*, Havelock Ellis, for example, pleads for his own authority based on his own "sincerity."[107] This sincerity acquires greater value against the reticence of the public. Not only is the public itself generally shy, but also previous writers on sexual perversion, namely travelers, are "shy of touching subject, ignorant of main points for investigation." Travelers, moreover, speak "vaguely about crimes against nature without defining the precise relationship involved."[108] He began his *Studies in the Psychology of Sex* with an essay on modesty. He argues,

> modesty—while common to both sexes, is more peculiarly feminine, so that it may almost be regarded as the chief secondary character of women on the psychical side. The woman who is lacking in this kind of fear is lacking, also, in sexual attractiveness to the normal and average man. The apparent exceptions prove the rule, for it will generally be found that the women who are, not immodest (for immodesty is more closely related to modesty than the mere negative absence of the sense of modesty), but without the fear which implies the presence of complex emotional feminine organization to defend, only make a strong sexual appeal to men who are themselves lacking in the complementary masculine qualities.[109]

Only a sexologist like Ellis can cut through the Gordian knots of modesty, for modesty is at once common to both sexes, but instinctively feminine, a fear that is concealed yet nonetheless instinctive, and present even in immodesty. In keeping with his uncanny ability to read both modesty and immodesty as symptoms of perversion, and to distinguish between 'mere negative absence of modesty' and immodesty, Ellis has one lack (fear) name another lack (sexual attractiveness and even femaleness) and proves his claims on the basis of exceptions rather than rules. This ability to perceive the modesty within immodesty, rendering both into symptoms of a diseased psyche and body, gives Ellis authority. Notions of an omnipresent instinct, along with an idea of modesty that does not allow for its absence, make absence present. Ellis further defines desire itself as essentially heterosexual since the only man who would be attracted to a woman lacking in modesty is a man who lacks masculinity. Ergo, only a lack of masculinity can find a lack of femininity attractive. Because modesty is what demands the "complex emotional feminine organization" to defend itself, women without modesty are not really women.

Like Havelock Ellis's insistence on modesty's essentially sexual core, the German dermatologist Iwan Bloch defined English prudery as the "surface veneer under which the darkest depravity may often lurk."[110] Bloch's metaphors of surface and depth grant his peeling off of the skin of prudery both legitimacy and necessary payoff. Such prudery, Bloch maintains, is heightened by England's geographical insularity. Surrounded by "precipitous cliffs and an angry sea," England is Bloch's Galapagos Islands.[111] While insularity promotes self-assurance, awareness of his own "inborn brutality" makes the English highly squeamish about propriety. Because a human being is "most completely himself" in the sexual sphere, sexuality thereby becomes the key to both the character and individuality of a people.[112]

Another way to promote the sexologist's authority was to control the traffic between sexual acts and identities, and to rethink the role of biological determinism. Concepts like "pseudo-homosexuality" or "accidental homosexuality," which severed acts from identity, or Bloch's and Freud's universalizing of perversion, which distributed perversion across the species and thereby made problematic the relation of perversion to individuality, helped to minimize the costs of a sexual identity. A congenital condition seemed at odds with criminal responsibility. Far from acceding to psychology the ultimate power to tell the truth about homosexuality, Bloch argued in *The Sexual Life of Our Time* that "to do justice to the whole importance of love in the life of the individual and in that of society, this particular branch of inquiry must be treated in its proper subordination as part of the general science of mankind, . . . constituted by a union of all other sciences—of general biology, anthropology and ethnology, philosophy and psychology, the history of literature, and the entire history of civilization."[113] Even Freud claimed that "an instinct is an urge inherent in organic life to restore an earlier state of things which the living entity has been obliged to abandon under pressure of external disturbing forces; that is, it is a kind of organic elasticity, or, to put it another way, the expression of the inertia inherent in organic life."[114] Instinct itself was partly somatic, "inherent in organic life." Hence sexologists would call upon anatomy to buttress their medical authority, even when they could not draw definitive conclusions, as did Ellis when he lavished attention on the abnormality of women's sexual organs in his case studies without specifying criteria for abnormality.[115]

The notion of an in-born sexual instinct was critical to the sexologist's authority.[116] The instinct provided an elaborate shell game, one that enabled the simultaneous grounding of and yet deferral of authority. When pressed to localize the instinct, the sexologist could move it from the body to the brain, from one region of the brain to another, from the brain to the mind, and then back again. Hirschfeld, for instance, used "brain" and "mind" as if the two were interchangeable when one refers to a physical organ and the other refers to an idea.[117] Although Krafft-Ebing located the instinct in the cerebral cortex, he had to admit that no part of the cortex had been identified as the seat of this instinct.[118] Despite the fact that sexuality was now the province of psychiatrists, this did not mean psychiatrists were willing to relinquish the moral authority that perversity provided. Because it refers to an evolutionary trajectory in which form atrophies from the organism's tending to abstain from function, hereditary degeneration provided another key means by which the somatic basis of perversion could be offered and then deferred or projected onto the future.[119] One could claim degeneration without proof. According to Bloch, one's congenital heterosexuality, moreover, did not preclude one's "becoming homosexual," as was the case of female prostitutes who were influenced by lesbian associates.[120]

Sexologists too recognized the value of disconnecting sex from identity. Having dwelled upon how sexology anchored sexuality in psyche, historians have paid less attention to how sexology undermined connections between sex and identity when identity proved too costly. Krafft-Ebing's concept of "periodical insanity" acknowledged that male sexuality was cyclical and that "while in the intervals the sexual instinct is neither intense nor perverse."[121] Krafft-Ebing, professor of psychiatry and Neurology at the University of Vienna, moreover, treated homosexuals who wanted to be cured; yet, he began to acknowledge that the unhappiness of homosexuals was due to societal condemnation rather than any inherent degeneracy.[122] Such a realization hinted that society needed curing, not his patients. Although his concept of "congenital urning" ("urning," a term coined by Karl Ulrichs, referred to a feminized male that sought other males) insisted upon sexed identity, Krafft-Ebing expressly noted that in some urnings their anomaly was limited "to the *vita sexualis*, and does not more deeply and seriously affect character and mental personality."[123] If sex life was separable from the rest of the psyche, how could sexuality capture identity? The "congenital" status of the urning was further undercut by the fact that sexuality did not rise to consciousness until puberty.[124] How to overcome the gap between sexuality and consciousness? He attempted to explain this gap away through concepts of "neuropathic predisposition" and "hereditary degenerate conditions." In the end, he had to admit that the "abnormal psycho-sexual constitution . . . is absolutely unknown."[125] Of course, such predispositions and hereditary conditions require the later manifestation of abnormal desire to become predispositions. Perversity is transformed into perversion when effect can be traced back to congenital cause; perversity is simply perversion before a case history. Although he believed female urnings had masculine souls, he thought that "the majority of female urnings do not act in obedience to an innate impulse." Krafft-Ebing's "cultivated pederasty" and "acquired contrary sexual instinct" acknowledges that when normal sexual activity is thwarted, as in prison or on shipboard, desire takes on whatever is at hand.[126] While acquired cases could be cured through the prevention of perverse conditions and through therapy, in congenital cases, "the benefit is probable only within the category of the psychical hermaphrodites, though possible . . . in that of urnings."[127]

Havelock Ellis claimed that sex was "all-pervading, deep rooted, permanent,"[128] yet he also argued that "sexual attraction between persons of the same sex, due merely to the accidental absence of the natural objects of sexual attraction, is . . . of universal occurrence among all human races."[129] Linking the accidental to the universal allowed Ellis to limit congenital inversion while universalizing homosexual acts. The idea of congenital homosexuality provided an origin that could look scientific, with the added benefit of not having to show proof. Perhaps because congenital inversion was empirically impossible to prove, Ellis returned to the climate's role in creating "proclivities" to "homosexual passion," and from there to race and gender. "This proclivity seems more common in the hotter regions of the globe."[130] He warned that although he felt female emancipation was a "wholesome and inevitable" movement. it would lead to more lesbianism because marriage is decaying and independence promotes "hereditary neurosis" the "germs" of which would result in women's "spurious imitation" of men.[131] He could thus claim progressiveness while clamping down on progress. The sexologist and British champion of birth control, Marie Stopes, likewise recognized identity's limits. She urged married couples to experiment with "the many sexual positions that suit them both," and she thereby weakened the connection between certain sexual positions and identities.[132]

Iwan Bloch similarly fudged the relation of sexuality to identity. Bloch protested that "the character of the people is revealed nowhere more clearly than in the realm of sex."[133] Simultaneously, he argued that much that passed for congenital homosexuality was acquired. Something acquired can only tell the truth of character if there is a predisposition toward the acquisition. Bloch warned that obscene photographs depicted perversions of every kind, and that these were coming into the possession of "persons of thoroughly normal sexuality or decidedly not subject to a specific perversion."[134] Panic over exposure thus intensified his authority; making sexual preference contingent, furthermore, gave him the power to prevent it. Bloch's concept of "pseudo homosexuality" described many homosexual practitioners who turned to same-sex partners for lack of heterosexual opportunity (his examples included sailors, athletes, and members of the English club).[135] He argued that childish impulses were not congenital dispositions because whereas a child's imagination is "permanently possessed" by accidental impressions, adults can easily free themselves of them.[136] Under the aegis of the child's imagination, congenital homosexuality is really a delusion. Prostitutes, moreover, were inclined to homosexuality not because of "somatic conditions," but rather by their "growing repugnance against intercourse with men."[137] Bloch would later confess that homosexuality was a "riddle" because "the more I have endeavored to study it scientifically, the more enigmatical, the more obscure, the more incomprehensible, it has become to me." He then sheepishly offered his only certainty: "but it exists. About that there is no doubt."[138]

Notwithstanding this confession, sexuality's ambivalent role in identity allows Bloch to leverage his authority. Where acquired pseudohomosexuality allows him control over the social environment, congenital homosexuality helps him to narrate the other's identity. When he thought most homosexuality was acquired, Bloch sought to depathologize the sexual perversions, locating them under physiology and not pathology because they were so ubiquitous. Nonetheless, Bloch urged the state to repress homosexuality, and he warned that a repeal of the criminality of homosexuality would "contribute tremendously to the ruin of youth."[139] The only cure was to guard children from exposure to harmful influences. While later operating under the assumption that most homosexuality was congenital, Bloch hoped to decriminalize homosexuality and to make it "compatible with complete mental and physical health." Of course, compatibility is not identity. Formerly, he thought that "true homosexuality was only a variety of pseudo-homosexuality—in a sense a *larval* pseudo-homosexuality." Conceding that "true homosexuality as a congenital phenomenon is far greater than" he had once assumed, Bloch normalizes it because it is contained within individual psyches.[140] Only after there are fewer larvae can Bloch stifle his homosexual panic.

The Munich physician Albert von Schrenck-Notzing thought suggestion therapy could cure all forms of inversion, and he believed one could inherit a predisposition to normal or perverse sexuality. Heredity, however, was a limited explanation because "every hereditary disposition, to become active, requires its specific excitant," and education is what teaches us to select some and not others.[141] To minimize the role of heredity, Schrenck-Notzing spoke of it in pejoratively as in "hereditary weak-mindedness."[142] And he argued that the idea of congenital homosexuality justified homosexuals' obedience to impulse, and allowed them to regard themselves as not responsible for their actions.[143] By thinking about what the language of perversion did—it permitted the expression of certain forms of sexuality without any responsibility for them—instead of assuming that perversion was based on impulse, Schrenck-Notzing demonstrated the very power of suggestion therapy. Seeing innate impulse as an excuse allows perversion to evaporate into thin air. Ellis nonetheless tartly retorted that because homosexuals imagined their own sex even in the act of heterosexual coitus, all Schrenck-Notzing had done was to teach male homosexuals to masturbate with a vagina.[144]

If psychology can simultaneously individualize and universalize, then, its role in identity formation has been oversimplified. How can what belongs to the species also be individual? By shuttling between the individual and the species, perversion acquires the ability simultaneously to deliver the self and to explain humanity, even as psychiatrists widen their pool of potential patients. Bloch insisted that the sexual anomalies were "universal human, ubiquitous phenomena" and that "extravagance in the manner of increasing the libido sexualis is characteristic of the human race."[145] Freud argued that since "no healthy person, . . . , can fail to make some addition that might be called perverse to the normal sexual aim, . . . the universality of this finding is . . . enough to show how inappropriate it is to use the word perversion as . . . reproach."[146] Hirschfeld, a homosexual who urged the decriminalization of homosexuality, coined the term "transvestite."[147] He claimed that a diagnosis of homosexuality was decisive when a "homosexual psyche" could be adduced.[148] But the absolute difference of this homosexual psyche was undermined by the fact that he considered humans to be universally bisexual, with either the homosexual or heterosexual component being predominant. Difference of kind—homosexual psyche—becomes one of degree, predominance, and the fluctuation of kind to degree implies that both otherness and identity have their strengths and limitations. He continues, "homosexuality is neither a disease nor a degeneration, but rather is a part of the natural order, a sexual variant, such as numerous analogous sexual modifications in the animal and plant kingdoms."[149] By enabling sexual perversion to become sexual modification by stripping away both disease and degeneration, at least for the homosexual, Hirschfeld allowed perversion to dissolve into mere sexual variation.

In sum, from 1800 to 1920, perverse sexuality is a capacious dialectical structure, one that continually repositions the self in relation to its other. Such repositioning looks like fixed identity as sexuality reincarnates itself into gender, class, nationality, species, and race. The British middle class needed the perverted other to consolidate its class position and to maintain empire, whereas sexologists turned to racialized others so as to deflect attention away from their sympathy toward perverts. To the extent that historians of sexuality have not yet come to terms with the role of the other in the history of sexual identity, then they have missed its role in the formation of empire, both at home and abroad, not to mention its crucial role in the formation of identity. Because identity has its costs and benefits, writers on sex have long recognized that detaching sex from identity facilitates the foisting of the self onto the other. If sexologists helped to consolidate rather than invent sexuality as identity, they were also surprisingly open to thinking about the implications of sex beyond identity. Such flexibility partly explains why readers of sexology have found such varying reasons to take comfort in it.

CHAPTER FOUR

Medical Science, Technology, and the Body

CHANDAK SENGOOPTA

In 1920, the Viennese physiologist Eugen Steinach (1861–1944) introduced an operation that became wildly popular with men of a certain age. In a strictly surgical sense, the operation was far from novel. It was no more than a simple vasectomy, but its results, Steinach claimed, were revolutionary. Cutting and ligating the vas deferens (which normally carries semen) stimulated the testicle to produce more sex hormones, which revitalized the entire organism, leading to virtual "rejuvenation."[1]

The first patient was a coachman, only forty-four years old, but presenting "a typical picture of premature senility without organic disease." He had lately been unable to work for long hours and had lost weight and appetite. His skin was dull, his hair gray and scanty, and his muscles weak. The patient also had a hydrocele, and the vasoligature was performed without his knowledge during an operation for the hydrocele performed on November 1, 1918. There were no dramatic consequences for the first three months. Then gradually, the patient's appetite increased, he gained weight, and his appearance became hale and hearty; a year later, his hair had grown thicker, and he reported that he now carried "loads up to 220 pounds with ease." His hair grew so luxuriantly that he had to shave twice as often as earlier. Eighteen months after the operation, the patient "with his smooth, unwrinkled face, his smart and upright bearing" looked like "a youthful man at the height of his vitality."[2]

The news spread like wildfire and reports of successful rejuvenation by the Steinach operation began to pour in from all over the world.[3] The *New York Times* reported in 1923 that every major American and European city already had a number of surgeons specializing in the operation.[4] By 1926, New York alone reportedly had more than a hundred surgeons who performed it "especially in cases of acute premature loss of vitality."[5] One of them, Steinach's admirer and tireless publicist Harry Benjamin, made sweeping claims for the operation. "The point that so far has not been brought out with necessary clearness," he observed,

> is, that by the Steinach operation the patient is given a more or less massive and continuous dose of his own gonadal hormone . . . All symptoms due to senility, including sexual impotence, as a rule improve after the operation . . . A few other possible indications would be: beginning arteriosclerosis, hypertrophy of the prostate, eunuchoidism, mental depression, and cases of dementia praecox, where we suspect gonadal deficiency as cause.[6]

Senile and decrepit men became energetic, muscular, virile, and sexually active after the operation, but its magical effects were not confined to the body. The creatively inclined regained their former imaginative vigor—the poet W. B. Yeats underwent the operation during a barren period and claimed to have regained his sexual and his literary abilities.[7] Writing in 1940, Steinach himself referred to more than 2.000 operations performed by surgeons in Vienna, New York, London, St. Petersburg, Copenhagen, Chile, Cuba, and India.[8] By the time Steinach wrote these words, however, the operation had already gone out of vogue and Steinach himself was almost a forgotten man.

At one level, the story of the Steinach operation may seem familiar, even banal—a medical innovation is launched with dazzling claims, proves popular for a while, but bites the dust when its results are shown to be less than impressive. Approached from a different perspective, however, the story illustrates a range of interesting and important issues, from the growth of laboratory medicine to the rise of medical consumerism, not to mention changing cultural attitudes toward aging, sexuality, and gender. Above all, however, Steinach's work demonstrates the scientific and technological transformation of the nineteenth-century body into a new entity that, despite its apparent solidity, was "produced" by a dynamic interplay of chemical forces and that could be significantly modified by influencing those forces.

THE NEURAL BODY

How does the organism work as a whole? What organizes and coordinates the functions of the diverse parts of the body? The ancient answer was sympathy, especially for situations where organs with no obvious connection affected one another, such as the characteristic flush of the cheeks in tuberculosis of the lungs. Sympathy could, obviously, be mediated through the nerves or the blood. Galen had allowed for both, but by the seventeenth century, the nervous connections came to be accorded paramount importance. The concept of sympathy remained of importance until the nineteenth century, and the nervous system—the brain, of course, but especially the spinal cord, the ganglia, and the innumerable nervous fibers traversing every tissue and organ—represented the communicative and integrative network of the body.[9]

The early nineteenth-century body was governed by the nervous system. The nerves carried messages, connected different viscera, and modulated their functions. Many of these nervous links remained assumptions rather than verifiable anatomical features—the supposed connections between many organs and the brain or the spinal cord were never really identified in clear, structural terms. The network of nerve fibers, ganglia, and plexuses, governed ultimately by the brain, was assumed to connect the vast and diverse masses of organs, cells, and tissues, enabling them to coordinate their functions. The notion of a neurally governed body seemed especially plausible because, as was established by the mid-century research of Emil Du Bois-Reymond, the nervous system communicated by electrical signals.[10] Electricity was the object of extraordinary scientific and popular fascination at the time, and analogies between the nervous system and the telegraph were frequent.[11]

As streets and houses began to be lit with electric lamps and telegraph wires hummed across the planet, physicians began to represent the body as an electrical battery. If the body was driven too hard, then, like any battery, it could run down, resulting in what the New York neurologist George Miller Beard (1839–1883) called neurasthenia, a debilitating condition characterized by a galaxy of symptoms ranging from headache, pervasive fatigue, and sleeplessness to blushing, tooth decay, and dyspepsia.[12] The nervous energy of the body, Beard claimed, was being drained by the relentless pressures and vicissitudes of modern life. The obvious treatment, of course, was to replenish the depleted electrical charge.

Electrotherapy, which had long been practiced in various forms, received a great boost thanks to the popularity of the neurasthenia diagnosis, and a whole range of electrical devices and methods were introduced for recharging depleted nerves.[13] The body, in other words, was redesigned as symbol, victim, and triumph of industrial modernity—driven by electricity, vulnerable to the cultural acceleration brought about (in large part) by that same magical fluid, and redeemable through a relatively simple technological fix.[14]

The neurocentrism of nineteenth-century physiology was hitched to contemporary ideas of femininity in explanations of ovulation and menstruation. In a mid-century essay, the young Rudolf Virchow (1821–1902) argued

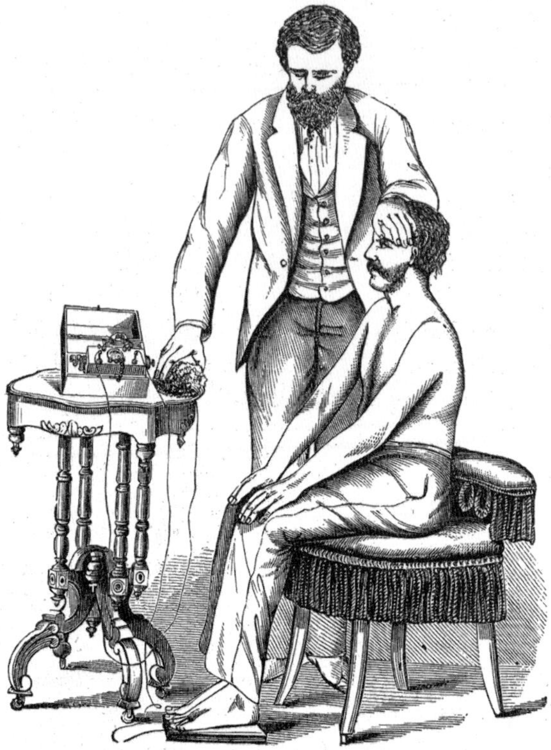

FIG. 126.—General Faradisation.

FIGURE 4.1: Doctor administering electrotherapy. From Julius Althaus, *A Treatise on Medical Electricity* (London: Longman, 1873). Wellcome Library, London (L0002393)

that the cardinal function of the ovary was the production of the ova, which caused menstruation.[15] What explained ovulation, therefore, would also explain menstruation. The ova ripened and were expelled only periodically from the ovary—but what determined that cycle? Was the ovulatory stimulus communicated through the nerves or the blood? This was the crucial question and to answer it, he emphasized, one had to choose between humoralism and solidism. Virchow, of course, was renowned for his solidistic doctrine that disease originated in cellular structures.[16] Solidism, however, was more than a matter of attributing the origin of *disease* to solid structures. Virchow's views on the nature of the body and its functions, too, were solidist to the core, and he heaped scorn on humoral concepts in pathology and physiology.[17]

If it was indeed the blood that carried the ovulatory impulse to the ovary, then, he argued, one would have to decide whether the periodicity of ovulation

FIGURE 4.2: Hydroelectric bath, ca. 1910. From J. A. Riviere, *Equisses Cliniques de Physicotherapie* (Paris: Bouchy, 1910). Wellcome Library, London (M0015161)

(and menstruation) was caused by periodic changes in the composition of the blood itself or by some mysterious substance produced by an unknown organ and carried to the ovaries by the blood. Nothing, he asserted, was known so far about the first alternative, and the second was implausible. Virchow cited a case of female conjoined twins who, despite living for twenty-two years with what was essentially a single circulatory system, menstruated at different times. Ovulation, therefore, was likely to be brought about by periodic changes that were intrinsic to the ovary. What could be more natural than to assume that those periodic, rhythmic changes were mediated by the nervous system? The rhythmic processes of respiration and heartbeat and the contractions of the uterus during labor were all instances of neurologically regulated periodicity—it was simply logical to assume that ovulation and menstruation, too, were similarly regulated.[18]

But the ovaries did much more than control ovulation and menstruation—they generated femininity itself and all its physical, mental, and behavioral attributes:

> The female is female because of her reproductive glands. All her characteristics of body and mind, of nutrition and nervous activity, the sweet delicacy and roundedness of limbs . . . the development of the breasts and non-development of the vocal organ, the beauties of her hair and the soft down on her body, those depths of feeling, that unerring intuition, that gentleness, devotion and loyalty—in short, all that we respect and admire as truly feminine are dependent on the ovaries. Take the ovaries away and we get the repulsive, coarsely formed, large-boned, moustached, deep-voiced, flat-breasted, resentful and egoistic virago (*Mannweib*).

Woman, in other words, was woman because of her ovaries—*propter ovarium solum mulier est quod est*—and the ovaries exerted their influence over body and mind through the nervous system.[19]

The ovaries were often surgically removed in nervous disorders, because an irritative focus in the ovaries was held to endanger the entire nervous network; conversely, an irritation developing in other parts of the neural network could implicate the ovary. These ideas were rooted in the theory of spinal irritation, which held that impulses from an irritative focus in the reproductive system could radiate to the spinal cord and subsequently to the entire body via the nerves, and in the allied notion of reflex neuroses introduced by neurologist Moritz Romberg in the 1850s, according to which the female reproductive system was the ultimate source of hysterical manifestations such as headaches, cardialgia, and epilepsy-like convulsions.[20] All bodies, in short, were neurally governed, but female bodies were vulnerable to particular nervous disturbances that could commence in or spread to organs that were not part of the nervous system.

The testicles had long been supposed to be the seat of virility, and the consequences of castration had been remarked upon since the days of Aristotle. It remained conventional wisdom for centuries that castrated males became feminized. The British naturalist William Yarrell (1784–1856) provided a classic account of the feminization of the castrated cock, which was approvingly quoted by Charles Darwin, August Weismann, and many other nineteenth-century scientists and physicians.[21] The discussions of the consequences of castration contain much material worthy of analysis, but for our purposes what is most important is the mechanism whereby the removal of the testicles led to the loss of masculine attributes. Just as in the case of the ovary, the influence of the testicles on the body was supposed to be mediated through the nervous system. Reports of localized consequences of unilateral castration—such as the loss of, say, the left antler in reindeer who had lost the left testicle—or the fact that only certain regions of the body sprouted hair at puberty seemed to exclude even the possibility of a blood-borne agent. Although the connections between the testicles and the central nervous system were never used to justify the removal of testicles in nervous or mental disorders, the testicles, too, were held to be crucial nodes in the neural network governing the body.[22]

THE EMERGENCE OF A "NEW HUMORALISM"

With hindsight, one can see that not all nineteenth-century scientists endorsed the concept of the neurocentric body. In a series of imaginative experiments, the Göttingen physiologist Arnold Adolph Berthold (1803–1861) showed in the 1840s that castrated cocks could be swiftly revirilized by reimplanting their own testicles on their intestines. Because the reimplanted testicles could not possibly have developed nervous links to the organism in such a short period of time, the virilizing influence of the testicles, Berthold argued, could only be mediated through the blood. Unfortunately, Berthold's experiments proved hard to replicate, and they did not lead to any reformulation of neural theories of gonadal function.[23] Even less influential than Berthold's work was the 1874 report of Strasbourg physiologist Friedrich Goltz (1834–1902) that a bitch with her spinal cord transected at the level of the first cervical vertebra had gone into estrus, mated (with a male toward whom she had previously been antagonistic), and became pregnant with triplets.[24]

In an intact animal, Goltz argued, one could assume that estrus was produced by nervous impulses traveling from the sex glands to the brain; that hypothesis, however, was obviously inapplicable when the spinal cord had been completely separated from the brain, as no nervous impulse could reach the brain without going through the spinal cord. Goltz, therefore, concluded that the sex glands exerted their effect on the brain by releasing specific chemical substances into the blood. Despite its acuity, Goltz's hypothesis was virtually ignored. This, Hans Simmer has observed, may have been because of the absence of a theoretical framework that would permit a proper evaluation of Goltz's finding. Only after the conceptualization of a chemical system of regulation akin to the nervous system could one perceive the true significance of Goltz's experiment.[25] That thoroughgoing reconceptualization of physiology began only in the late 1880s.

On June 1, 1889, the distinguished seventy-two-year-old French-American-Mauritian physiologist Charles-Edouard Brown-Séquard made a sensational announcement at a scientific meeting in Paris. For the past few years, he told his august audience, he had been feeling increasingly frail and decrepit but instead of accepting it as natural for his age, he had tried to "rejuvenate" himself by injections of testicular extracts from dogs and guinea pigs. His experiments, he claimed, had been resoundingly successful—he now felt many years younger and could work with the full vigor of youth.[26] Although Brown-Séquard attracted more than his fair share of derision (and his death within a few years of his supposed rejuvenation certainly did not help his cause), his apparently misguided experiment was to prove enormously fruitful.

Merriley Borell has emphasized that Brown-Séquard's obsession with the testicles was based less on endocrine notions of testicular function than on a more traditional faith in the energizing qualities of semen. The nineteenth century, it is well-known, was the century of the spermatic economy, when any inordinate loss of semen, whether through masturbation or excessive venery, was supposed to cause immense damage to health and morals.[27] Because the loss of semen was harmful, the absorption of semen was likely to be of some benefit. Hence, testicular extract could be seen as a likely source of health and renewed vitality in old age, when the production of semen was on the wane. Brown-Séquard may have been mocked by his fellow scientists, but he inaugurated a clinical revolution. Following his announcement and his subsequent reports, the medical marketplace was flooded with extracts and various preparations of virtually every organ—from the brain to the ovaries, from the testicles to the spleen—which were prescribed for virtually every conceivable disorder that did not respond to orthodox treatment. "Organotherapy, opotherapy, or the Method of Brown-Séquard as it was often called, came," says Merriley Borell, "to be the therapeutic hope of physicians from Cleveland to Bucharest."[28]

But Brown-Séquard had not merely introduced a new therapeutic fad; he had opened a whole new perspective on physiology. Benjamin Harrow of Columbia University, an eminent researcher on vitamins and hormones, observed in the 1920s that Brown-Séquard had revived the old humoralism "in quite a modern form, and with reasons for its revival drawn from the knowledge of the nineteenth and not the ninth century. One may truly say of him that he is the founder of the conception of ductless glandular function as we understand it today."[29] Harrow was exaggerating the modernity of Brown-Séquard's own physiological views, but he was right about the consequences of the elderly scientist's sensational experiment.

Partly to investigate the rationale of organotherapy, there was much research in the 1890s on the secretions of the so-called ductless or endocrine glands, such as the thyroid, adrenals, ovaries, and testicles. Thyroid extract was shown to be of real therapeutic value, and a few years later, it was found that extracts of the adrenal glands could raise blood pressure. By 1903, the British physiologists Ernest Starling and William Bayliss proved that an important step in the digestion of food was governed by a chemical secreted by the small intestine. Brown-Séquard and the organotherapists may have been wrong about many things, but they had been right in their belief that certain organs produced chemical secretions that served important functions in the body. In 1905, Ernest Starling, on the suggestion of a Cambridge classicist, named these chemicals "hormones" (from the Greek "I excite").[30] The first half of the twentieth century was to be the heroic age of hormone research. Over this period, hormone after hormone was identified, and the complex workings of the endocrine glands elucidated at a furious pace.

The testicles remained prominent in this research. After Brown-Séquard, it was no longer possible to dismiss the latter associations as mythical, and many scientists devoted their careers to the study of testicular functions. The elaboration of modern concepts of testicular endocrinology was greatly dependent upon the elucidation of testicular histophysiology. In 1850, anatomist Franz Leydig (1821–1908) had identified that the testicles comprised at least two distinct groups of cells: the germinal cells, which produced spermatozoa, and the interstitial cells, which came to known as Leydig cells. It was noticed that the destruction of the sperm-producing cells—which occurred naturally in undescended testes—did not interfere with the development of the secondary sexual characteristics. Boys with undescended testicles usually grew up to be fairly typical men.[31] Similar findings were obtained after experimental destruction of the germinal tissue by other investigators. The virilizing secretions of the testicle, therefore, were not produced by the germinal cells.[32] The experiments of Pol André Bouin (1870–1962) and Paul Ancel (1873–1961), anatomists at the University of Nancy and then at Strasbourg, provided persuasive evidence that male sexual characters depended entirely on the secretions of the interstitial (Leydig) cells.[33]

MODIFYING THE HUMORAL BODY

The scientist who was to build most extensively on these findings was Eugen Steinach (1861–1944). Through a series of imaginative experiments on rats and guinea pigs, Steinach established that the physical and behavioral characteristics of sex—body build, fur, breasts, internal and external genitals, sexual orientation, mating rituals—were all generated and maintained by the testicular and ovarian hormones. Moreover, Steinach claimed, the *degree* of somatic and functional masculinity was related to the degree of secretory activity of the testicles. With partially functioning testicles, one obtained partly masculine subjects. As far as sexual behavior was concerned, the development of an immature male animal into a sexually active male depended on the action of the testicular secretions on the central nervous system. Steinach called this effect "erotization" (*Erotisierung*). Once achieved, erotization lasted for a long time, even in the absence of the testes, a claim supported by his own earlier studies of rats castrated *after* puberty and by reports on the persistent sexuality of human eunuchs.[34] All the masculinizing effects of the testicle, Steinach emphasized, were brought about entirely and exclusively by the Leydig cells, which he collectively named the "puberty gland" (*Pubertätsdrüse*).[35]

Sexual development, then, was under the control of the internal secretions produced by the gonads and could be modified at least to some degree by

FIGURE 4.3: Eugen Steinach. Photograph by J. Scherb after a painting. Wellcome Library, London (V0027219)

eliminating or modifying those secretions. Forgotten by endocrinologists and just beginning to interest historians, these experiments were accorded high importance by some of Steinach's most prominent scientific contemporaries. Nobody in early twentieth-century central Europe exercised greater authority on endocrinology than Artur Biedl, and in the English-speaking world, Francis Hugh Adam Marshall's *Physiology of Reproduction* was the most authoritative conduit for information on new endocrinological approaches to reproductive biology. Both Biedl and Marshall made considerable space for Steinach's experiments, without necessarily agreeing with all of Steinach's claims.[36] Other scientists of the time attempted to replicate Steinach's results, and their experiments, too, were widely discussed and debated in the contemporary medical literature.[37] As a stimulus to research and discussion, Steinach's work was simply nonpareil. Although he never did receive the Nobel Prize, he was nominated for it six times between 1921 and 1938.[38]

Steinach's science, respectable as it was by the standards of the time, was obviously shaped by contemporary ideas of gender, but in spite of its roots in traditional, polarized views of masculinity and femininity, its results served ultimately to undermine some of those notions. Steinach's experiments showed above all that masculinity and femininity were not immutable givens, but chemical phenomena that were malleable at least up to a point. Gender and its attributes, therefore, were always potentially unstable. By repeatedly demonstrating how easily masculine or feminine sexual characteristics could be distorted, accentuated, or attenuated, Steinach ultimately subverted the fundamental concept of sexual polarity. This subversive dimension was reinforced by Steinach's lack of interest in the genetic determination of sex. Although there was growing interest in the genetic basis of sex in the early twentieth century, Steinach never even mentioned it in any of his published writings. In one private letter, however, he conceded that sex *was* indeed determined by genetic factors but insisted nevertheless that—and this, for him, was crucial—the *characteristics* of sex could always be modified by modulating the functions of the sex glands.[39]

Neither the physical sexual characters nor the complex psychological and behavioral manifestations of sex were governed by the chromosomes. With the inappropriate kind of internal secretion, an organism's sexual characters and behavior might be the reverse of what was determined by the chromosomes. The message was clear and simple: even if sex itself was determined in some metaphysical sense by genetic factors, its actual, outward manifestations depended on hormones, and by manipulating the latter, one could, for all practical purposes, manipulate sex.[40] What had emerged, then, from Steinach's apparently bizarre sex-change experiments was not simply a new concept of sex but a whole new understanding of the body. The solid, neurocentric body was now shown to be largely under the control of chemical secretions, and as

Steinach and his followers emphasized time and again, the actions of the sexual hormones did not merely concern copulation or reproduction. They were responsible for endowing the organism with vigor and vitality. The perennially adventurous Steinach was so convinced of the potential of the new science that he could not wait to apply his findings to humans.

The world was more than ready for him. "Glands" and their secretions had exploded into popular consciousness after the World War I. Scientists and journalists, ordinary doctors and elite specialists, novelists and scientific popularizers talked endlessly about those enigmatic organs. The American biochemist Louis Berman, author of the phenomenally popular 1921 work *The Glands Regulating Personality*, predicted that by studying "the chemical conditions of his being," man would "climb to those dizzy heights where he will stretch out his hands and find himself a God." In 1922, the medical correspondent of the *New York Times* complained that "a war-ridden world" had been succeeded by "a gland-ridden world"—despite the uncertainty of the scientific evidence, more and more people were coming to believe that the endocrine glands and their secretions held the key to the secrets of life, sex, and death. Because sex was the basis of life, and because the secretions of the testicles and ovaries controlled sex in all its protean manifestations, it seemed self-evident that life itself be enriched and ennobled by glandular means. Or, to use the words of Steinach himself, "just as the sex glands are situated in the middle of the body, so too are they at the centre of life itself."[41]

Steinach's attempt to apply his experimental findings to human problems began with male homosexuality. Collaborating with a urologist and a clinical sexologist, he introduced a treatment for homosexuality that involved the removal of a testicle and its replacement by one from a heterosexual donor. The first subject, a thirty-year-old homosexual with an effeminate body, began to have heterosexual erotic dreams soon after the operation. Within a few weeks, he had had sex with a female prostitute, and his body was becoming leaner, hardier, *and* more masculine. In less than a year, he married, and wrote to his doctors: "I am disgusted to think of the time when I felt that other passion." No less a figure than Sigmund Freud greeted Steinach's results as "very important" but other doctors found Steinach's results hard to replicate, and the operation quickly went out of vogue. By then, the irrepressible Steinach had moved on to what would be seen as his greatest achievement—and later, as his greatest folly.[42]

In his experiments with rats and guinea pigs, Steinach had been struck by the fact that the physical characteristics of masculinity or femininity were gradually effaced in old age. Because those characteristics were determined by the hormones of the testicles or the ovaries, it was likely that senile individuals suffered from a deficiency of sex hormones. If their gonads could be encouraged to produce more sex hormones, then the sexual characteristics would regain their pristine sharpness, and—this was the key point—they might also regain some of the general vitality of youth. One simple way of raising the hormone output of the testicle, Steinach claimed, was a vasectomy. Subjecting senile rats to this minor operation, he found that the previously lethargic, underweight, and almost lifeless animals became active, gained weight, developed a glossy new fur, and regained sexual interest. Some even turned into tireless sexual athletes, copulating as many as nineteen times in fifteen minutes. Would the same technique work in human beings, however? Teaming up again with his urologist colleague, Steinach tried out his technique on the coachman we met at the beginning of this chapter, with spectacular results.

The news, as we saw, spread like wildfire. Steinach's very name soon became a verb; people no longer had the Steinach operation—they were Steinached. The operation attracted some very high-profile clients. One was the poet William Butler Yeats. Passing through a period of bad health and creative barrenness in his late sixties, Yeats resorted to it with spectacular results. "It revived my creative power," he wrote, "it revived also sexual desire; and that in all likelihood will last me until I die." It was not just his lust that was stimulated

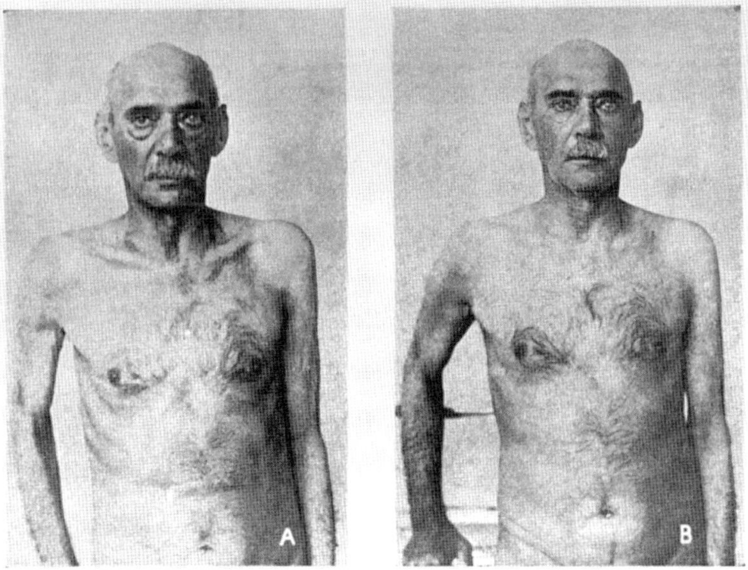

(a) SEVENTY YEAR OLD PATIENT BEFORE THE STEINACH OPERATION
(b) TWO MONTHS AFTER THE STEINACH OPERATION
(*Ufa Steinach-Film*)

FIGURE 4.4: Before and after picture of a rejuvenated man. From Paul Kammerer, *Rejuvenation and the Prolongation of Human Efficiency* (London: Methuen & Co., 1924). Wellcome Library, London (L0025163) by the operation, but also his art, resulting in a crop of late poems ranked with his best work. Innumerable case histories of less famous (or at least, less identifiable) patients were published, reporting remarkable results ranging from weight gain, improvement of skin, and hair growth to greater interest in sex, enhanced energy, and even, in applicable cases, a flowering of creativity. One admirer of Steinach hoped that the operation would soon become as mandatory for older men as vaccination was for children.

The Steinach operation was soon joined by an even more colorful rival—the monkey-gland operation, which involved grafting testicular tissue from chimpanzees or baboons into aging men. The use of chimpanzee testicles in man was the innovation of the Paris surgeon Serge Voronoff. "In the manifestation of his physical and intellectual qualities, man himself," declared Voronoff, "is worth whatever his sex glands are worth." Aging and senile debility were due to the exhaustion of the testicles, and the solution was to supplement the failing glands with a transplant. Because French law prohibited the removal "of any part or parts of the body, even in death," using human testicles was out of the question; in a post-Darwinian age, the obvious solution was to use ape testicles instead, and by the mid-1920s, Voronoff had performed about a thousand monkey-gland transplants. (He had set up his own monkey farm on the Italian Riviera, employing a former circus animal-keeper to run it.) And in America, perhaps the most enthusiastic land for glandular treatments of all kinds, monkey glands soon became all the rage. Voronoff's American acolyte, Chicago surgeon Max Thorek recalled that soon, "fashionable dinner parties and cracker barrel confabs, as well as sedate gatherings of the medical élite, were alive with the whisper—'Monkey Glands'." Poet E.E. Cummings sang of a "famous doctor who inserts monkeyglands in millionaires," and indeed, there were many such doctors and much money to be made with the Parisian miracle cure.[43]

Thanks to monkey glands and the Steinach operation, the 1920s turned into the decade of the testicle. One can hardly think of another era when one bodily organ featured so consistently and so prominently in the popular media and the specialist medical press. From *Good Housekeeping* to the *Journal of the American Medical Association*, there was no escape from the latest news and debates on the secretory functions of the testicles, even though the media referred prudently to "glands" rather than to testicles or gonads. There were doubts and concerns aplenty about the rejuvenation operations, but no accusation of quackery or fraud. In Britain, for instance, Ernest Starling told the Royal College of Physicians in 1923 that although further observations were necessary before the results of the Steinach operation could be accepted unreservedly, "it must be owned that they are perfectly reasonable and follow, as a logical sequence, many years' observations and experiments in this field."

Most of the hope and the hype about rejuvenation centered on men, but procedures were available for women, too. One, introduced by Steinach himself, involved irradiation of the ovaries by X-rays. The American novelist Gertrude Atherton was a famous recipient of this treatment. After years of churning out novels and journalistic pieces, Atherton had succumbed to a barren phase around the age of sixty-five, and she approached an American associate of Steinach's for rejuvenation. After the treatment, "torpor vanished. My brain seemed sparkling with light," Atherton exclaimed in her memoirs. "I almost flung myself at my desk. I wrote steadily for four hours." The fruit of this creative ferment was *Black Oxen*, a novel about the rejuvenation of a woman, which turned out to be the biggest success of her career, earning more than $30,000 in royalties in one year.

The later history of the Steinach operation cannot be covered here, but briefly, with more research on sexual physiology, the principles on which Steinach had built his operation came to seem quite simplistic. The idea that the testicles or the ovaries somehow generated the biological characteristics of sex all by themselves was replaced by far more complex theories stressing the interconnected nature of endocrine glands—the gonads were powerful organs, to be sure, but they did not act in the kind of splendid isolation that Steinach and his generation had imagined. Also, by the time of World War II, a veritable revolution in biochemical techniques had led to the availability of potent, standardized hormone extracts—even if procedures like the Steinach operation or the monkey-gland transplant worked, there was no longer any need for them. The same effects could be obtained more conveniently by injections of sex hormones. The Steinach operation and most of Steinach's theories, therefore, were obsolete even before his death in 1944.

A NEW BODY FOR A NEW AGE

One could, of course, look upon the history of the Steinach operation as an episode in the early history of endocrinology. I would suggest, however, that the story offers bigger, deeper insights. What sustained the whole Steinach saga was the interaction of concepts, techniques, and imperatives that either did not exist in 1800 or were not regarded as important. The complex history of Steinach's research illustrates many of the fundamental transformations in medical concepts of the body over the long nineteenth century.

The most obvious place to begin is, of course, with anatomy and physiology. From ancient times until the early nineteenth century, physicians and patients thought of the body as a solid assemblage of organs traversed and governed by fluids called humors. Disease was caused by the imbalance of the four humoral fluids—blood, yellow bile, phlegm, and black bile—and the aim of treatment, therefore, was to restore the balance.[44] In 1800, this notion of the body was only beginning to crumble, but by 1920, as we have seen, it had been swept irrevocably into history. Thanks largely to the well-known focus on anatomy and pathology in post-Revolutionary Paris hospital medicine, the body came increasingly to be regarded as a conglomerate of different kinds of tissue, and disease, far from being the result of the imbalance of humors, was seen to be located in the tissues.[45]

Anatomy became the medical student's foundational science and the doctor's scientific sheet anchor; it also spread into popular discourse, and the anatomical picture of the body and its workings helped construct new, purportedly more scientific images of the body and, indeed, new ideas of selfhood. The anatomical body, with its precisely delineated parts and organs, all governed by rational laws, provided an ideal foundation not only for the new scientific medicine but also for the domesticated, civilized, bourgeois self. In premodern times, the body had doubtless been important, but the true motor of the self was regarded to be the spirit; in the new scientific world of the nineteenth century, despite fears of materialism, the body became integral to, and nearly synonymous with, the self.[46]

And that body did not remain the assemblage of complex but still largely macroscopic parts and tissues that René Laennec's generation knew so well. The later nineteenth-century body became a cellular one, whose structure could no longer be ascertained by the naked eye. Neither microscopy nor the concept of cells was new—both can be traced back to the seventeenth-century research of the British scientific polymath Robert Hooke (1635–1703) and the Dutch microscopist Antonie van Leeuwenhoek (1632–1723)—but it was only over the long nineteenth century that microscopic research, powered by technological, conceptual, and institutional innovations, produced a grand new vision of the microstructure of the body and the world.[47] At the center of this revolution was the cell. From the 1830s, the cell was increasingly seen as the fundamental unit of plant and animal life, and in the 1850s, the celebrated German scientist Rudolf Virchow (1821–1902) enshrined the cell as the ultimate locus of disease and healing. Research on cellular physiology and pathology proceeded all over Europe at a dizzying pace, and the body was revealed to be a complex patchwork of cells as diverse in their structure as in their functions, and by the end of the century, the microscope had also illuminated the myriad forms of life that shared our environment and caused some of our most serious ailments.[48]

The impact of the cell theory was not confined to the internal domains of biology and medicine—it was, as Paul Weindling has shown, shaped by and in turn influenced social, political, and cultural ideas of body and society. Just as the individual organism represented a "society of living cells, a tiny well-ordered state," the ideal society, too, was now portrayed as an organic consortium of unitary individuals—a "cell-state." The survival of an organism depended on cooperation and coordination between its cells, and the survival of society, likewise, depended on cooperation between the individuals composing it. Such organicist political metaphors, which cannot be reduced to a monolithic "social Darwinism" figured in complex intellectual and political debates. Some could be republican in spirit, while others could be quite authoritarian. The body could, for instance, be seen as a hierarchy of cells ruled by a monarch (the brain) or an aristocracy—the nervous system.[49] Diverse as their social and political connotations might have been, in strictly intellectual terms, what these theories were striving to explain was the coordination of the diversely constituted body. What, in other words, held the cells, tissues, and organs together, enabling the body to act as one unitary organism? As we have seen, the universal answer up until the late nineteenth century was the nervous system. We have also seen how that neurocentric view was displaced by a "new humoralism," which began with the experiments of scientists like Berthold or Goltz but attained prominence only with the coming of Brown-Séquard's organotherapy at the end of the century. Steinach's generation built on those foundations.

The concept of hormones led to a radical overhaul of previous notions of bodily communication, integration, and regulation. In 1931, the doyen of British physiologists and pioneering endocrine researcher Edward Schäfer (1850–1935) declared that "the old physiology" had been based on the concept that bodily functions were regulated by the nervous system, whereas the "new physiology" emphasized the chemical regulation of the organism.[50] The body was no longer a solid, relatively unchanging entity—the stability of the organism was real, but it depended on a balance of secretions. Change that balance and the characteristics of the organism could change. The humoral body was always potentially malleable, offering unlimited possibilities for what one may call physiological engineering.

This is where physiology merged with therapeutics. It is well-known that for all its scientific glories and despite the revolutionary changes in explanations of disease, nineteenth-century medicine did not possess many efficacious remedies—hence the vogue for electrotherapy, hence the popularity of organotherapy. Although many doctors scoffed at Beard's faradic coils and, subsequently, at Brown-Séquard's testicular fluid, many more hailed them as dramatic new advances. One could at last *do* something for ill people—and thanks to the novelty, relative scarcity, and apparent sophistication of the agents, one could also charge substantial fees. Today, those interventions might seem like quackery, but in their time, they were regarded as magic bullets. From this perspective, the Steinach operation was merely the latest in a line of so-called scientific miracle cures. And, although we cannot go into details here, its popularity was also influenced by the sexual and gender anxieties of the fin de siècle, and the promise of nineteenth-century biology and medicine to resolve them rationally and scientifically.[51] By 1920, these issues had been joined by even graver ones. Reeling from the ravages of World War I and in quest of regeneration, the Western world welcomed the idea that the secret of regaining youth and vitality had been found within the body. Rejuvenation by glandular surgery, one could argue, was the electrotherapy for a new, darker age—a way to reactivate bodies (and perhaps cultures) that were now facing challenges far more onerous than those of the late nineteenth century.[52]

Although the Steinach operation was a relatively simple procedure in technical terms, it was the fruit of a century of innovation in medical technology. It required anesthesia, antisepsis (or asepsis), and, as far as the rejuvenation of women was concerned, X-rays. The former two were well-established by Steinach's time, but X-rays were still very new (Röntgen's serendipitous discovery was announced in late 1895), and their use still almost experimental. Although it is true that by 1914, "most hospitals had an x-ray outfit, many had an x-ray department, and in many places x-ray diagnosis was becoming routine," the potential and risks of the new rays remained very imperfectly known.[53] It was only after World War I that X-rays entered the popular domain with a vengeance—shoe shops used fluoroscopes to X-ray feet, some doctors used the rays to treat acne, the removal of unwanted body hair by X-rays became popular with many American women, and Steinach and his followers, as we saw, used the rays to stimulate aging ovaries.[54] Far more enduringly valuable therapeutic effects of the rays—for example, in cancer—also came to be appreciated. Röntgen's discovery not only produced new images of the body's interior—the "skull beneath the skin" had never been so evident to all and sundry—but also generated dramatic new ways of destroying, stimulating, or modifying parts of the body that even the surgeon's knife might find difficult to reach.[55]

Steinach, to be sure, occupied only a moment in the long and tortuous history of transformations in concepts of the body between 1800 and 1920. Nevertheless, focusing on his research and using it as a reference point, we can form a tolerably clear conception of how changes and shifts within medical science and technology affected professional and public understandings of the body. In 1800, surgery was a last resort and elective surgery a dream. By 1920, surgery had become the heroic exemplar of a new, impeccably scientific medicine that not only saved lives but, through procedures such as Steinach's or those of the emerging field of aesthetic surgery, could correct our physical deficiencies, improve our appearance, and enhance our functionality. The new surgery and the new endocrine science came together in Steinach's career, illustrating the emergence of new concepts of the body and, more importantly, the new confidence, shared by wide sections of the public, that medicine now had the knowledge and the technical expertise to reengineer our bodies, our lives, and, indeed, our very selves.

CHAPTER FIVE

Popular Beliefs and the Body

"A Nation of Good Animals"

PAMELA K. GILBERT

> To be a good animal is the first requisite to success in life, and to be a nation of good animals is the first condition of national prosperity.
>
> —Herbert Spencer

Herbert Spencer's dictum has often been quoted as an irreverent take on theories of society derived from evolutionary theory.[1] But long before the eugenic policies of the fin de siècle, nineteenth-century Britons were already engaged in an active debate about how best to create a nation of healthy bodies and bodily practices—and avoid the opposite. Nineteenth-century Britain, responding to the cholera epidemics of 1832–1867, as well as to the newfound visibility of other urban epidemics as statistical science became a part of the techniques of government, developed a focus on public health as hygiene that was to penetrate every aspect of urban and village planning, schooling, housing, literature, and public discourse of all kinds. This new focus implied not only a prescriptive effort but also a new scrutiny of existing bodily practices. In this chapter, we will examine how the construction of the body changed in medicine and popular understandings of medicine. We will also explore how hygienic practices affected understandings of the urban and rural poor and the physical conditions under which they lived. The new focus on hygiene meant that, beyond interventions in working and living conditions, physical culture came to be emphasized, both for the middle to upper classes and for the working classes. Finally, other leisure-time activities also emphasized the pleasures (and, occasionally, the perils) of cultivated flesh.

The emergence of a public health authority in response to epidemic diseases enabled sanitarians such as Edwin Chadwick to disseminate their ideas about health. As public medicine became a function of the state, however vexed, the doctor emerged as a heroic public figure, gaining a place in the public sphere along with legislators and clergy. Additionally, competing quasi-scientific knowledges, such as homeopathy and phrenology, sometimes endorsed by medical professionals, vied for space in popular understandings of the body. In describing history before the late eighteenth and the nineteenth centuries, the term "popular" has often been often used to mean "folk." However, the increase in literacy and emergence of mass print media in this period, along with rapid urbanization, problematizes this simple equivalence. There were many popular audiences in this period, including the middle class, the small trading and skilled laboring classes, and the literate working classes. And in these groups, popular beliefs were diverse and changed radically over the century. Finally, regional communal beliefs—often founded on older humoral modes of medical knowledge—coexisted with all of these understandings and practices. These knowledges varied widely by geographic area, level of urbanization, and ethnic group.

Popular is a slippery term—it has often been used pejoratively: for example, the "merely popular" versus high culture, or "popular opinion" versus reasoned discourse. As these examples suggest, it has been associated with the masses, and the masses have generally been seen as unwashed. Although liberal and Marxist scholars have rehabilitated the masses as "the people" or "the folk," and popular "wisdom" as "common sense" as opposed to suspect "elite knowledges," the popular retains a whiff of dirt and danger—whether the exciting danger of subversion or the less thrilling danger of submersion in mass-marketed mediocrity. In cultural studies, popular aesthetic forms could be seen as a manipulable form of false consciousness or a liberatory expression of authenticity, although the best work has refused this simple binary. In the area of knowledge production, liberal readings of popular or common knowledge opposed these knowledges as authentic and organic against professional elite knowledges used as social control—a tradition that continues in interesting ways through many Foucauldian studies. In the nineteenth century, Britons thought of the popular ambivalently.[2] The advent of changes in the English franchise, beginning with the famous Reform Bill of 1832, brought exaggerated fears of a dictatorship of the masses, as the merits of democracy were debated in the United States, France, and Britain, and the continent faced a succession of revolutionary events that often seemed in danger of spreading across the channel. Popular leaders were feared as demagogues and popular beliefs were often thought to rest on inadequate knowledge and easily manipulated emotionalism.

Yet, at the same time, the spread of literacy and leisure time downward through society as industrialism progressed meant that the vast gap between popular/folk and elite/lettered knowledges began, slowly, to narrow and change in nature. And the advent of liberal beliefs about the role of representation in government, in addition to the availability of cheaper print media to a wider reading public, meant that both private and public interests increasingly sought to change popular opinion through propaganda and education, rather than relying on more direct uses of power to enforce their views. Finally, the increase in middle-class power and size meant that the "Philistines," as Matthew Arnold dubbed them, could no longer be either left to their own devices by elite "Barbarians," nor could the middle classes themselves rely on being unaffected by the behaviors and beliefs of the working classes or the poor. Widely shared popular beliefs and practices came into being, then, not only accidentally, as culture was homogenized through the development of national media, national and international trade and transportation, and communication networks such as mail, classified ads, and telegraphs. They were also shaped through compulsory education and the often intentionally didactic influence of literature—sentimental, religious, sensationalist literature, the sporting press, women's magazines, the illustrated weeklies, and the medical press—as well as through theater and sporting events, music halls, ballads and broadsheets, circuses, "rational amusements," exhibitions, mechanics institutes, and public lectures.[3] By the mid-nineteenth century, describing a novel, event, practice, or fashion as "popular" might be intended as a scornful condemnation or a positive description of widespread consumption, depending on context. In this chapter, which will largely focus on the mid- to late nineteenth century, the term "popular" will refer to beliefs and practices widely shared across the social spectrum and promulgated in publications and activities, especially among middle-class and working-class people.

THE DANGEROUS BODY

The development of the medical vision of the body over this period moves in two directions: the first does not significantly differ from views widely held early in the century, relying on a theoretical understanding of the body as multiple systems—mostly intake and excretion and, to a lesser extent, chemical systems—which required balance to be healthy. Although incorporating new material and perspectives, these concepts are not incompatible with earlier ways of thinking about physiology. The second development has to do with emerging medico-surgical work on the body, in which there is increasing attention to structures and seats of disease. Although the second set of ideas came to be dominant within medical professional writing over the course of the mid-century, the sanitary perspective was still shaped by a general and fairly easily grasped notion of the body as systemically regulated by intake and excretion, especially through respiration. This notion was compatible with popularizations of the new understanding of the body, based in part on early nineteenth-century theories of pathology, as always dying, always death producing in all of its processes. One sees this understanding emerging in mid- to late nineteenth-century British medicine and popular health discourse, in which it harmonized with evolutionary theories of the survival of the fittest. As popular health discourses seek to harmonize these ways of envisioning the body, three trends emerge quite dramatically in the 1860s: a flattening of the body into surface, a tendency to think of the body's multiple systems (respiratory, alimentary) as having differences in substance (tissue) more important than differences of overall function, and an overriding concern with bodily output. Intake is characterized less in terms of type and quantity than of purity, and the ability to rid the body of—and distance it from—its excreta came to be of urgent interest.

The beginning of the period posited a body that was, in itself, healthy as long as it was balanced, using an economic model focusing on intake of "fuel" to power the body without overstraining its ability to digest. The later model is one in which the body is essentially unhealthy—death creating, death infested. The essential task of health is to excrete that poisonous or degenerate—literally decaying—material as quickly as possible. The earlier model is exemplified in the Society for the Diffusion of Useful Knowledge's (SDUK) 1832 instructions to the working man regarding the cholera and temperance: "if a man's natural spirits and strength are habitually exhausted by artificial stimulants, his stock of spirits and strength will be so taken up beforehand, that if the cholera makes a sudden demand upon this stock, even his life must go toward the payment."[4] To avoid "depleting one's stock," the reader is advised that, "Coarse sour food; spoiled vegetables . . . [etc.] are all unwholesome, and produce all the uneasinesses and evils of indigestion. Food that is too rich or too nutritious will produce the same kind of mischief . . . It is moderation that is everything."[5] This point is repeated several times, and in italics at the end of the section. Moderate intake—neither too much nor too little, but just enough to keep all running properly—ensures a body that is healthy and therefore resistant to external enemies. Later, however, the enemies of the body are most significantly internal. Whereas the body in the 1830s was not itself dangerous to others, especially in health, by the 1860s, the body is essentially dangerous—if it excretes healthily for itself, it poisons the environment for contiguous bodies. As William Farr says, "Men . . . are always surrounded, in air and water, by an atmosphere of decaying matter, which is given off from their own bodies and from the animals by which they are surrounded."[6] In fact, even as medical knowledge moves toward germ theory (as the theory that disease was caused by specific living organisms later came to be called after Louis Pasteur's discoveries in the late 1870s), popular medical discourse seems to compensate with an ever-intensifying suspicion of the body itself. Early sanitarians wanted to clean acquired dirt off the body; later, they were concerned with cleaning the body's own perspiration off to keep the pores open for continuous excretion of the body's internally generated filth.

Norbert Elias has argued that the modern body emerges as a body concerned with closure and regulation of its openings. Pierre Bourdieu has also traced the emergence of the middle class body, especially that of the petit bourgeois, as a body concerned with "narrowness," regulation, and the diminution of excess.[7] Certainly the gendering of the body, during the emergence of the rhetoric of separate spheres and the consequent reclassification (and reanatomization) of female (and less obviously male) sexuality, confirms this increasing focus on the disciplining of the body's openings.[8] This is particularly true of what Bakhtin describes as the grotesque body celebrated in carnival—a body defined by its openness, especially of the lower body, and its incontinence, publicness, and impropriety.[9] The grotesque body was increasingly identified with improper femininity or indiscriminate sexuality, pauperism, drunkenness, and the foreign (and, secondarily, with the decadent aristocracy). It is this body that is envisioned in polemics directed at legislating control of the body, and it is the "abject" products of this body, including disease, that must be eliminated from public view.

It is this sense of the body that informs the targeting of working-class women by the Contagious Diseases Acts in the 1860s, which allowed authorities in port cities to arrest women suspected of being prostitutes, forcibly examine them for venereal disease, and incarcerate them for treatment until they were cured—that is, until the visible signs of infection disappeared. This was not aimed at controlling venereal disease throughout the kingdom, but specifically in areas wherein sailors of the British navy might become infected; the danger was to Her Majesty's military. (This also led to the establishment of governmentally controlled brothels in various colonial areas for Her Majesty's troops, using native women.)[10] Women's bodies, in particular, were the weak link in the social body; closer to nature and the animal, women were thought to be more susceptible to sensory impression, leading them to be more easily corrupted. Yet, as the mothers of the nation, the replication of the healthy social body depended on women's bodily continence. Mothers (especially working-class and poor mothers) were the targets of hygienic education, both in Britain and on the continent. And the failure of working-class and poor men or children to live up to middle-class ideals of hygiene—not drinking, washing regularly, staying at home instead of the pub or the street—was largely blamed on their wives and mothers. As the women of a family were the first (and often only) providers of medical care, from first aid to nursing, midwifery, and death-bed attendance, they were the practical purveyors of medical knowledge within the family, and so the most important target for educational intervention.

However, despite the increasing homogenization of beliefs about health, these interventions were by no means wholly uncontested. Women's unique role in the family not only made them appropriate targets for intervention but also gave them also unique authority to resist. Based on their supposed closeness to animals and resulting "mother instinct," women were respected within the working classes (as in the middle classes) as the sources of intuitive knowledge about health generally and especially the well-being of children. As Nadja Durbach points out, this "feminine authority" was used in the rhetoric of resistance to vaccination.[11] Additionally, the working classes as a group were by no means entirely convinced that the danger to society emanated from their bodies, rather than ill-considered and invasive interventions from above. As Durbach elegantly illustrates in *Bodily Matters: The Anti-Vaccination Movement in England, 1853–1907*, "Anti-vaccinators saw their bodies not as potentially contagious and thus dangerous to the social body, but as highly vulnerable to contamination and violation."[12] They resisted a dangerous procedure, in which fluids from the blisters that developed on a vaccinated child were transferred directly into cuts made on the arm of another child, often resulting in illness, disfigurement, and infection. Although this was a cross-class movement involving men and women, it also became a specifically feminist issue.[13]

Like the body of the woman, the body of the racial other was often considered weak—an avenue for infection and danger to the nation. From the mid-century on, theories of evolution, anthropology, and race dovetailed to focus attention on race as an essential biological trait rather than primarily an effect of geography and climate. Still, for a very long time, race as it was popularly understood remained a slippery combination of biology, ethnic heritage, and geographic location. Geography and culture were still thought by many to change biology itself, which might then in turn become heritable. Broadly, darker peoples were considered to be more emotionally labile, more sexual, lazier, less intelligent, and more childlike than lighter complected northern Europeans. (The Irish and French, because of their Celtic heritage, were assimilated to a southern European model.) This was related to a warmer climate, thought to speed sexual maturity and inflame the passions, while reducing the disposition toward intellectual or physical effort. By the end of the 1860s, the ideal body is male, English, imperial, disciplined, and active. This is the good animal of Spencer's successful nation.

Sanitary writings and other popularizations of medical models in this period tend to focus on the surface of the body as a dangerous point of contact between self and not self. One most important tissue was the membrane that both divided and connected inside and outside, in mediated communication. An 1871 pamphlet titled *Sanitary Rhymes*, intended as an educational piece for the working class, conveyed current medical understandings of sanitation and the body in verse, beginning, significantly, with "The Skin" (the entire body is summarized in three poems on the skin, the blood, and the nervous system, all tissues which mediate communication between the body and its environment).

> There's a skin without and a skin within,
> A covering skin and a lining skin;
> But the skin within is the skin without
> Doubled inwards and carried completely throughout.
>
> The palate, the nostrils, the windpipe and throat,
> Are all of them lined with this inner coat;
> Which through every part is made to extend—
> Lungs, liver and bowels, from end to end.
>
> The outside skin is a marvellous plan
> For exuding the dregs of the flesh of man;
> While the inner extracts from the food and the air
> What is needed the waste in his flesh to repair. . . .

Verses follow on proper food, water, and hygiene. The poem concludes:

> All you, who thus kindly take care of your skin,
> And attend to its wants without and within,
> Never of Cholera need have any fears,
> And your skin may last you a hundred years![14]

The body here is flattened out into a two-dimensional surface—a skin—in which the inside and outside of the body are not really differentiated as the difference between that body and its environment is. Instead of eating being a taking of materials inside through an opening of the body—through a clearly demarcated boundary between outside and inside—it becomes a placing-in-contiguity of certain, potentially dangerous (impure) materials with the porous surface of the body. The body is thus, in one sense, sealed (it has no openings, only a continuous skin—even materials taken into the body are outside this skin—that is, the stomach becomes a space exterior to the body proper), yet radically permeable (the whole surface is now a potential place of ingress and egress). Systems that would earlier have needed separate accounting—the digestive system, for example—can be absorbed under this model. Eating becomes a form of respiration, just as breathing is tissue nourishment.

The focus on skin and elimination of wastes emphasized perspiration as a method of cleansing, added to long-standing interests in purges and vomitives. In George Meredith's novel *The Ordeal of Richard Feverel*, when the pragmatic Mrs. Berry wants to warn the hero of the sexual dangers of being away from his bride in the early days of his marriage, she has no better recourse than the analogy of sweating: "we all know what checked perspiration is. It fly to the lungs, it give ye mortal inflammation, and it carries ye off. Then I say checked matrimony is as bad. It fly to the heart, and it carries off the virtue that's in ye, and you might as well be dead!"[15] Although this analogy provokes the young men she is addressing to laughter, they immediately take her meaning (and the plot proves her correct).

VISIONS OF THE POOR: URBAN HOUSING

The focus on the body as essentially diseased and the skin and lungs as a purveyor of the body's uncleanliness to the outside world legitimated existing bourgeois practices of privacy and separation as matters of health and survival rather than simply of taste. It was thought that 700 cubic feet of air per person was an appropriate environment in which to live and sleep. Long-standing representations of the poor as teeming, aggregated bodies dovetailed with these medical theories; now the congregation of the poor in ill-ventilated, overcrowded buildings was not merely morally untoward but a source of disease. The masses were, precisely, massed bodies, filthy and insufficiently individuated. They were too close—sharing space, water, air, sexual congress. These bodies, being too contiguous, became continuous. This was potentially a nation-within-the-nation, one comprised of bad animals.

Both poor environment and disease itself were held to have bad effects on massed bodies that would accrue also to the body social. Disease attacked persons, but the moral effects of disease were deleterious to populations. Famous sanitarian and architect of the New Poor Law of 1834 Edwin Chadwick and his employee, physician Thomas Southwood Smith, argue, "Depression is the normal condition of the residents in all unhealthy districts. . . . The offspring of people in this enfeebled condition are puny and sickly . . . and the physical deterioration goes on increasing with each successive generation." Further, by reducing the proportion of working adults to children and the aged in a population, "this destruction of the heads of family produce[s] . . . pauperism." Not only does the community then suffer economically, but also morally, as "the steadying principle of the community is lessened, the acquisition of productive skill is obstructed, the difficulty of extending moral culture and forming moral habits is increased . . . there is substituted a population always young, inexperienced, ignorant, credulous, passionate, violent, and proportionately dangerous, with a perpetual tendency to moral as well as physical deterioration."[16]

Crowding and unsanitary environments were also thought to create racial degeneration. In 1852, William Farr, collator of abstracts for the Registrar General's Office, wrote, "history . . . exhibits this great fact—the gradual descent of races from the high lands, their establishment on the coasts in cities sustained and refreshed for a season by immigration from the interior; their degradation in successive generations under the influence of the unhealthy earth, and their final ruin, effacement, or subjugation by new races of conquerors."[17] He thus justifies imperialism, staking Britain's racial claims to leadership on its public health and its geography. Farr explains that the periodic cholera epidemics that had swept through Britain (and indeed, the entire globe) beginning in 1832, were God's warnings that the British community was also in danger of this elevation-related degeneration because so many had settled near the Thames: "the pestilence speaks to nations, in order that greater calamities than the death of the population may be averted. For to a nation of good and noble men Death, is a less evil than the Degradation of Race."[18]

FIGURE 5.1: *Lodging House in Field Lane*. Lodging houses were places of particular concern, not merely because of their poor sanitary arrangements but also because of the so-called promiscuous association of the sexes in sleeping areas. From Hector Gavin, *Sanitary Ramblings* (London: Churchill, 1848). Wellcome Library, London (L0009845)

By the 1850s, then, the social problems of hygiene were no longer defined solely in terms of nuisance removal; they involved also the people who lived in what were considered problem environments. However, an underlying concern about overcrowding was not only the amount of space, but its uses. Descriptions of persons huddled together in one room often explicitly define incest as an inevitable result of such crowding. "Talk of morality!" says sanitarian Edward Bickersteth, in a lecture quoted by philanthropist and editor of *The Builder*, George Godwin, "amongst people who herd—men, women, and children—together, with no regard of age or sex, in one narrow, confined apartment! You might as well talk of cleanliness in a sty, or of limpid purity in the contents of a cesspool . . . the first token of moral life is an attempt to migrate, as though by instinct of self-preservation, to some purer scene."[19] By the 1860s, the housing movement itself, although legislatively concerned with sanitation—the destruction of slums, the repair of drains, the construction of new housing up to a certain code—was just as concerned with the inculcation of domesticity and the performance of privacy, particularly among the lower working classes.

Overcrowding, which had first become important in medical literature in the 1840s, was central in the legislative and philanthropic literature of the 1860s. In 1866, Mr. Bruce, MP, warned, "The House had already dealt with two great causes of disease . . . [water and drainage of nuisances]. But the source of evil the most difficult of all . . . was the overcrowding of houses . . . in every large town thousands of persons were brought up in a state of moral degradation, which could only end in a great national danger."[20] Only after the body was properly separated from other, contaminating bodies could other interventions take place. Even those who opposed legislation agreed that hygiene was important. Herbert Spencer argued that legislation was simply not the best way to achieve these changes. After all, he argued, "now that the laws of health are beginning to be generally studied; now that people are reforming their habits of living; now that the use of baths is spreading; now that temperance, and ventilation, and due exercise are getting thought about—to interfere now, of all times, is surely as rash and uncalled-for a step as was ever taken."[21]

BATHING: PUTTING THE SKIN TO WORK

It is perhaps not surprising, with all of the focus on the body as producer of filth, that mid-Victorians should have been particularly interested in bathing.[22] Indeed, frequent bathing, especially in cold water, came to be seen as a particularly English obsession. However, there was also a long-standing recreational and medical bathing culture to which was added in the second half of the century a new emphasis on sea bathing across classes down to the artisan

FIGURE 5.2: *Dudley St., Seven Dials*. Overcrowding and lack of privacy were central issues for reformers faced with slums like this famous London neighborhood. Engraving by Gustave Dore, from Gustave Dore and Blanchard Jerrold, *London: A Pilgrimage* (London: Grant, 1872). Wellcome Library, London (L0000881)

class. Spa or bath towns were popular resorts for the healthy middle to upper classes in the nineteenth century who came for change of air, sea bathing, and relaxation. This continued an older tradition of using the baths medicinally; both the sick and the well came to fast or participate in special supervised diets; "take the waters" internally and externally; and receive massages, enemas, purges, cold showers, hot showers, and so forth either as preventive maintenance or to treat existing illnesses. Over the course of the nineteenth century, as the free pleasures of the beach were emphasized over those of the costly spa waters, more lower-middle-class workers and artisans came out either for short stays or as "day trippers" after cheaper excursion trains started to run in the 1840s, for example, to Brighton.[23]

In the 1850s and 1860s, dry air sauna (Turkish) and vapor (sometimes called Roman) baths became popular in northern England, and later in the south. The objects of such baths were at least in part to induce sweating to clean the body and the skin. George Elson, who wrote of beginning a second career (he had been a sweep) as a shampooer and swimming instructor at a Turkish bath, wrote of his early experience,

FIGURE 5.3: *The Queen's Landing* depicts a queen, possibly Queen Victoria, being driven through the sea in a bathing machine. The upper classes enjoyed sea bathing in elaborate bathing machines like this that ensured the privacy of bathers (especially women) entering the water. Wood engraving by John Leech. Wellcome Library, London (V0020078)

> At the beginning I thought I should be dried up entirely with the extreme heat; I lost a stone weight in the first fortnight, which caused me to fear I should fail in my new venture; but the theory and system were right. I had need to get rid of the superfluous matters—not that I was stout. I required a rectification of the physical powers ere I could recuperate and get back my weight, which was returned to me in the course of a few weeks.[24]

Turkish baths were used mostly by middle-class men, and to a lesser extent women, though some artisans did use them, and especially in the north, many were opened to the public with the intention of drawing working-class people. Depending on what one spent and where one went, one could be sweated, plunged, shampooed, and massaged in a Turkish bath establishment, and then enjoy coffee, tobacco, and other refreshments, but regular (wet) public baths were also available for a bit less money. A Turkish bath at Faulkner's hotel in London could be had for as little as 1 and 6 on a Saturday in the late 1880s.[25]

An astonishing number and variety of bath vessels were made and sold for private use in this period as well.[26] Many of these uses were therapeutic for specific ailments, but they were also used for general mental and physical hygiene beyond mere cleanliness. One lady wrote of her experience that,

> In future when life ceases to be interesting, and I feel rigid and miserable, I shall get into my Cabinet Turkish Bath and set my skin at work instead of my brain. . . . After sitting in it for half an hour, a tepid cold sponge bath and brisk rubbing with rough towels should follow, and the sensation of relief and comfort is experienced which only those who attend to their skins, as important factors in good health, can realise.[27]

Those taking Turkish baths were convinced that they were ridding themselves of noxious substances through the skin's ability to excrete those materials in sweat, an important step beyond the way other baths simply washed away surface dirt. But even regular baths were held to stimulate the skin's abilities, especially if they were taken cold. Although the truly poor did not have frequent access to full baths, even the poorest people probably bathed in cleanish water at least occasionally in the second half of the century—the children were often bathed and deloused if they went to charity, or later state schools, and those sleeping in casual wards bathed and sometimes deloused before they dossed down. Antibathing beliefs—that bathing was dangerous, that dirt protected the health of children—seem to have become outmoded among the middle and upper classes by the mid-nineteenth century, but are said by various writers to have persisted for a while among some working class and rural people. By the end of the century, the holdouts seem to have gradually given way under the onslaught of hygienic haranguing.

Cold-water bathing was considered manly and very English; it was also, like athleticism, considered a specific against venery, as well as more general moral laxity. Charles Kingsley admonishes his little boy reader in his children's book, *The Water-Babies*, that dirt leads to degeneration,

> efts are nothing else but the water-babies who are stupid and dirty, and will not learn their lessons and keep themselves clean; and, therefore . . . their skulls grow flat, their jaws grow out, and their brains grow small, and their tails grow long, and they lose all their ribs (which I am sure you would not like to do), and their skins grow dirty and spotted, and they never get into the clear rivers, much less into the great wide sea, but hang about in dirty ponds, and . . . and eat worms, as they deserve to do. . . . Meanwhile, do you learn your lessons, and thank God that you have plenty of cold water to wash in; and wash in it too, like a true Englishman.[28]

And the novelist Ouida, in *Under Two Flags*, writes of a French soldier and a French *vivandiere* in Africa trying to identify the hero's nationality; they find "overwhelming proof" of his Englishness in the fact that he "bathes—splash! Like any water-dog."[29] Although middle-class women also attended baths, they were much more popular among men, and the baths could be, like clubs, spaces of homosocial bonding for business and pleasure. They could also be perceived as something of an equalizer, at least within the ranks of those who could afford the entry fee. Anthony Trollope's "The Turkish Bath" (1869) humorously writes of being trapped into an interview by an indigent writer, whose scruffy clothing on the street attracts the editor's pity, but whose superb physique and grand manner commands respect in the bath, where his sartorial disadvantages are neutralized by nudity. He spends his last few shillings to enter, as it is the only way to gain access to the editor, who would never have agreed to see him under normal circumstances. (The writer turns out to be insane, unfortunately.)[30] No doubt these circumstances did not often occur outside of fiction, but the uneasy perception that the nudity of the bath concealed, rather than revealed, identity was part and parcel of the general understanding of the city as a heterogeneous space in which social contact was much less mediated than in the highly controlled social rituals of access that were common in the country or in clubs. In the baths, as in stores, parks, omnibuses, and streets, contact with the social other was unavoidable. If epidemic disease was the tragic story of this contact, then social discomfort was its comic counterpart.

Frequent bathing was considered very English (at least by the English themselves) despite the fact that many bathing practices popularized in Europe and Britain in the eighteenth and nineteenth centuries were borrowed from the colonies or from the East (as Turkish bathing was copied from the Ottoman Empire). In the broader context of the British Empire, views of native bodies and their health changed dramatically over the course of the century. In India, for example, although many attitudes can be found throughout the period, there is a developmental arc wherein the preponderance of beliefs shift from a more constructionist perspective, in which native land and peoples are seen as in need of colonial guidance to improve their situation, but not as substantially different in kind from the urban British poor (an outlook prevalent in the first part of the century), to a belief that Indians are fundamentally biologically different (and inferior). Alan Bewell notes that in the early nineteenth century, British slums and factories are often described as tropicalized environments that foster disease and are similar to colonial settings.[31] However, by the mid-century, the environment of India was seen as essentially different from that of Britain, and as essentially productive of disease.[32] Although Mark Harrison points out that Anglo-Indian representations of India were much more nuanced, in Britain itself, India came to be viewed as a fairly simplistic and negative totality by mid-century: "guarded optimism about acclimatization and the colonization of India prior to 1800, gave way to pessimism and the alienation of Europeans from the Indian environment; a shift which was closely related to the emergence of ideas of race and the consolidation of colonial rule."[33] This follows the track of the writings on other Indian cultural practices, such as Hindu and Muslim bathing habits, which were in the Romantic period held up as exemplary of civilized cleanliness. By the mid-century, however, despite such borrowings, Eastern habits and health practices were more likely to be thought filthy and disease producing.[34] In part this attitude derived from the association of disease with contaminated water after Snow's controversial discovery that cholera was waterborne, which was publicized in 1857. Though Hindu bathing habits were admired in the early nineteenth century, by 1870 George Johnson, MD, in an address to a missionary group, locates danger in Indian social practices, especially bathing in the Ganges. He described these bathing practices as immersing clothing and the body in water and taking water into the mouth, then spitting it out for the use of other bathers. He continues, "we may reasonably hope that the time is not far distant when the natives of India may learn that the only sure preventive of cholera is through cleanliness in their persons, their dwellings, their food, and above all in their drink. . . . [or] the scourge of cholera will continually recur."[35]

Whereas cold-water bathing seemed manly, perhaps because there was little temptation to linger over it, various hot-bathing practices evoked more ambivalence. Turkish baths were tarred with the brush of oriental luxury and

FIGURE 5.4: Turkish bath in London. Notice the humorous commentary on anxieties about racialized or foreign influence. *Turkish Bath in Latherington Street, W.1, London*. From *Punch* 50 (1866). Wellcome Library, London (L0005394)

the effeminacy popularly associated with the East. Because the hot bath was largely passive, and was historically associated with excess and sensuality, men worried about relaxing too much. Roman or steam baths were associated with tropical climates that Britons believed had enervating effects: "There never perhaps existed a people so thoroughly fitted by habit for a foreign yoke. The physical organization of the Bengalee is feeble even to effeminacy," Macauley sneered, "He lives in a constant vapour bath."[36] Australian humorist Joseph Furphy (1843–1912) facetiously derides the Australian and English cult of bathing—and imperial pretensions—in his novel *Such Is Life*: "how few of our later novels or notes of travel are without that bit of description [of bathing]; generally set-off by an ungainly reflection on the dirt of some other person, class, or community. The noxious affectation is everywhere." However, he suggests, the notion that bathing is manly is of recent provenance:

> The Spartans (so ran my reflections) were as much addicted to dirt as the Sybarites to cleanliness; and just compare the two communities. The conquering races of later ages—Goths, Huns, Vandals, Longobards, &c.—were no less celebrated for one kind of grit than for the other. It is the Turkish bath that has made the once-formidable Ottoman Empire the sick man of Europe. . . . Bathing did the business for Italy, as it does the business for all its victims. . . . Indeed, the most sweepingly appropriate bestowal of the title, 'Great,' is made when we refer to the adherents of the dirt-cult, collectively, as the Great Unwashed. . . . we will entice external leakage of such incipient greatness as we have—soaking ourselves in water, as if we were possums, and our virility a eucalyptus flavour that we sought to dissipate.[37]

Furphy is joking, but using a logic to which those suspicious of spas and hot baths often had recourse without a trace of humor. The common rejection of the dirty, unsophisticated working classes sat uneasily with the valorization of supposedly working-class traits, such as physical strength, anti-intellectualism, anti-aestheticism, and simplicity, as masculine. Good animals were to be clean but were not to enjoy their cleanliness too much, lest it become a form of effete fastidiousness. Masculinity had to be carefully contained, as a leaky body could compromise its virility.

BRITISHNESS AND ATHLETICISM

In addition to new emphases on techniques of pleasurable and hygienic care of the body, such as bathing, the nineteenth century was the great era of athletics, especially in the second half of the century in Britain and her domains. Like other hygienic practices, athletics were thought to be not just healthy but also

FIGURE 5.5: *Outer Cooling Room of a Turkish Bath-house*. The Turkish bath evoked fantasies—and distrust—of oriental luxury and passivity. Wood engraving by Thomas Abiel Prior after Thomas Allom. Wellcome Library, London (V0020026)

character building. In rural areas, festivals and market days were occasion for traditional games of strength and prowess—wrestling, fighting games, hurling stones or heavy weights, and so forth, as well as spectacles such as pig wrestling, cockfighting, and dogfighting. Betting usually added to the entertainment value, as did drinking alcohol. Like most working-class entertainments, such games came in for a certain amount of middle-class scrutiny and disapproval, though they had their partisans, such as Thomas Hughes and Charles Kingsley. For these apologists, such activities represented a folk tradition on which a concept of national and perhaps racial identity could be founded. From the mid-nineteenth century on, however, such games increasingly came under the control of middle-class organizers who attempted to replace them with what they considered more rational amusements. Some of these rustic pleasures—dogfighting and cockfighting for example—were also popular town entertainments and were the bêtes noires (so to speak) of animal rights activists and advocates of rational amusements. But rural entertainments that became of interest to hygiene advocates included fighting (middle-class hygiene advocates preferred boxing, with padded gloves and three-minute rounds, but spectators of whatever class often preferred the much more brutal spectacle of bare-knuckle fighting, in which a round continued until one man was felled or both were too exhausted to continue without a rest). In Lancashire, working-class men practiced a form of kickboxing in heavy boots known as "purring." This was used to settle disputes but also as a formalized entertainment with betting. No doubt there were many such local variations.

For participants, such sports sometimes served as ways of making money, gaining prizes and recognition, and settling disputes. They also provided ways to affirm masculinity through the demonstration of strength, competitiveness, and the ability to endure pain stoically. None of these were sports approved for women, and although spectators might gather and cheer when women fought, such displays were usually impromptu and seem to have been motivated more by anger than the desire to display prowess. Thomas Hughes approvingly describes a rural "backswording" match, in which men armed with large ash sticks compete to be the first to draw blood from the other man's head "above the eyebrow": "Whack, whack, whack, come his blows, breaking down the gipsy's guard, and threatening to reach his head every moment. There it is at last. 'Blood, blood!' shout the spectators, as a thin stream oozes out slowly from the roots of his hair, and the umpire calls to them to stop."[38] It was, of course, the fighting and drinking, as well as more sexual entertainments such as peep shows, that gave such rural fairs a bad name among middle-class reformers. In the cities as well, working-class (and cross-class) sporting events were often organized by tavern keepers who provided refreshments and some-

FIGURE 5.6: Two Men Boxing. Boxing was considered a gentlemanly sport. Eadweard Muybridge's photos appeal not only because of the audience's interest in the intricacies of the sport but also because of the aesthetic appeal of athletic male bodies in motion. Photogravure after Eadweard Muybridge, 1887. Wellcome Library, London (V0048677)

times held the purse.[39] At all levels of society, parties and festivals, and even weekends, were occasions for dancing and games that involved some level of physical exercise, often doubling as courting play. Blind man's bluff, for example, was played by middle-class and working people alike.

Women in the nineteenth century distinguished themselves—or were supposed to distinguish themselves—from both men and girl children by quiet, decorous movements. Still, concerns about hygiene did eventually push public opinion toward a cautious acceptance of moderate exercise for middle-class women. In addition, increasing concern about factory girls and other urban workers' unhealthy, cramped, or repetitive working conditions led to both a movement to provide cheap fares for day trips to the country and an increased emphasis on urban green spaces, so that workers might walk and enjoy the pleasures of nature without going to the country. Housing and social work activist Octavia Hill (1838–1912) championed the open spaces movement in the last third of the century that resulted in the preservation of many permanent public spaces in working-class areas; she advocated the use of cemeteries as recreational spaces in order that workers might have a location to congregate outside for activities in locations other than stairwells and stoops. (It also made their activities easier to observe and police, supposedly discouraging drinking and gambling.)

Open spaces advocates emphasized the need for a healthy body—male or female—to walk, breathe clean air, and rest from the sights, noise, and dirt of the city. The focus on respiration is particularly important. Otherwise, rural and servant working women, it was assumed, generally got enough exercise. As far as participation in competitive sports, women lagged behind men. Rural working women at mid-century tended to compete at fairs and market days more through domestic arts (cooking, baking, plain sewing) than athletic competitions. However, by the end of the century, working women, or at least girls, were drawn a bit more into school gymnastics and an educational emphasis on the virtues of walking and other moderate physical activity. Their middle- and upper-class sisters had always walked and ridden horses, and by the mid-century enjoyed decorous outside games like croquet and sports like archery. Increasingly, "fast" middle- and upper-class young ladies, as well as their married sisters, even participated in hunting. By the 1870s, middle- and upper-class women adopted such sporting crazes as roller skating (as well as ice skating, which had long been a rural pleasure of the young), and later, bicycling and tennis. Gymnastics and calisthenics became somewhat popular for middle-class girls on parts of the Continent; some British families did adopt the practice, but it seems not to have become very popular. Such individual athletic activities were encouraged more for girls than boys in England until late in the century, when calisthenics began to be employed for boys as part of military drill. However, such activities had long been popular on the Continent for boys as well.[40] Britain distinguished itself by the depth of its commitment to team sports and its explicit connection of such sports to national identity.[41]

Upper- and middle-class men rowed, ran, rode, hunted, shot, and increasingly played organized ball sports well into manhood. But in the mid-1860s, the curricula of public schools increasingly focused on organized sports (cricket, rugby, soccer, footraces, rowing, swimming). Many "old boys" went on to organize weekend teams in the city once they began office work; additionally, some philanthropically organized teams for working class and even street children and young men. Organizations like the YMCA meant to provide a healthy alternative to the temptations of the pub and street for young male clerks and shopboys, as well as an outlet for copious male vital energy, which, it was believed, should be dissipated through physical exercise and proverbial cold showers lest it be expended in debauchery, whether alone or in the company of prostitutes. Although blood sports other than hunting were considered declassé, a certain amount of pain and violence were considered a normal part of masculine life, and it was considered manly to be able and indeed eager to endure pain. Rugby, the form of football popularized by the Rugby school, allowed shin kicking, and Thomas Hughes's famous chronicle of *Tom Brown's Schooldays* (1857) shows a young player's instruction to a first-time player to wear white trousers so the "hacks" or kicks would show, as badges of courage.[42] The same player proudly explains that public school football is a much more serious business than football played in a village school: "it's no joke playing-up in a match, I can tell you—quite another thing from your private school games. Why, there's been two collar-bones broken this half, and a dozen fellows lamed. And last year a fellow had his leg broken."[43] However violent such sports were, however, the goal was to win, and injury was supposed to be simply a by-product. (British authors often express disgust at the popular German practice among middle- and upper-class boys and young men of participating in dueling clubs that resulted in dramatic facial scars highly prized by their wearers.)

Kingsley and Hughes's portrayals capitalize on trends toward athleticism that already existed, but popularized them in new ways, giving them a coherent philosophical and theological base that came to be referred to—at first slightingly, but then admiringly, as "muscular Christianity." Muscular Christianity included a general sanitary theory of physical health as fundamental—and prior—to moral well-being, which built on existing prejudices toward the able-bodied and codified anti-intellectual tendencies in notions of both national and masculine identity. Further, it posited a physical, and in shifting but still insistent terms, a racial profile for healthy English manhood. This body was based less on class than on other attributes seen as quintessentially British, and it insisted on a body toughened either by physical labor or physical sport (preferably some of both), and a keen delight in the physical pleasure of exer-

FIGURE 5.7: *A game of cricket with Lord Morpeth and Lord John Russell as the two batsmen.* Sports in private schools were preparation for networks of privilege among older boys, as well as metaphors for martial and political life, as seen here in this humorous portrayal of parliamentary politics. Colored lithograph by H. B. [John Doyle], 1840. Wellcome Library, London (V0050277)

tion and strength, which militated a bit against some earlier ascetic notions of gentlemanliness.

In reality, of course, a laborer's body was often shaped by exhausting and repetitive labor, which rarely resulted in a physique of the classical Greek ideal of athleticism so generally praised in this period. Urban laborers, in particular, were likely to be stunted by illness and developmental vitamin deficiencies resulting in shorter, smaller bodies than either their middle-class or agricultural laborer counterparts. Their bodies were sometimes marred by the distinctive bowleggedness of rickets or by exposure to industrial hazards—the testicular cancer characteristic of sweeps, for example, or the degenerative "phossy jaw" of matchgirls exposed to phosphorous.[44] Rural laborers, though often better fed, taller, and more muscular than their urban counterparts, often aged rapidly and suffered from back and joint problems in middle age.[45] In their paeans to labor, middle-class writers could be appallingly blind to the conditions under which real workers worked. Still, attention to the differences between the middle- and working-class body did increasingly lead to legislation on workplace safety, and later in the century, when the need to recruit soldiers made evident the effects of poverty on the vast majority of potential soldiers, adequate nutrition and childhood fitness became widely discussed matters of public health.

Ball games were supposed to inculcate the martial virtues; Hughes's narrator describes the sixth-form captain of the football team this way: "over all is old Brooke, absolute as he of Russia, but wisely and bravely ruling over willing and worshipping subjects, a true football king. His face is earnest and careful as he glances a last time over his array, but full of pluck and hope—the sort of look I hope to see in my general when I go out to fight."[46] They were also supposed to inculcate discipline and a team ethic, while fostering leadership skills—the qualities of gentlemen. But not all ball games were equally susceptible of gentrification. By the end of the century, cricket or rugby-style football were considered gentlemanly sports, but regular football (soccer), once the subject of great hopes as a carefully controlled rational amusement for the masses, had lost its middle-class adherents. Even with the imposition of common rules and time limits with which reformers had hoped to tame the folk elements of the game, it retained its status as a fiercely partisan local sport, in which (mostly male) spectators supported particular local heroes, drank and betted copiously, and brawled or threw garbage if they didn't like the referee's calls.[47] Working men's teams were generally named for the locality, often the street, of their origin; teams named for factories or church groups were comparatively rare.[48] In the second half of the century, sporting papers (often printed on colored paper as weekend supplements) provided a way to link the fortunes of local teams with a national identity, as readers could follow a favorite team after they moved away from the neighborhood or talk about especially strong teams with their own amateur teammates at the local pub.[49] In this sense, they did supply the sense of loyalty and identity that reformers hoped, though not always, perhaps, to the values reformers hoped to inculcate.

Just as Englishness was identified with sport, the empire was a realm in which sport could be used to socially bind the local British ruling class as they practiced sports together (tennis, hunting, cricket, croquet) in clubs that excluded natives.[50] For the so-called white colonies, such as Canada or Australia, sport provided continuity with the values, practices, and fashions of Britain. But British values could also be inculcated in colonial subjects through sport. In India, sons of ruling families were brought into public schools and taught British games, especially cricket, in order to consolidate their ties with British rulers and to inculcate what were thought of as British values of sportsmanship and self-control. Lord Curzon set up Chiefs' Colleges in the 1870s and 1880s, which offered a curriculum based on that of the English public school, to the sons of leading families. (On the other hand, polo, a game native to Asia, was adopted by the British, even though it was also used by local princes as a nationalist/nativist practice considered superior to imported British games.)[51] Games were also often used as part of evangelical muscular Christian training of natives. Theodore Leighton Pennell, a medical missionary to Afghanistan, proudly wrote that "simpler native games" like wrestling and tent-pegging lost

FIGURE 5.8: Photograph of RAMC Football Team, Howick, Natal, 24/10/1900. From the playing fields to the combat zone, team sports were thought to inculcate discipline and the martial virtues. By 1900, however, football had lost much of its association with gentlemanliness. From photograph album of Sergeant George Charles Maidment, RAMC on the Boer War 1899–1900, including No. 12 Field Hospital and No. 1 Stationary Hospital at Ladysmith. Wellcome Library, London (L0028920)

favor among Afghan boys in his school to the "superior attractions of cricket and football," and they developed a "fascination for these sports which have done so much to make England what she is."[52]

As with bathing, then, attention to the body through athletics was potentially two-edged. If being a "nation of good animals" was the ground of "national prosperity," such attention was often considered successful—Britons were proud of their reputation for healthy, boisterous, athletic males fed on British beef and beer. Yet others feared that such an emphasis on the body would lead to a neglect of intellect and spirituality, which would in turn lead to a different kind of individual and national degeneration—good animals, but not very admirable human beings. The ideal of the happy Hellenic athlete, in uncomplicated and pleasurable possession of sublime beauty and health, also entailed the suspicion that such a man would be at best childlike and amoral, at worst a dissipated, evil Dorian Gray. As is suggested by this analogy, one underlying suspicion of athletics, again as with bathing, was that a homosexual impulse underlay its worship of the male body and privileging of body-oriented homosocial practices. The body was dangerous, not only in its physical production of unhealthy substances, but even in its vitality, which might be a source of moral evil. Thus, the new emphasis on physical culture in Victorian Britain celebrated the healthy body as never before, not only as the result of moral and physical hygienic practices but also as the source of morality itself. Yet, at the same time, longer-standing suspicions of the body and embodied experience as a guide to moral decision making remained, simply taking a new form as the body was seen not simply as in conflict with the spirit but as productive of it.

CHAPTER SIX

The Normal, the Ideal, and the Beautiful

Perfect Bodies during the Age of Empire

MICHAEL HAU

> Ancient Greek art created its motives from life. Neither rough weather, nor physical ailments moved the population of Greece to cloak their beautiful figures with clothes which is why the first basic condition for the creative artist, the daily study and comparison of different forms of the nude body in its most perfect shape, was given.[1]
>
> —Carl Heinrich Stratz, 1899

Between 1800 and 1920, processes of class formation, changing gender roles, and the emergence of popular mass culture profoundly transformed people's ways of self-perception and self-presentation. Body ideals changed in tandem with social identities. There was no single generally accepted normative ideal of physical beauty. Before the late nineteenth century, the portly belly of a French bourgeois signified wealth and social status and was therefore an acceptable body image.[2] Slender, pale, and frail, almost consumptive men's and women's bodies were the ideal of the respectable American middle and upper classes in the first decades of the nineteenth century. In the second half of that century, such ideal body types were superseded by images of heavier men, and plump, buxom, voluptuous women who emulated the stars of burlesque shows and musicals. Around 1900, athletic men and sportive women became the ideal.[3] During the 1920s, the cigarette-smoking androgynous flapper with her boyish figure—representing an independent-minded woman who defined herself in opposition to traditional gender roles—contrasted sharply with supposedly wholesome and feminine ideals of women's beauty.[4]

Despite this diversity of ideal body types across time and place, many people believed there were absolute and timeless ideals of beauty that had their origins in the sculptures of Greek antiquity. In the eighteenth century, people such as the German writer Johann Joachim Winckelmann argued for the transcendent normativity of Greek representations of human bodies.[5] During the French Revolution, classical female bodies served as allegorical representations of the virtuous nation triumphing over the grotesque and disproportionate bodies of the aristocracy. Republican body imagery invoked Hercules facing the royal monster and striking blows against enemies of the revolution, whereas classical images of Liberty represented a return to stability and order after the excesses of revolutionary terror.[6] Although classical ideals were not universally accepted during the modern period, they played a prominent role in the arts, sciences, medicine, and popular discourses. Their appeal to various groups within society and shapers of social thought depended on specific cultural, social, and political contexts within which they emerged and circulated.

BEAUTY AND CONCEPTS OF THE IDEAL BEFORE THE LATE NINETEENTH CENTURY

> The human body springs from an ancient ideal type (*Uridee*) in all its parts and dimensions, structured according to one and the same basic proportion and is, within all its endless manifestations of individual forms and liberty of its movements, an organism permeated by the most perfect harmony and eurythmy.[7]
>
> —Adolf Zeising, 1854

This is how, in 1854, the German scholar Adolf Zeising described the natural law that governed the proportions of the human body. Before Zeising became a renowned writer on aesthetics, he had taught at a German gymnasium, an elite secondary school that prepared students for university studies through a rigorous humanistic education in Greek, Latin, and the arts. He believed the basic principle of aesthetic design occurred in nature and in art. The ancient beginnings of all forms (*Formbildungen*)—"the cosmic as well as individualizing, the organic as well as the anorganic, the acoustic as well as the optical"—were based on a single aesthetic principle that was "their highest goal and ideal." This ideal found "in the human form its most perfect realization."[8]

The notion that natural forms were more or less perfect realizations of basic ideas is an ancient one. Plato, for example, saw in natural phenomena the imperfect realization of ideal forms. Zeising's own understanding of nature and beauty owed much to the writings of the poet-scientist Johann Wolfgang von Goethe and Alexander von Humboldt's *Cosmos*, a comprehensive work of science that attempted a holistic synthesis of natural history. Humboldt, the famous explorer, geographer, and natural scientist, proposed an aesthetic understanding of nature as a "living whole" that was regulated by universal principles. Like Humboldt, Zeising believed the quantitative exploration of such principles contributed to an emotionally deepened understanding of nature and, by extension, human beings.[9] For Goethe, beauty was the manifestation of secret laws of nature. Morphology was the study of the forces of organic development (*Bildungstrieb*). As a science of form, it examined organisms as a whole and in relation to their parts. Natural forms were individualized expressions of a basic type, and any form that was found in the real world was the modification of a morphological archetype. According to Goethe, plants and animals developed from imperfect and primitive states toward perfection; humans were the epitome of perfection among animals.[10]

In Zeising's view, beauty was the natural expression of the "golden proportion," a mathematical relationship that was embedded in normal natural forms. The golden proportion was the division of a line in such a way that the length of the shorter portion of the line related to the longer portion of the line in the same way as the longer portion related to the line as a whole. From this principle could be derived all morphological proportions of natural phenomena, including the human body, because these proportions were the convincing proof of "how the world-creating force accomplished with the simplest means the most august and grandest effects and found from the singular the transition into the endless plurality and diversity."[11]

For Zeising, beauty was embodied in the forms of nature. The subjectivity of the observer played no role in the judgment of beauty. Beauty was thus not in the eyes of the beholder. This was in contrast to the claims of philosophers and psychologists like Theodor Lipps, Friedrich Theodor Vischer, and Robert Vischer who interpreted the perception of beauty as a psychological phenomenon arising from the subjectivity of the observer and not as a quality embodied in objects and organisms.[12] Zeising's attempt to codify the laws of natural beauty was just one of several similar efforts by artists, art critics, physicians, and biologists during the early and mid-nineteenth century. Most tried to explain and validate the aesthetic norms that found expression in the art of Ancient Greece. The Berlin sculptor Gottfried Schadow, for example, attempted to reconstruct the lost canon of the ancient sculptor Polycletos by measuring human bodies and antique statues. A canon was a set of basic principles from which ideal proportions for human bodies could be derived.[13] Schadow's work not only influenced the attempts by late nineteenth-century scholars to codify natural laws governing beauty, but his effect was also felt in art schools across

Europe. In France and England his work was well received, which demonstrated how classicism promoted transnational aesthetic orientations among artists and cultural elites.[14]

In the 1850s, Carl Gustav Carus created one of the most influential systems for the judgment of human beauty. Carus, an accomplished painter, had served as professor of obstetrics, medical councilor at the royal court, and personal physician to the king of Saxony. In his *Symbolism of Human Form*, Carus argued that human beauty was a reflection of a "beautiful godly idea" (*eines schönen Gottesgedankens*). In their purest forms, male and female beauty was represented in the art of Ancient Greece. There, the creative spirit of "divinely inspired artists" (*Gott-begabter Künstler*) had produced the most magnificent representations of Jupiter, Apollo, and Venus. According to Carus, these ideal forms represented the "pure middle" (*reine Mitte*) from which the physician and artist could develop an understanding of the numerous atrophies, deviant states, and aberrations that characterized the appearance of the great mass of humanity. For Carus, the middle or norm meant an ideal. It did not refer to modern notions of a statistical mean or median derived from anthropometric measurements of a large group of people.[15] This does not mean Carus ignored nineteenth-century statistical notions of the normal and average. He relied, for example, on measurements of the Belgian statistician Quetelet, whose "middle numbers" he considered an approximate if not fully reliable guide for what he considered to be normal.[16] As Theodore Porter has shown, Quetelet's own thought was infused with aesthetic assumptions regarding the normal, ideal, and average. He identified greatness and beauty with the mean and maintained that "an individual who epitomized in himself, at a given time all the qualities of average man, would represent at once all the greatness, beauty and goodness of that being." Deviations from the normal or mean signified vice, sickness, and "ugliness in body."[17]

Carus thought that the "normal proportion" of the human body could be derived from the length of the spine, which encapsulated the "highest formation of life." He maintained that the spine of the adult was three times the length of the spine of the newborn. Therefore, its length could serve as a "primal measure" (*Urmaß*) of the entire human organism. This measure he called a "modul" from which he constructed the ideal proportions of the human body (Figure 6.1). The length of the head without lower jaw was equal to one modul, as was the length of the collarbone. All proportions of the human body could be derived with reference to this modul. In an ideal body, the length of the arm corresponded to three moduls and the length of the thighs to two and a half moduls. But this construction was an ideal that rarely existed in nature.[18] Real people deviated more or less from these proportions depending on their age, sex, tribal origin (*Volksstamm*), or constitution. All in all, Carus distinguished between eighteen constitutional body types, with different affinities to six different temperaments and four main intellectual states: geniuses, talents, idiots, and the great mass of what he called "elementary mankind."[19] According to Carus, deviations from the "pure middle" were the keys to physiognomic science, because they symbolized sexual, racial, and individual differences in temperament and character. Women had, for example, proportionally smaller heads compared with the ideal, and male heads were larger, which was taken as a reflection of different mental and emotional traits in men and women. Although men were more energetic, women tended to be more phlegmatic, and their intellectual powers were weaker than men's.[20] Based on this diagnosis, Carus claimed that men with refined dispositions could become geniuses, whereas men in a raw state (*roher Ausbildung*) were part of elementary mankind. By contrast, women were more limited in their natural endowments. Refined women could have individual talents, but unrefined women often displayed greater mental limitations.[21] Carus's categories were shaped by class and gender perceptions. As a physician of the elite, he appears to have been disturbed by the coarseness of elementary mankind.

Carus also regarded non-European physical features as deviations from the normal/ideal body. He broke down humanity into three different races: Caucasians, who he called "peoples of the daylight" (*Tagvölker*); negroes, who he termed "peoples of the night" (*Nachtvölker*); and the peoples of the eastern or western dawn (*östliche und westliche Dämmerungsvölker*), including Mongols, Malays, and Americans.[22] According to Carus, these "tribes" (*Volksstämme*) had different mental abilities and characters, because their brain sizes differed. He supported his claims with the dubious data produced by the American physician Samuel Morton, whose research results supposedly demonstrated that there were significant differences in skull sizes in different races.[23] The peoples of the night were characterized by low intelligence and a lack of emotional depth, in conjunction with strong drives they could not control. Their psychological makeup was thus similar to criminals, who Carus sometimes described in terms of their pronounced "Negroid" and Jewish features.[24] According to Carus, the physical characteristics of black people, with their flat brows and "animal-like form" of the nose, lips, hands, thighs, and feet, allowed for an accurate assessment of the entire tribe (*Stamm*). Their mental development was low. Whatever *Bildung* they did possess had been learned from the "peoples of the day."[25]

Racial taxonomies that attributed aesthetic perfection to European whites were not particularly new. In the eighteenth century, the Enlightenment scholar Johann Friedrich Blumenbach discussed the bodies of non-Europeans in terms of deviance from classical European norms of beauty. In his 1854 essay on "Personal Beauty," Herbert Spencer made similar claims about the relationship between race and beauty. Both Carus and Spencer stand in the middle of a European tradition that equated beauty with racial purity and attributed

Bestimmungen des gesammten Körperbaues, seinem Skelet nach, durch den Modul = ⅓ des freien Rückgraths.

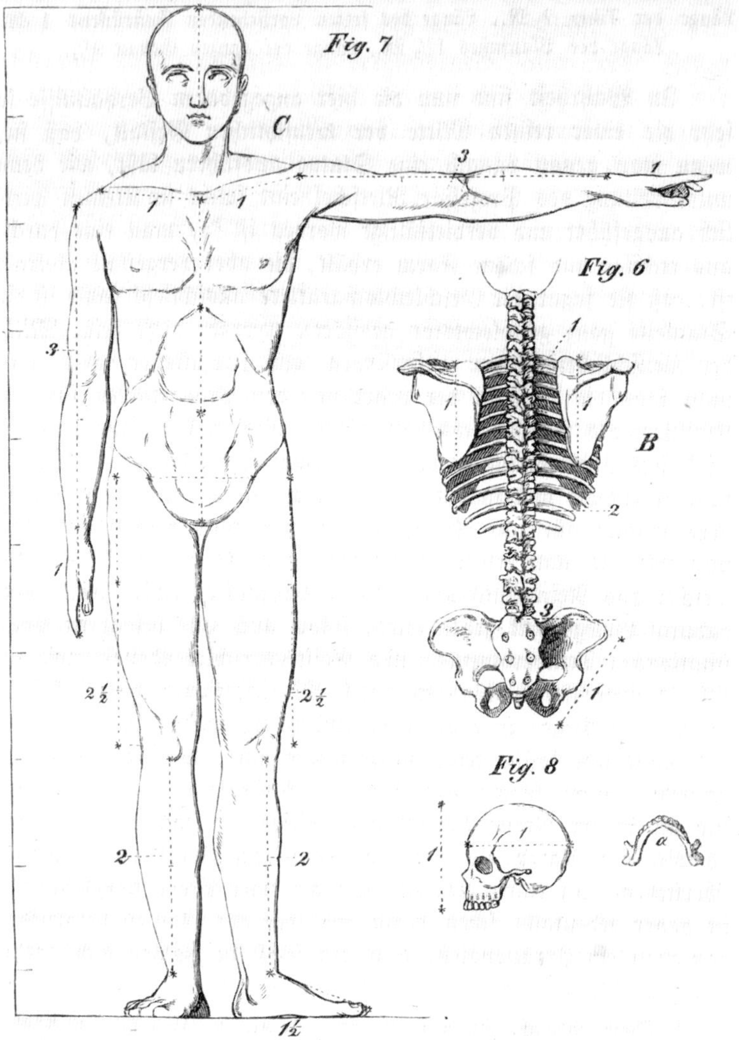

Kopf (Fig. 8) Längendurchmesser 1 M., Höhe (ohne Unterkiefer) 1 M., größter Umfang 3 M., Bogen der Unterkieferäste (Fig. 8 a) 1 M.; Stamm (Fig. 6 und 7) freies Rückgrath 3 M., jede halbe Schulterbreite längs des

FIGURE 6.1: A drawing of the human form from Carl Gustav Carus, *Symbolik der menschlichen Gestalt. Ein Handbuch der Menschenkenntnis*, 2nd ed. (1858; repr. Hildesheim: Georg Olms, 1962), 59. National Library of Medicine.

ugliness to unharmonious and unbalanced constitutions resulting from racial miscegenation.[26] During the latter part of the nineteenth century such views would be recast within a Darwinian evolutionary framework. Darwin himself saw human beauty as an outcome of sexual selection. Men selected women for their beautiful appearance, which led to an aesthetic improvement of humankind over time. In 1864, Darwin claimed that sexual selection was the most important factor for the diverging development of different races, because different races had "widely different standards of beauty."[27] In subsequent decades, this natural historical account of the development of beauty and racial differences proved to be quite compatible with the timeless normativity of the classical ideal. Darwinian anthropologists argued that the outcome of human evolution found its most perfect expression in the white race of Ancient Greece and thought that modern Europeans could recapture this perfection—which distinguished them from non-Europeans—through physical exercise and healthy living.

LIFE SCIENCES, ART, AND MASS CULTURE

From the mid-nineteenth century onward the growing prestige of scientific medicine and the life sciences provided scholars from these fields with opportunities to make authoritative statements about religion, philosophy, and art. For example, Ernst Haeckel, an eminent biologist and the most important popularizer of Darwinist thought in Germany, advocated the aesthetic appreciation of natural phenomena. In his lavishly illustrated *Art Forms of Nature*, Haeckel demonstrated to an educated lay public the beauty of living organisms.[28] Haeckel argued that the aesthetic appreciation of nature was not only a source of pleasure and an inspiration for artists, but combined with a scientific understanding of nature, it also contributed to a higher veneration of "god-nature" and culminated in a new monistic religion based on a unified understanding of science and art.[29] Haeckel believed science-based knowledge of natural forms was a prerequisite for reliable aesthetic judgment. This was also the assumption of physicians who wrote anatomical manuals for art students—a more general European phenomenon that attests to the fluidity of boundaries between art and life sciences throughout the nineteenth century.[30]

Similar exchanges between the arts and sciences were evident in controversies about the fine arts and mass culture. In the 1890s, the Berlin physiologist and anthropologist Gustav Fritsch tried to reaffirm the validity of transcendent aesthetic norms, which he saw threatened by an emerging mass culture and modern art.[31] An outspoken critic of contemporary trends in art, Fritsch rejected works of art that perverted classical ideals by appealing to sexual instincts. He accused the Munich secessionist artist Franz von Stuck of indulging his dirty fantasies. About three-quarters of Stuck's paintings involved the erotic.

One of Stuck's most famous works, titled *Sin* (1893), showed a seductress with partially exposed breasts and an orange-eyed snake around her shoulder. Such works can be interpreted as protest against the repressions of civilized society or alternatively as a sophisticated, lustful play with taboos—probably one of the reasons for Stuck's financial success.[32] Fritsch did not spare the impressionists and naturalists either: without mentioning names he claimed that for such artists "no subject was frivolous or nauseating . . . if only the dirt (*D . . . k*) had been represented faithful to nature." He berated the symbolist artist Max Klinger for painting a bloated "beer nymph" disfigured by ugly wrinkles and "angular breasts." The figure of a woman in Klinger's painting "The Blue Hour" was for Fritsch an example of an artist's intentional rejection of the ideal body in favor of the repulsive—an alarming symptom for the impending decay of good taste.[33]

Given the contemporary state of art, Fritsch considered it a matter of urgency to enlighten the public about the ideal normal proportions of human bodies from the medical point of view. Although outstanding artists like Leonardo da Vinci had created works of beauty because they had a reliable "pair of compasses in the eye" that followed ancient ideals of beauty, most contemporary artists had abandoned such lofty aesthetic orientations.[34] To determine which human forms were natural, normal, and beautiful, Fritsch picked up on Carus's notion of an ideal normal body. He considered "ideal-normal humans" people who were normal in the sense that they were free from "generally widespread individual flaws and imperfections."[35] Such imperfections were deviations from the aesthetic norms embodied in the great works of Greek art. They were also aberrations from the ideal of modern, cultivated man (*Culturmensch*) who had achieved "a certain maximum of perfection" in accordance with ancient ideals.[36] The classical ideal thus represented a lost state of physical perfection that modern cultivated humankind could regain by pursuing a moral and hygienic life in accordance with human nature.

In defending the normativity of Greek ideals, people like Fritsch, saw themselves as spokespersons for cultivated people (the so-called *Gebildeten*). Wilhelmine debates about the validity of Greek aesthetic norms reflected conflicts about the nature of higher cultivation (*Bildung*), good taste, and art. People with *Bildung* saw their cultural hegemony threatened through innovations in mass culture and art. They hoped to stem the tide by giving advice about the proper care for the body, fashion, and consumption in accordance with good taste. The architect and cultural critic Paul Schultze-Naumburg, for example, admonished women to adopt healthy living in order to live up to the standards of beauty and body care of the Ancient Greeks. As a negative counter-image to the classical ideal emerged the body type associated with the corset, French fashion, and decadence. The French ideal—promoted through fashion magazines and erotic photography—emphasized women's sexual nature at the expense of their health. For Schultze-Naumburg this vulgar ideal of beauty was the "ideal of the mass"—people who were ignorant in matters of health and beauty because they lacked *Bildung* and *Kultur*.[37] From a feminist point of view, the excesses of French fashion—the corset and high heels—accentuated women's breasts and hips and made them objects of men's sexual pleasures. The health reformer Anna Fischer-Dückelmann counseled women to avoid "the frivolous French spirit" of women who traded motherhood for the pleasures of consumption and entertainment. But the homely German housewife, the "mother animal" (*Muttertier*) who produced ten children in twelve years was not her ideal either. Instead of ruining herself physically and mentally through too frequent pregnancies, the German woman had a right to develop her own personality and cultivate her body according to classical ideals. As a warning, Fischer-Dückelmann's popular health manual for women contrasted the Medicean Venus with the grotesque bodies of contemporary women who had neglected their health.[38] Like Schultze-Naumburg, she associated natural beauty with *Bildung* and artificial means of beautification with the fake promises of mass culture, cosmetics, and jewelry: "The uneducated [woman] or the woman standing on a lower cultural stage paints herself in the most extreme manner, [she] adorns herself with rings and the most colorful bric-a-brac; the cultivated woman of higher ethical standing will always conserve a noble reserve, avoid everything that is gaudy and instead exercise natural body care."[39]

CLASS AND MASCULINITY

In nineteenth-century Europe, attitudes and practices related to the body and symbolic contests about the meanings of body types and body ideals became a field for the confirmation and contestation of class and gender identities. In Germany, lower-middle-class propagators of physical culture who lacked *Bildung* claimed that the drinking rituals and immoral lifestyles of university students contributed to the effeminacy and "degeneracy of the educated classes." They maintained that true cultivation could not emerge from the neohumanistic education in elite secondary schools, the *Gymnasien*. Instead, it had to be based on the cultivation of the body through physical culture and exercise as in the ancient Greek gymnasium. In denouncing the cultural capital of the educated elite as irrelevant for practical life, people without *Bildung*—in the sense of higher education and cultivation—turned the well-trained body into an alternative source of social esteem.[40]

For late nineteenth-century France, historian Christopher Forth has examined how body ideals of healthy masculinity were deployed in the political and cultural conflicts engendered by the Dreyfus affair. In 1894, Alfred Dreyfus, a young Jewish officer of the army's general staff, was accused of selling artillery secrets to the Germans. His subsequent trial, public humiliation, and deportation to Devil's Island in French Guiana, along with the 1899 retrial that led to his pardon, was the greatest political scandal of the Third Republic. These events reinforced and recast stereotypes about deficient Jewish masculinity and honor that were widespread in many parts of Europe. According to detractors of Dreyfus, the affair showed that Jews were not loyal to the French state. In their view, they also lacked masculine strength, courage, and honor, which made them unfit for military service. By contrast, affirmations of Jewish manhood entailed the renunciation of a scholarly lifestyle and the cultivation of physical strength and health in accordance with the manly ideals associated with military discipline.[41]

Both detractors and supporters of Dreyfus resorted to medical imagery, in nearly identical terms, to denounce the deficient masculinity of their respective opponents. In this way, the affair expressed anxieties about the degeneration of the male physique and psyche as a result of modern civilization and sedentary lifestyles of the urban population in general and intellectuals in particular. According to Forth, issues of the body and gender provided a common idiom for conceptualizing and discussing political, social, and cultural issues. Sports and their concomitant valorization of physical strength challenged men with higher education and sedentary lifestyles. It made intellectuals aware of their distance from more physically heroic and martial forms of masculinity.[42] At the same time obesity (which earlier had signified social status and health) became increasingly stigmatized, because it signified the lack of virility, self-control, energy, and will in men. The obese were considered outside the sphere of heroic male activity and volition. Obese and weak puny men were both represented as negative counterimages to a new ideal of muscular masculinity.[43] The writer Emile Zola, famous for his girth and for his prominent role in the campaign to rehabilitate Alfred Dreyfus, therefore, tried to change his public image by adhering to a strict diet and exercise regimen.[44] Physical culturists like Edmond Desbonnet denounced obese and puny men for their lack of character, energy, and will power. The sedentary lifestyles of state bureaucrats and intellectuals and the pleasure-seeking ways of dandies were held responsible for the degeneration of French men.[45] Such distinctions between different male types also found their way into the terminology and taxonomy of clinical medicine. The physicians Claude Sigaud and Leon Vincent, for example, constructed a typology of male constitutional types: Their *type musculaire* with a sportive muscular body, contrasted sharply with the *type cerebral*, a puny body that was found most often among intellectuals and solitary researchers (Figures 6.2 and 6.3).[46]

In countries such as France and Germany, where public intellectualism and privileges of higher education carried considerable social esteem, the degenerate physical state of intellectuals attracted the malicious criticism of physical fitness advocates. In the United States, middle- and upper-class advocates of

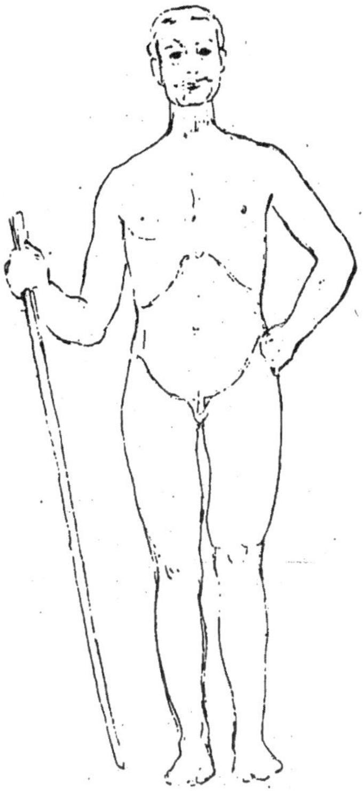

FIGURE 6.2: "*Type Musculaire.*" From Claude Sigaud and Leon Vincent, *Les Origines de la Maladie. Essai sur l'Evolution de la Forme du Corps humain*, 2nd ed. (Paris: A. Maloine, 1912), 57. National Library of Medicine.

physical exercise were concerned about the sedentary way of life of the business and professional classes. To counter the perceived emasculation and feminization of upper- and middle-class men, they promoted sports as an important part of the educational experience of male college students—necessary for the building of moral character as well as physical vigor.[47] For people like Theodore Roosevelt, sports, physical culture and strenuous outdoor activities were central to a narrative of self-transformation: hard and determined work on his body transformed him from a sick, weak, and nervous child to a healthy, strong, and dominant male who had overcome the debilitating effects of civilization.[48] In contrast to formal higher education, "body work" was not confined to the middle and upper classes. The physical culture regimens propagated by fitness entrepreneurs like Bernarr MacFadden and Eugen Sandow offered lower-class men a body-centered form of self-discipline and respectability. In this process, they redefined social aspirations and success in terms of physical fitness: As MacFadden put it: "If you have health of the supreme sort, it makes you throb with superabundant energy that gives you ambition, endurance, determination. Then you are indeed rich, regardless of the size of your bank account."[49]

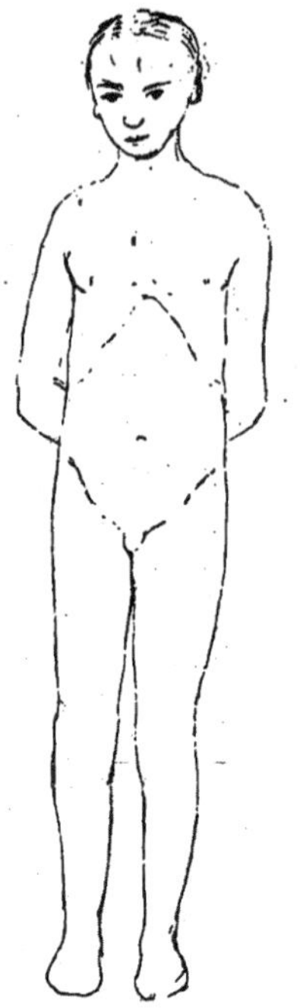

FIG. 4. — *Type cérébral.*

FIGURE 6.3: "*Type Cérébral.*" From Claude Sigaud and Leon Vincent, *Les Origines de la Maladie. Essai sur l'Evolution de la Forme du Corps humain*, 2nd ed. (Paris: A. Maloine, 1912), 59. National Library of Medicine.

NATIONAL ORIGIN MYTHS AND RACIAL BEAUTIES

During the second half of the nineteenth century, representations of classical bodies became increasingly central to European fantasies about the racial ancestry of nations. European intellectuals and anthropologists developed visions of Greek racial origins for their nations. After the Franco-Prussian War, for example, French anthropologists liked to depict brown-haired Frenchmen as representatives of the Ancient Greek type of human perfection, which contrasted with degenerate Prussian blonds.[50] Such claims to Greek racial origins often had particular national inflections. Although French anthropologists and artists favored constructing common Gaul, Celtic, and Greek racial origins for the French, German promoters of the classical ideal preferred to see themselves as the racial descendents of Tacitus's *Germanen*, who shared common racial origins with the Greeks. In the 1920s, the German movie *The Golden Path to Health and Beauty* (*Wege zu Kraft und Schönheit*) invoked this origin myth by showing scenes of Greek athletes competing in a gymnasium and of ancient Germans trying to jump over several horses. The movie promoted national fitness through modern sports and physical culture as compensation for the degenerating effects of sedentary lifestyles and modern work.[51]

Around 1900, colonialism and the conflicts between imperialist powers changed the ways in which Europeans imagined themselves and others. These developments had a profound impact on the meanings people attributed to ideals of physical beauty during the age of empire. The emergence of the vigorous, healthy, and muscular male as a masculine ideal in late nineteenth-century France was tied to concerns about the vitality of the nation and its ability to compete with other imperialist nation states. In France, the memory of the defeat in the Franco-Prussian war merged with concerns about the declining birthrate to exacerbate fears about the debilitating and feminizing aspects of modern civilization. In this context, physical culture was promoted as a collective redemptive project for restoring French vitality.[52]

In Germany, concerns about military fitness and the national body had already merged in the nineteenth-century gymnastics movement. During the late nineteenth century, German gymnasts (*Turner*) based their claims to citizen rights on the contribution they made to the military fitness of the nation.[53] After 1900, physicians and health reformers advocated physical fitness as a way to systematically increase the vitality of the nation. The medical professor Ferdinand Hueppe supported physical culture and sports in an effort to improve national health and military fitness. This was in line with his engagement as a eugenicist who propagated racial hygiene in order to prevent degeneracy and improve the hereditary fitness of the nation over the long term.[54] Like many contemporary health reformers, he was worried that the sedentary way of life of the urban population contributed to the degeneracy of modern humankind. In Hueppe's view, physical fitness, beauty, character, and mental abilities could be improved through exercise. Geniuses like Leonardo da Vinci or Goethe, he claimed, were also exceptional in their physical prowess and beauty.[55]

Hueppe was a Darwinist who believed cultured humankind represented the apex of human development and perfection. He rejected the view that so-called natural peoples (*Naturvölker*) were physically superior to modern cultured humankind. Instead, the Nordic race represented the highest stage ever achieved by humankind. This perfection was also expressed in the beauty of the race, which corresponded with the ideals of physical beauty represented in Greek art.[56] This conviction he shared with other Darwinist anthropologists. Writing during the high point of European expansionism, Gustav Fritsch claimed that the beauty of cultivated modern man and woman provided an ideal from which the difference and distance between human races could be assessed. Because human diversity was threatened by European expansionism, Fritsch urged that photographs be taken of as many naked bodies as possible, before it was too late to establish scientifically different racial types.[57]

The motivations for anthropological and popular interests in race were complex. On one level, the construction of a hierarchy of racial types certainly served to legitimize European colonialism. One historian has argued recently that racial science was fundamental to colonial projects because it objectified colonial subjects and denied them any subjectivity as a basis for human rights.[58] However, debates about race are not only interesting in terms of the history of colonial relations, but because racial constructs intersect with understandings of social class, gender, and sexuality, they also provide idioms for the discussion of social relations. This is why discussions of social class were sometimes cast in racial terms. In Germany, there was an assumption among sectors of the educated upper middle-classes that only people of the Nordic race had the innate ability to acquire true *Bildung*—in the sense of a deep inner cultivation of the self—and express their inner beauty in their physical appearance. The ideal Nordic woman, according to this line of thought, was beautiful and free of the base sexual instincts that were commonly attributed to lower-class women.[59] In popular culture, the sexuality of lower-class women became associated with racial otherness. Around 1900, commercial ethnographic exhibitions (*Völkerschauen*) of Africans and other colonial subjects became the focal point for concerns about hedonistic mass entertainment and the sexuality of lower-class women who might be tempted by the exotic bodies of some of the performers in these shows.[60]

The interest in exotic bodies combined scientific interest in racial bodies with erotic interests in the diversity of exotic nudity. Fritsch's photo album on nude racial types showed exotic women in erotic poses that were incompatible with the standardized requirements of anthropological photographs that he himself proposed.[61] The gynecologist and anthropologist Carl Heinrich Stratz—an avid collector of photographs who contributed to Fritsch's collection—showed a similar interest in the erotic (Figures 6.4, 6.5, and 6.6). He published photographs of nude women from all over the world in a book on the racial beauty of women in order to demonstrate to his readers the timeless validity of ideals of classical beauty. According to Stratz, white Nordic women represented timeless aesthetic norms, but black women deviated from these proportions because their extremities were too long, and Asian women deviated from these norms because their heads were too large and their limbs were too short. White women thus represented the ideal middle. Stratz did not deny that exotic women could be erotic. Their distinctive racial beauty might be pretty (*hübsch*), and their erotic attraction might tempt white men, but they were quite different from the chaste white women respectable white men preferred as their wives.[62] The discourse on body aesthetics and racial and national diversity legitimated heterosexual interests, which fueled a large market for photographs of French and Hungarian artist models.[63] The German photographer Wilhelm von Gloeden published images of young southern Italian male nudes in the journal *Die Schönheit*, which had more than a subtle subtext of the homoerotic.[64]

FIGURE 6.4: "Seduction of the Exotic: Japanese Woman." From Carl Heinrich Stratz, *Die Rassenschönheit des Weibes*, 5th ed. (Stuttgart, Germany: Ferdinand Enke, 1904), 119. National Library of Medicine.

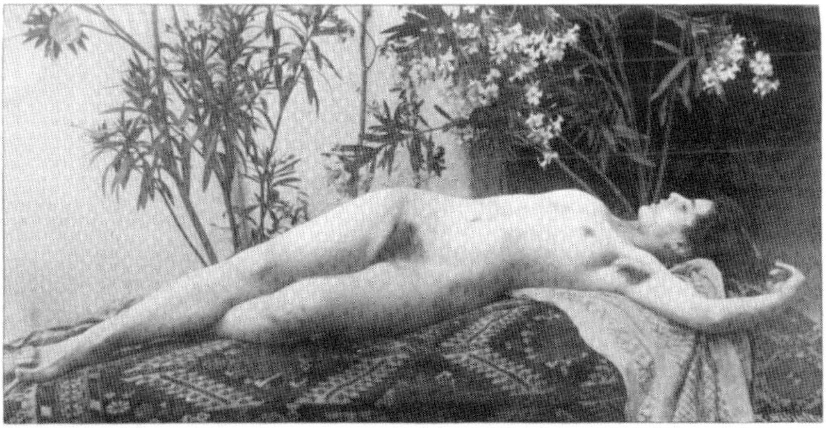

Fig. 202. Römisches Mädchen von 18 Jahren, reiner Nationaltypus. (Phot. von Plüschow.)

FIGURE 6.5: "Italian Woman." From Carl Heinrich Stratz, *Die Rassenschönheit des Weibes*, 5th ed. (Stuttgart, Germany: Ferdinand Enke, 1904), 304. National Library of Medicine.

Fig. 218. 20jähriges Mädchen aus Paris.

FIGURE 6.6: "French Woman." From Carl Heinrich Stratz, *Die Rassenschönheit des Weibes*, 5th ed. (Stuttgart, Germany: Ferdinand Enke, 1904), 326. National Library of Medicine.

IDEALS, VARIABILITY, AND HUMAN TYPOLOGIES

Scientists' attempts to establish universal rules for the assessment of physical beauty concealed underlying epistemic tensions in the discourse on the normal and beautiful. The authors often had difficulties reconciling the reductionism of their systems with the facts of human variability. Stratz and Fritsch thought that the only human proportions that were beautiful were those that corresponded to Fritsch's canon of ideal proportions. They claimed that the same canon applied to men and women but never reconciled this conviction with the assumption that men and women ideally had different masculine and feminine proportions. If there was only a single standard for beautiful proportions, how could feminine women and masculine men represent this same ideal? Ferdinand Hueppe tried to account for human diversity by diagnosing different body types among male athletes and Greek representations of beautiful bodies. Beautiful males were either a muscular Heracles type or resembled a leaner Hermes type (Figures 6.7 and 6.8). The former was the patron of wrestlers and

Fig. 28.

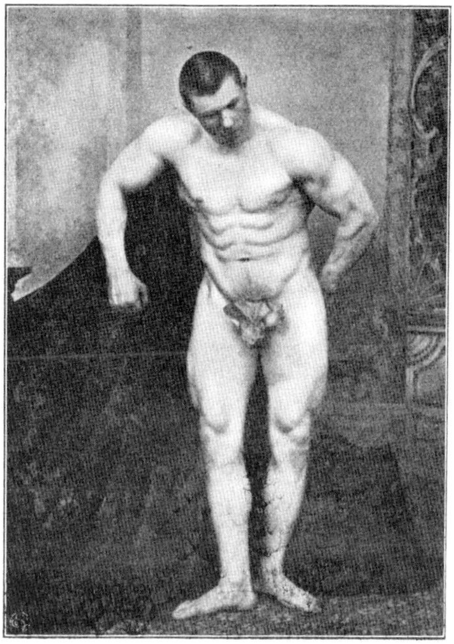

Hackenschmidt, um die Wende des Jahrhunderts der beste Ringer und Gewichtsathlet, am 21. Geburtstage; typische herkulische Muskulatur; Größe 1,75 m. Entwicklung besonders durch deutsches Turnen, Radfahren, dann durch Schwergewichte und Ringen.

FIGURE 6.7: "Wrestler and Weight Lifter Hackenschmidt as Example for the Herculean Type." From Ferdinand Hueppe, *Hygiene der Körperübungen* (Leipzig, Germany: S. Hirzel, 1910), 53. National Library of Medicine.

heavy athletes (*Schwerathleten*). The latter represented track and field, and pentathlon athletes. For Hueppe, both types were beautiful, because they were adapted to different kinds of physical performances.[65] This followed from his assumption that a fit and well-performing body was inherently beautiful. Exercise, therefore, had to improve the physical performance of the body to improve its appearance and make it truly beautiful. Horse breeders, he claimed, no longer selected horses for breeding based on their physical appearances. Instead they selected the highest-performing horses, but the result was that contemporary race horses were more beautiful than ever.[66]

As an advocate of specialized sports that favor different body types, Hueppe moved away from the notion of a single aesthetic ideal toward a notion of different physiological and morphological constitutional types that were suited for different sports. During the 1920s, such approaches would culminate in research programs about sports and work types that were suited for different

Fig. 31.

Unger gen. *Lionel Strongfort*, ebenmäßig entwickelt, nur Kapuzenmuskel etwas zu stark; schöner Fünfkämpfertypus; Größe 1,72 m; ohne besondere Methode aufgewachsen, dann mit Attila-System oder 5-Pfd.-Hantel-Methode planmäßig entwickelt, dann Schwergewichte und Boxen.

FIGURE 6.8: "Lionel Strongfort Max Unger as Beautiful Pentathlon or Hermes Type." From Ferdinand Hueppe, *Hygiene der Körperübungen* (Leipzig, Germany: S. Hirzel, 1910), 56. National Library of Medicine.

kinds of sports and physical labor. The sports physician Wolfgang Kohlrausch, for example, measured the bodies of athletes during sporting competitions in order to establish the physical characteristics that determined success in different sports. At the 1928 Amsterdam Olympics he took more than ten anthropometric measures from more than 300 athletes, almost all of them men. Kohlrausch, who directed the anthropometric laboratory of the German University for Physical Exercise in Berlin, hoped to develop a taxonomical system of sports types in order to be able to assess the aptitude of people for different sports.[67]

Such typologies were not confined to sports medicine. From the 1920s onward, physicians and psychologists developed typological norms for bodies to account for different physical and mental characteristics and abilities. The psychiatrist Ernst Kretschmer, for example, distinguished between weak leptosome, muscular athletic, and short and stocky pyknic body types. The first two types had a constitutional predisposition toward schizophrenia, the third a predisposition toward manic-depressive disorders.[68] Inspired in part by Kretschmer, the American psychiatrist William Sheldon proposed his own taxonomy of human constitutional types to diagnose what he considered the hereditary physical and mental potential of people. During the twentieth century, such efforts to create individualized norms for different body types did not however fully erase the normativity of the classical ideal. In America and Europe, a muscular athletic type remained an important reference point for many males.[69]

SELF-REALIZATION AND HEDONISM

In most Western countries, the spread of physical culture and sports contributed to the growing prominence of ideals of physical beauty in twentieth-century mass culture. This signified a profound change in the ways people related to their own bodies. Physical fitness and attractiveness played an important role in consumer culture. Exercise regimens and physical culture, but also methods of artificial rejuvenation such as cosmetic surgery, all point to a growing self-reflexivity centered on the body. In Nikolas Rose's terms, one can see such body-centered practices as technologies of subjectivation. Physical culture, sports, dieting, aesthetic surgery, and other forms of body work aimed at changing the self. They were instruments for creating self-knowledge and self-mastery in a world that lacked binding value orientations based on traditions.[70]

Work on their bodies enabled people to develop a sense of agency. By focusing on those aspects of their lives that they could control—their bodies—men and women could acquire social and cultural capital by shaping their bodies according to cultural ideals. Because success was seen to depend on physical attractiveness, health entrepreneurs in Europe and the United States claimed that their exercise systems could help people to become more successful in their professional and personal lives.[71] As a result of an individual's own body work, the well-shaped attractive body signified health, vitality, and strength of character. In Germany, promoters of exercise and physical culture claimed that a well-developed natural body was a reliable representation of an authentic self that contrasted with the false selves or "masks" of people who artificially enhanced their bodies through cosmetics or cosmetic surgery. Although such natural beautification promised to break down privileges of social class because it was (at least in theory) accessible to everybody and did not depend on a person's wealth, the distinctions between artificial and natural cemented new distinctions between the authentic and inauthentic body.[72] This is also evident from the post–World War I discourses on rejuvenation that were recently examined by Heiko Stoff. Stoff contrasted debates about the natural rejuvenation of male and female bodies through exercise and physical culture with controversies about artificial rejuvenation through surgery. Influenced by new research on sex glands and hormones, physicians like Eugen Steinach and Serge Voronoff promised to rejuvenate men and women through sterilization, organ substitution, or transplantation of sex glands. The therapeutic measures advocated by such rejuvenation physicians were quite diverse, but they all tapped into the widespread desire of people to regain healthy productive bodies and attractive and youthful appearances.[73]

Some historians have pointed to the liberating effects of modern body culture. The emphasis on physical health and beauty indeed had liberating aspects for men and women who felt constrained by a sedentary way of life at work or at home. Feminists, for example, often resorted to hygienic arguments when they argued that women should have greater professional opportunities and should be less restricted in their social lives. At least in urban areas, many restrictions on the social mingling of the sexes ceased to exist after World War I. In major cities, young people went without parental supervision to movie theaters and dance halls. Young women, inspired by movies and illustrated magazines, wore sexually enticing clothes. New attitudes toward women's sexuality began to emerge. Some people openly defended premarital sexuality as an important part of personal self-fulfillment.[74] Sports and physical culture were part of a hedonistic postwar culture that stressed the enjoyment of physical activity of both sexes.

However, while sports and physical activities legitimized people's interest in the sexual body, social conservatives promoted physical exercise as a means to curb the sexual desires of youth. They hoped that sports as a wholesome character-building activity would lure young people away from the enticements of urban consumer culture: drinking, partying, and premarital sex. The wholesomeness of exercise was one of the messages of the movie *The Golden Path to Health and Beauty*. A scene in which an alcohol-drinking dissolute young woman amused herself with men in a bar served as a contrast to healthy young women engaged in various sports and exercise. However, the movie is also a good example of the general ambiguity of contemporary attitudes toward the sexual body. The condemnation of the sexually enticing went hand in hand with the legitimization of the erotic through the presentation of nude or scantily clad bodies that evoked sexual desires and wholesome notions of Ancient Greece.[75]

Hedonism and self-realization were important promises of the popular culture after World War I.[76] Yet the work of several historians suggests that the transition to the consumer and mass society of the 1920s, with its promises of pleasure, was not a simple story of a hedonistic liberation of desires. Rejuvenation and aesthetic surgery, as well as physical culture, held out the promise that people could refashion their own selves in accordance with notions of ideal bodies that referred back to normative ideals of classical antiquity. However, the ability to refashion one's body and identity in accordance with accepted norms could prove elusive for many people who could not approach the ideal. As Elizabeth Haiken has shown for early twentieth-century America, the psychological construct of an inferiority complex served to justify the nontherapeutic use of plastic surgery. By ridding people of physical stigma, they would be able to overcome their inferiority complexes, develop self-confidence, and become successful in life. However, by emphasizing physical attractiveness as a key to social and professional success, health entrepreneurs and aesthetic surgeons also fostered personal insecurities.[77] The marketing of the Lionel Strongfort system of physical culture that was promoted by the German health entrepreneur Max Unger in Europe and the United States demonstrates this mechanism. The advertisements deliberately encouraged men's awareness of their individual deficiencies. Describing Strongfort's system as the "Science of the Normal," they denounced men who suffered from aesthetic deficits and lack of vigor as "abnormal"—a state they could only overcome by following a training regimen based on the principles of "Strongfortism."[78]

By asserting the normativity of body ideals that many people could not live up to, physical fitness and beautification advocates promoted the very feelings of inadequacy that their interventions promised to cure. One need not go so far as an American historian who saw in modern body discourses "not a hedonistic quest for pleasure but a pervasive anxiety and . . . fear of the biological processes in one's own body that might give offense to others and bring the precarious self to social ruin."[79] Yet the insistence of popular health discourse that one's own body could be perfected and that every one had the potential to achieve ideals of physical beauty was certainly a source of anxiety as well as of pleasure and self-realization.

CHAPTER SEVEN

Empire, Boundaries, and Bodies

Colonial Tattooing Practices

CLARE ANDERSON

In recent years the body has become a fashionable mode of inquiry into the nature of colonial societies. Historians have used a Foucauldian or Saidian framework to focus on the relationship between power and the body (notably within colonial institutions, but also those embodied in cultural practices) and/or on representations of the body (for instance in ethnographies of colonial difference). As Abby Schrader puts it, "the body itself constitutes a central text of cultural history."[1] India and Africa in particular have proved fertile ground for explorations of the colonial body, for it was peculiarly central to colonial understandings of societies in which the organizing principles of caste and tribe—with their seemingly incomprehensible array of ritual practices and taboos—seemed so important.[2] In examining colonial tattooing practices and their representations, this chapter draws upon this empirically and theoretically pertinent set of historiography. It presupposes the surface of the skin—and its apparently permanent and always potentially visible inscriptions—as an important element of the embodied practices and representations that historians and anthropologists have described.[3]

Underlying this chapter are three assumptions. First, though tattooing leaves permanent marks, those marks can be read in multiple ways and so acquire multiple meanings. Even the most certain physical mark of identity—the apparently immutable tattoo—is subject to debate and (re)interpretation. As Ann Laura Stoler and Frederick Cooper argue in their theoretical consideration of the *Tensions of Empire*, the "otherness" of colonized persons is neither inherent nor stable.[4] And yet—and this is my second point—tattoos were (and still are) nevertheless an important means of establishing identities. As I will show, these could be individual or collective in nature—and sometimes a rather messy and conceptually inconsistent combination of both. Third, and perhaps most importantly, tattoos transcend the geographical boundaries that are commonly associated with state formation in the nineteenth and early twentieth centuries. Indeed, they represent the complex slippages in the relationship between Britain and its empire, and between colonies themselves. They also speak to the multiple cultural connections of European colonization in North America and, as such, might be considered one of the intimate domains that figured in the making of both cultural categories and imperial rule.[5] As we will see, during the long nineteenth century of this volume, marked bodies are viewed most fruitfully through a global lens. Though tattooing is the main focus of the chapter, I will also refer to other modes of bodily marking that break the surface of the skin—scarification (or cicatrization), piercing, and branding—recognizing too that nineteenth-century practices sometimes included aspects of each.[6]

TATTOOING AND THE COLONIAL ENCOUNTER

Tattooing is commonly believed to have arrived in Europe after James Cook returned from the South Seas in the late eighteenth century, giving rise to a derivative description from the Polynesian word *tatu*, or *tatau*. That tattooing in Europe in fact dated from medieval times has led Jane Caplan to describe it as a "promiscuous travelling sign," always represented as having arrived from somewhere else.[7] Though Cook's voyages did not introduce the practice into Europe, they certainly reinflected it at a time when the continent was open to its enthusiastic reception.[8] Early-modern encounters between Europeans and tattooed Polynesians—both in Europe and the South Seas—and their multiple artistic and discursive representations are beyond the temporal scope of this chapter. However, I would like to note two things. First, that Pacific practices involved on occasion the tattooing of symmetrical lines and patterns on the face. This was different from the position and form of European tattoos, which did not usually incorporate facial designs and were more image or object-based forms of representation. Second, that because by the beginning of the nineteenth century tattooing in Europe was largely associated with the overseas travel of sailors and soldiers, it might best be described at that time as a product—or a symptom—of the colonial encounter.[9]

FIGURE 7.1: *Mouina, Chief Warrior of the Tayehs, from Ua Huka*. Engraving by David Porter and William Strickland. From David Porter, *Journal of a Cruise Made to the Pacific Ocean* (Philadelphia: Bradford and Inskeep, 1815). Wellcome Library, London (L0040399)

James Bradley, Hamish Maxwell-Stewart, and Ian Duffield show that during the nineteenth century tattoos were a means through which marginalized Europeans made sense of a fast-changing world. Religious or individual mementos allowed sailors, soldiers serving overseas, and convicts transported to penal colonies in Australia to cope with dislocation, to remember kin networks, or to mark significant moments in their journeys overseas. One British jockey-turned-soldier court-martialed in Bombay and sent to Van Diemen's Land (Tasmania) in 1845 embodied the prospect of his transportation through tattoos of a kangaroo, a man with a spear, and a dog, each inscribed above the words *advancée australia*.[10] Tattoos like these were intensely personal—representations often included strings of letters or dates; the sun, moon, and stars; flowers; and hearts, all of which held individual or social meaning. The repetition of particular designs, such as the crucifix or the hope and anchor, reflected religious beliefs or broader social relationships with occupation and the cultural dynamics of colonial journeying.[11] British soldiers for example commonly sported regimental crests, and those serving in Burma sometimes acquired elaborate Southeast Asian designs, indicating that tattooing was also a type of collecting.[12] Burmese tattooing had long aroused the fascination of European travelers, for young men were tattooed from the hips to the knees in "breeches" style. The complex and often colorful images included tigers, lions, demons, dragons, peacocks, fish, and flying animals.[13] One decommissioned lieutenant seen by an army medical officer in 1861 was tattooed from the neck to mid-calf, with "not the space of a pin's point left uncoloured" between the designs: red, black, and blue letters and pictures of birds, animals, fish, and scrolls. He told the officer that Europeans serving in Burma frequently acquired some tattoos, but most could not bear the pain associated with the lengthy sittings necessary and so the tattooing remained incomplete.[14]

The suggestion that large numbers of nineteenth-century British soldiers "collected" Burmese tattoos further suggests the importance of tattooing as a cross-cultural practice, and there was a strong relationship between tattooing and overseas travel. In Europe, until the middle of the nineteenth century, tattooing was viewed largely as an expression of cultural disassociation, for it was aligned most closely to marginal social groups like the soldiers, sailors, and convicts that Bradley, Maxwell-Stewart, and Duffield describe. This pattern holds for North America, too. One anthropologist estimated that even in 1908 95 percent of the U.S. Infantry, and 90 percent of sailors on men-of-war were tattooed. Alan Govenar describes how the association between tattooing and social marginality meant that military officers wanting to improve soldiers' image during World War I moved to discourage the practice. Tattoos common at the time included lucky numbers, important dates, and mottoes.[15] During the nineteenth and early twentieth centuries, tattoos were often represented through the colonial trope of "primitivism," and aligned

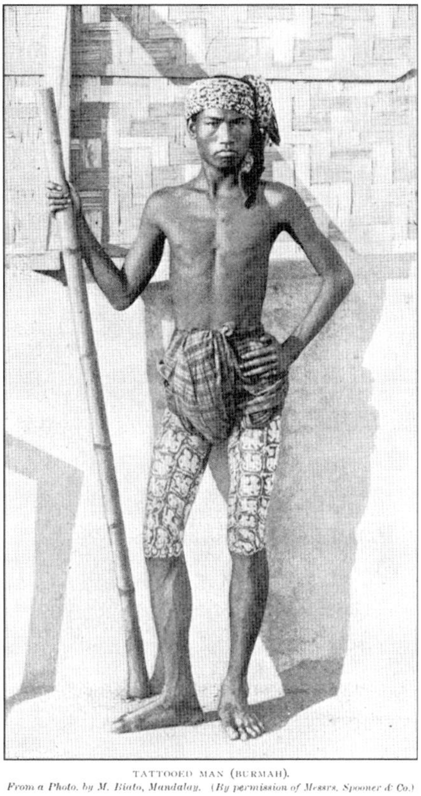

FIGURE 7.2: "Tatooed Man (Burmah)." From Gambier Bolton, "Pictures on the Human Skin," *Strand Magazine* 8 (1897): 427.

with premodern so-called savage practices. Enid Schildkrout describes this as a conflation of the European underclass with what they viewed as exotic African, Asian, and Native American bodies.[16] And yet the great paradox of tattooing is that this apparently impermeable sign marked the permeability of colonial borders. In this respect it also communicated a certain porousness in the discursive boundaries of representation and social understanding, for marked bodies in the age of empire often transcended the cultural categories (notably the tropes of "civilization" and "savagery" aligned with Europe and its others, respectively) tattooing was thought to express.

Joanna White offers further insights into cultural exchanges in a fascinating discussion of European beachcombers (deserters, shipwreck survivors, mutineers, escaped Australian convicts) in the Pacific islands during the late eighteenth and early nineteenth centuries. She shows how central tattooing was to their local acceptance and assimilation (or "going native"). Frequently, local practice included the tattooing of the face, which was quite different from the marking of the torso or limbs in Europe, and of course the removed potential to cover those marks from public view. A small number of beachcombers returned home, but their strongly visible tattoos proved both an obstacle to their cultural reintegration and a boon to their fame and subsequent wealth. Chelsea pensioner George Bruce, who returned from New Zealand to Britain at the beginning of the nineteenth century, recorded later the popular hostility he faced on account of his tattoos thus: "some calls me a Man Eater. another Says I am the Devil and another calls me a Traitor To my Country & this is all becas I don't sattsfie Them all. with the marks in my face." His petition for a land grant in New Zealand failed.[17] Unlike Bruce, a few, such as James O'Connell—the first tattooed European to work in American circuses—capitalized on the popular appeal of tattoos. O'Connell made a good living displaying his exotic marks—which included facial tattoos—through the second half of the 1830s to the 1850s. His act was part of the broader contemporary genre of captivity narratives through which the tales of Europeans captured by barbarian others—and in O'Connell's case were tattooed against their will—found a popular market.[18] As White shows, "the tattoos that had been customary marks of belonging in the Pacific context were deliberately transformed into spectacular symbols of exotic ornamentation within a new social context."[19]

Stephan Oettermann describes how in the later decades of the nineteenth century and during the years leading up to World War I, the "greatest showman on earth," P. T. Barnum put tattooed Europeans and North Americans on display in his circus sideshows, all of whom played to their audience in recounting outlandish colonial tales about how they had acquired their tattoos. Most famous of all during the nineteenth century was Greek Alexandrino, who was tattooed from head to toe with 388 images in Burmese style.[20] In the 1870s, contemporaries described his skin as like "a tightly-woven fabric of rich Turkish stuff," and only the soles of his feet were left unmarked. As well as Burmese script, the tattoos included both stylized and natural representations of sphinxes, apes, leopards, cats, tigers, eagles, storks, swans, frogs, peacocks, snakes, lions, elephants, crocodiles, salamanders, dragons, fish, fruit, leaves, and flowers. In Germany he was called "The Tattooed Man of Burma," and in America he was known as "The Turk" or "Prince/Captain Constantenus." He told various stories about his tattoos, claiming that he was a pirate, a fortune hunter, or an arms dealer who had been captured in China or Burma and as punishment tattooed forcibly. He enjoyed fame in European medical circles, too, and his tattoos were viewed, photographed, and discussed across the continent. One line of scientific enquiry was the damage tattooing potentially caused to the glands and pores of the skin.[21]

In Alexandrino's wake came John Hayes, an American who presented his tattoos within the same trope of captivity. He told stories of his imprisonment and forcible tattooing with 780 images during the Indian Wars. Later imitators included a woman named Irene Woodward, who was known popularly as "La Belle Irène." Her claim was that her father had tattooed her with blue and red images to prevent her abduction by Sioux Indians. The representa-

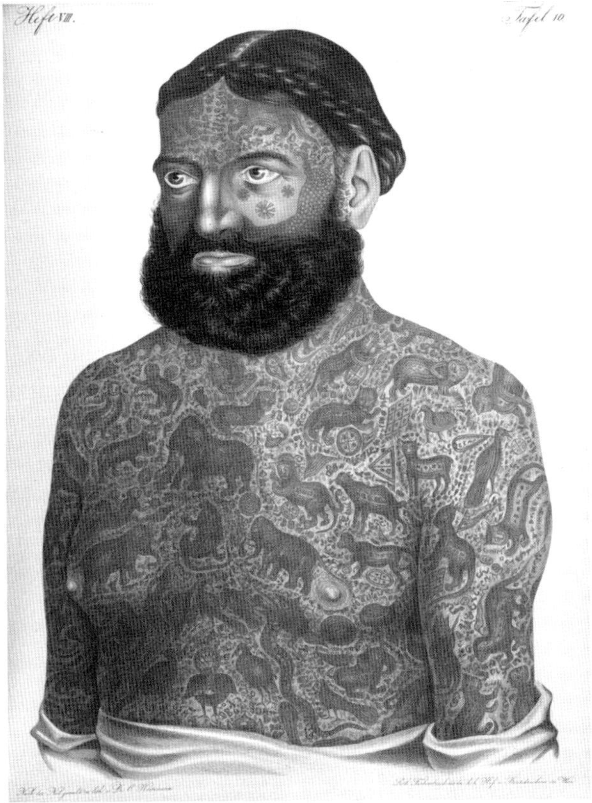

FIGURE 7.3: *Tattooed Figure*. From Ferdinand Hebra, *Atlas der Hautkrankheiten* (Vienna: C. Gerold's Sohn, 1856–1876). Lithograph by Anton Elfinger, after drawing by Carl Heitzmann. National Library of Medicine (67440930R)

tions included butterflies, flowers, suns, eyes, eagles, insects, snakes, and even the popular painting "A Sailor's Farewell." Tattooed entertainers remained popular in America until the end of the century, though increasingly tattoos alone were not enough to inspire awe. Instead, put on display were tattooed dwarves, fat ladies, wrestlers, sword-swallowers, and so-called Indians who performed conjuring, psychic, or juggling tricks. Tattooed families also appeared. Especially famous in Germany later on were tattooed couples, in the 1890s notably Frank and Emma de Burgh, whose tattoos included pictorial representations of flora, fauna, and their names. Performances by women were particularly exotic, for their show involved unpeeling clothing from usually covered skin. For Oetterman, this was something akin to a striptease.[22] Horace Ridler, the "Zebra Man" (also called "the Great Omi," after the Pacific islander famously brought to Britain by Captain Cook in the late eighteenth century), enjoyed enormous popularity in England, France, and on Broadway between the two world wars when he displayed new types of tattoos and body modifications. He reworked his old pictorial tattoos, which had been so popular at the end of the nineteenth century, into broad stripes and patterns that covered his face, scalp, and body. His teeth were filed to points, and his septum and earlobes were pierced and stretched. Though he told stories of capture and tattooing in New Guinea, just as John Hayes was tattooed by Samuel O'Reilly, the American inventor of the electric tattoo machine, Ridler's tattoos were in fact the work of the famous London-based artist George Burchett.

In Britain, as greater numbers of old elites and the growing middle class began to visit or work overseas, tattoos acquired a different meaning. Bradley shows that uniquely in Europe there was a late nineteenth-century fad for tattooing among the fashionable classes in Britain, coinciding with the return of the princes Albert and George from Japan sporting tattoos, apparently copies of the five crosses tattooed on their father, Edward, the Prince of Wales, on his tour of the Holy Land in 1862. This probably reflected rather than inspired the already growing popularity of tattooing in privileged social circles.[23] Indeed,

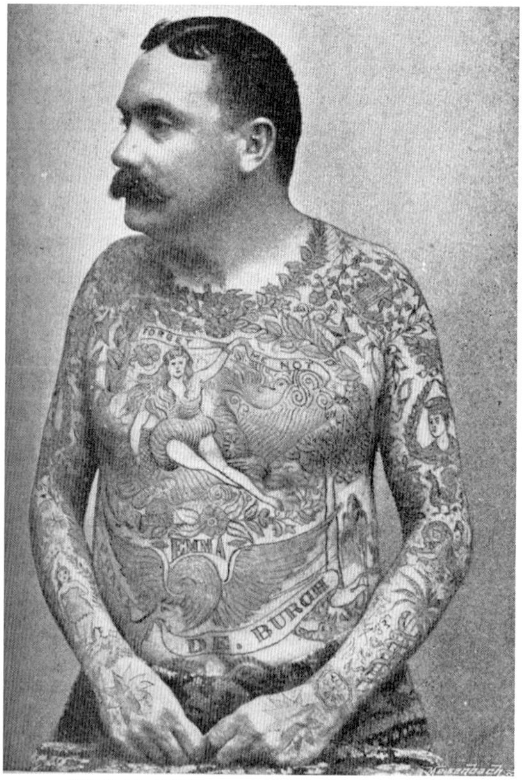

FRANK DE BURGH.

FIGURE 7.4: "Edward de Burgh." From Gambier Bolton, "Pictures on the Human Skin," *Strand Magazine* 8 (1897): 431.

evidence suggests that tattooing had spread to the officer class of the army in at least a limited way as early as the mid-1850s.[24]

Nevertheless, by the turn of the twentieth century the artistic merits of Japanese tattooing were widely and popularly known in Britain and its overseas colonies. There is no question that Japanese tattooing traditions made their way into the repertoire of the most skilful tattooists, like Sutherland Macdonald in London, himself famous in the colonies. In North America, too, tattooists drew upon overseas practices. Gus Wagner, who worked in Ohio at the turn of the twentieth century left America in 1896 and spent six years as a merchant seaman. He learned tattooing in Borneo and Java, and returned home sporting 264 designs. Although the electric tattoo machine was invented in about 1890—rendering the process quicker, less painful, and more precise—Wagner preferred to use hand-carved tools and needles, as used in the indigenous cultures from which he learned his art. In 1906, he began the process of tattooing his wife, Maude Stevens Wagner, all over her body, and when he had finished a year later they appeared together in circuses and carnivals, where they set up tattooing stalls.[25] No doubt the spread of tattoo designs was in part related to the technological innovations associated with the invention of the electric machine. Also important was the development and spread of photography and print technology, for tattoo flash (sheets of designs) could be captured and mass circulated between tattoo parlors and in illustrated books, magazines, and periodicals. The Wagners even displayed flash on temporary boards in their makeshift stalls. In Britain, these sources included *Strand Magazine* and *The Tatler and Bystander*.[26] The broadening cultural appeal of tattooing in Europe and North America was also reflected in the development and circulation of postcards that showed people with exotic tattoos, and sometimes tattoo artists at work, overseas in places like Burma.[27] However, as Govenar points out, in the North American context at least, mapping the precise relationship between the popular press and the demography of the practice is difficult.[28]

If there were distinctions within and between colonial, American, and European tattoos, in Britain there also remained strong class distinctions in tattooing practices. In Britain, upper-class tattooing cultures (performed in more socially refined tattoo parlors than those proliferating on seafronts and proletarian neighborhoods) and designs remained distinct from those of the working class. Each held what Bradley describes as "radically divergent meanings." He argues that working-class tattoos were a substitute for jewelry or other material possessions that articulated the meaning of relationships between the body, the self, and others. Those worn by their wealthy counterparts were, on the other hand, leisured displays of conspicuous consumption.[29] To a greater or lesser degree both embodied or expressed nineteenth-century global cultural and economic imbalances and exchanges.

Jordanna Bailkin has offered a uniquely gendered interpretation of the relationship between the nineteenth-century aristocratic tattooing fad described by Bradley (in which women also participated) and ethnographic representations of tattooed colonial populations in Burma (many of whom were women). For Bailkin, the British craze for tattooing is best understood in relation to its "colonial precedents," for it documented the insecurity of the British aristocracy and the instabilities and uncertainties of overseas expansion in places like Burma. Tribal women there enjoyed remarkable social and economic freedoms, and so although the patterns included the facial marks uncommon in Britain at the time, their tattoos could not be dismissed simply as primitive. Instead, the tattoos came to signify indigenous resistance to annexation and, as an expression of these women's cultural superiority, what Bailkin describes as the "hyper-modernity of the female body itself."[30]

COLONIAL ETHNOGRAPHIES OF THE TATTOO

The complexities of the performance and cultural representation of tattooing in Europe and North America suggest the importance of considering European and North American attitudes to non-European marked bodies more generally. British colonial ethnographies of the tattoo and, in the African context, of scarification, were largely concerned with uncovering origins, meanings, and significance—as ornamental or therapeutic marks, a religious expression, or an embodiment of the onset of puberty, the gendered social order, or ethnic affiliation. During the second half of the nineteenth and into the twentieth centuries, the pages of anthropological journals were full of discussions of tattooing and scarification, mostly written by colonial administrators or military officials. They shared the concerns presented in regional ethnographic handbooks, where tattooing was often a subsection of descriptions of other cultural practices relating to events like birth and marriage. In a general sense, contemporaries saw tattoos and scarification as markers of the otherness of whole societies. But where there was wide variation in practice they also used them to create cultural layers between communities. Tattoos and scars were seen as marks of apparent backwardness or primitiveness, a means of distinguishing between social groups, and sometimes a way of tracing relationships between them.[31] Moreover, although some Europeans collected tattoos during trips overseas, others collected tattooed skin for such research—or even for display in museums of natural history or anatomy.

Despite an important recent reminder of the theoretical complexities of revealing the origins of cultural practices or institutions,[32] as we have seen historical analyses have engaged with important questions about ongoing global movements of peoples and cultures. With relation to tattooing, Alfred Gell, in his landmark study of contemporary tattooing in Polynesia—*Wrapping in*

Images—describes its significance in relation to their revelation of the relationship between what is *on* and *under* the skin, speaking to the intertwining of tattoos and various—and contingent—cultural, religious, and political processes and values.[33] A nuanced consideration of colonial understandings of the origin, nature, and significance of tattooing during the nineteenth century raises other important questions, not only about the relationship between the skin's surface and broader social structures (or at least colonial understandings of them) but also about how visual representations were transformed into discursive ones.

Anne D'Alleva has engaged with some of these issues in her study of missionary responses to tattooing in the Pacific. During the nineteenth century, *tatau* came to the attention of missionaries, and they broadly equated them with apparently primitive cultures. Because they seemed to be associated with religion, missionaries further described tattooing as "a heathen and savage practice." The London Missionary Society's (LMS) repression of tattooing meant that by the mid-nineteenth century, it had almost disappeared in Tahiti and the Society Islands. In Samoa, however, when the LMS established stations on the three main islands (Upolu, Savai'i, and Tutuila) during the 1830s, it did not outlaw the practice, but rather "discountenanced" it. Station missionaries believed it was connected with "heathenism" or "abomination," or even performed in direct imitation of the gods, and therefore advancing European civilization (not least the gospel) would soon push it from favor. The Wesleyan Methodist Missionary Society in Tonga favored the earlier LMS approach, legislating against *tatau* in the island's first penal code (1839). Roman Catholic missionaries in Samoa did not on the other hand prohibit *tatau*—cannily providing an alternative place of Christian worship for those who continued to practice it. There were further layers to this inconsistent missionary approach, for as D'Alleva shows, their successful inscription of tattooing prohibitions into local penal codes largely depended on their relationships with powerful title-holding families. "[M]ultiple and complex practices and conceptions" therefore coalesced around *tatau*.[34] D'Alleva also raises the crucial question of the response of islanders to missionary attacks on tattooing, showing how their responses varied over time and place, and depended on factors like social status, gender, and age. Local communities were unwilling to give tattooing up, and in places like Samoa traveled overseas to have it done. This affected missionary efforts in profound ways.[35]

In colonial Mozambique, later on in the nineteenth century, indigenous women also responded to Portuguese attacks on tattooing (*tinhlanga*) in creative ways. The Portuguese authorities saw tattooing as a general mark of primitivism and a specific sign of ethnic affiliation, and its eradication became, therefore, one element in the promotion of civilization. In a fascinating survey, Heidi Gengenbach shows that tattooing and cicatrization became increasingly significant to women affected by colonial rule in negative ways. Tattoos and scarification marks were a means of forging "blood ties," and of mapping a gendered social world, but they also became a secret female language for commenting on social change. As in the Pacific, these practices continued under colonial rule, though Gengenbach suggests that changing images (by the 1920s incorporating images of keys, waistcoats, scissors, and watches) were one means through which women renegotiated the frontier between Mozambican and Portuguese culture.[36]

Historically, many African communities tattooed themselves, or practiced cicatrization, and to many European commentators these practices became markers of their general primitive otherness. Nineteenth-century European descriptions of the Makua, Yao, and Makonde—from the areas that now make up Tanzania, the Comoro Islands, and Mozambique—for instance, noted their "country marks," often on the face, at the same time commenting on the difficulty of establishing a precise correlation between types of tattoo/scarification (as well as forms of body modification like nose and lip plugs) and ethnic identities or tribal affiliation.[37] As Megan Vaughan has shown, a medical officer at work in Nyasaland during the first years of the twentieth century made similar references to the effects of migration, conquest, and the slave trade in the production of intermingled practices of scarification.[38] Indeed, even if it was possible to align bodily practices of this kind with precolonial ethnicities or tribes—and this is debatable—any simple reading of the social meaning of scarification would contradict the complex cultural roots of communities from places with long histories of movement and outside contact.[39]

Across the Indian Ocean in South Asia, colonial administrators viewed the tattoos of communities resistant to colonial incursion in the Naga Hills (between Assam and Upper Burma) as markers of backwardness or savagery, too. This was at least in part because in a practice almost unknown in Britain (though seen in parts of the Pacific)[40] they were tattooed on the face. Representations of their *ak* (tattoos) were of such curiosity that models of Naga heads were produced and then exhibited in the Colonial and Indian Exhibition of 1886.[41] Though the 1886 exhibition included representations of facial tattooing, colonial collectors who worked in or traveled through New Zealand acquired preserved Maori tattooed heads (*tā moko*), taking them back to Europe and elsewhere. As part of a more general trend toward the repatriation of human remains dating from the colonial era, in the past few years a number of *tā moko* have been sent back to New Zealand. An agreement was reached at the beginning of 2005, for instance, to return two from the museum in the Scottish city of Perth. They had been appropriated in a typically dubious way: a ship's surgeon sent them to the city's Literary and Antiquarian Society in 1825. About 200 Maori heads still remain overseas, most of which were no doubt donated in similar circumstances.[42] The largest known collection was at least in part purchased by the American Museum of Natural History in

New York from the collector Major-General Horatio Robley at the start of the twentieth century.[43] London's Wellcome Trust Library has photographs of others, including a picture of Robley with his collection and a set of plaster casts. These are ethically troubling objects and images; inflected as they are with the violence of British imperial encounters, and removed from their family, kin, and cultural context, they have become central to postcolonial controversies over the repatriation of human remains.

In India, officials like Superintendent of Ethnography Herbert Hope Risley believed tattoos provided a more nuanced means of understanding society. In his view, tattoos were unequivocal marks of the social classification of caste and—because for him caste was a racial endowment and not a social artifact—they could be used to unearth the racial origins of various social groups. There was also the possibility of using tattoos to establish the origin of European *roma* communities, who Risley, like many, believed had migrated from South Asia in the distant past. As part of broader ethnographic researches into caste then underway, in 1901 Risley commissioned a survey of tattooing in India's Bengal Presidency, asking district officers a series of questions about its practice, meaning, and significance. He called for details about the origins of tattooing, how it was performed, what type of designs were used, at what age it was performed, whether it was most common among men or women, and the significance of the designs and any associated ceremonies. His ethnographic assistant in South India, Edgar Thurston, gave the same instructions. Like the officials undertaking concurrent anthropometric surveys, district officers met with an unexpected degree of noncooperation, for the communities under the spotlight were fearful that their questions about tattooing represented some sort of threat to the practice. Officers complained that they had been unable to elicit more than the most basic information. In what did emerge, local communities challenged the notion that there was any sort of relationship between tattooing and social or cultural practice. This divergence from Risley's preexisting assumptions perhaps explains why he chose not to publish the report. It was clear that the origins of the practice were diverse, and there were few common styles. Tattooing was not a social sign as Risley understood it, except perhaps in its absence, for the most unequivocal finding of the district officers was that it was overwhelmingly confined to women, and that it was dying out among groups keen to emulate their (untattooed) social betters. The tattoo (or rather its opposite, bare skin) was therefore a sign of caste mobility, a form of social change with new meanings at least in part provoked by the multiple interventions of the colonial state into Indian social hierarchies.[44]

Despite the claim that by the beginning of the twentieth century tattooing was in decline, it is clear that during the same period the Indian subcontinent was permeable to overseas tattoo designs. During the nineteenth century, large numbers of male Tamil workers migrated to Burma in search of work, and like

British soldiers, many returned to India sporting elaborate tattoos. Tattooists in South India subsequently began to copy Burmese designs.[45] Perhaps the most interesting data in this respect emerged from an examination of 130 men of the Eurasian community in South India during Thurston's turn-of-the-century anthropometric survey. When they were instructed to strip themselves of clothing for the purpose of close measurement, they revealed various designs that speak to the cultural hybridity and ambivalence of what was a socially porous mixed-race community. A picture of the wife of King Edward VII, Queen Alexandra, was recorded, as was a royal coat of arms, a crown and flags, and a crucifix, but also a dancing girl, an elephant, a lizard, a scorpion, a boat, and a Burmese lady. All were of course viewed during an intensely colonial encounter. In the years following the social upheavals of the mutiny-rebellion of 1857–1858, though many Eurasians served in the army (hence the nature of some of the tattoo designs), they were uncovered in a cultural exchange that came about because white British officer Thurston viewed them primarily as natives.[46]

TATTOOING AND COLONIAL CONTROL

Though tattooing expressed something of global relationships (and representations) during the nineteenth century, its apparent permanence meant that it gradually also became encoded within metropolitan and colonial strategies of discipline and control. Tattoos were transformed from cultural signs to a means of establishing or confirming the individual or collective identity of criminal offenders. At this time, the tattoos or other distinctive marks of all convicts transported to Australia were carefully noted for surveillance purposes. Maxwell-Stewart and Duffield have written of the humiliation Australian convicts experienced as they were stripped and examined.[47] I have referred to this process in the nineteenth-century South Asian context, where Indian convicts were also made subject to close physical scrutiny, as the exercise of bio-power.[48] In Britain, the passing of the 1869 Habitual Criminals Act was accompanied by the establishment of a Register of Distinctive Marks to facilitate the identification of repeat offenders. In India, too, by the 1870s the police were compiling Old Offenders Registers.[49] As Simon Cole argues in a reading of identification practices in Europe and North America, such initiatives assumed that the body could be used to identify itself.[50]

Despite the strong association between tattooing and marginal communities like these through much of the 1800s, and the realization of the possibility that tattoos could be used to identify *individual* criminals, as Caplan demonstrates, by the end of the century some observers saw ornamental tattooing as a mark of *collective* criminality. However, there was considerable tension between two distinct approaches. One of which—the Italian criminological school headed by Cesare Lombroso (1835–1909)—favored inherited pathological (or biological) explanations for tattooing. The other—led by Alexandre Lacassagne (1843–1924) in France—stressed environmental causation. The reach of this debate was wide, and army, medical, and prison officers beyond Europe published theoretically derivative studies. In the 1920s, for instance, Megalos Caloyanni published a study of prison tattoos in Egypt, representing tattoos as inscriptions of external life rather than any pathological condition.[51] In mid-1920s Russia, interested parties examined the tattoos of prisoners in Moscow and the Ukraine. Their published study reported that between sixty and seventy percent of prisoners who had tattoos had been tattooed in jail. They used tattoos as a means of upholding tradition, displaying group solidarity, and communicating in a secret language. The fingers—constantly on display—were in this respect an important representational site. Some convicts were put under a great deal of pressure from fellow convicts to acquire tattoos.[52] Invariably, these studies glossed over national tattooing traditions (which as Caplan shows genuinely may have disappeared) and presented tattooing within the colonial trope of savage cultural practice.[53]

By the 1890s, tattooing had been transformed into a mark of hereditary criminality in South Asia, though representations of the practice were embedded in ambivalence and inconsistency. In a series of legislative measures dating from 1871 (the Criminal Tribes Acts), hundreds of thousands of people were placed in so-called criminal tribes and classified as hereditary criminals. These groups mainly comprised itinerant groups that existed at the margins of society, and the government of India forced many into industrial and agricultural reformatory settlements. During the late nineteenth and early twentieth centuries, some police officers drew on colonial ethnographies of the tattoo to argue that the criminal tribes could be identified collectively through distinct tattoo designs and particular types of clothing, jewelry, or disguise. The deputy inspector-general of police in Bombay, M. Kennedy, wrote *Notes on the Criminal Classes* in 1908, for instance, a handbook that carefully noted socially distinctive tattoo designs.[54] And yet it seems that the links between caste, social status, and tattoo designs was tenuous at best. The principal of the Police Training School remarked: "One might easily be deluded into the idea that every tribe has such marked characteristics that anyone could recognize them at sight. Nothing could be further from the truth."[55] Given the inconclusiveness of Risley's earlier surveys, the very general comments on tattooing in many contemporary ethnographic texts, and the widespread recognition that tattoos were easy to enlarge or alter, it is likely that far from being a mark of individual or caste identity, in practice the police viewed tattooed individuals generally with suspicion.[56]

And yet there was a significant gender dimension to tattooing practices in South Asia, for it was mainly women from marginal communities who wore tattoos and who worked as tattooists. With the transformation of tattoos into a sign of criminal tendency (specific or general), the construction of criminality itself became an integrally gendered process. As I have argued elsewhere, colonial observers were fascinated by female interventions into each other's bodies and constructed discursively the practice and meaning of tattooing and tattoos (which they rarely if ever saw) in erotic ways. Sexuality became a central trope for imagining the performance and design of bodily marking, and officials saw it variously as a signal of puberty, a means of enhancing sexual attractiveness, a curb on (women's) sexual desire, or even a disfigurement that prevented abduction and rape.[57]

THE PENAL TATTOO IN SOUTH ASIA

In precolonial South Asia, mutilation (amputation of nose, ears, or hands) and public shaming (head shaving, face blackening) were used as discretionary punishments.[58] There is suggestive evidence that facial penal marking was a further means of stigmatizing—and identifying—offenders. Penal tattooing and branding was common in Europe and its overseas colonies, too, and convicted criminals and army deserters were subjected to the tattooing needle or branding iron. Across the globe in the Mascarenes, southern Africa, the Americas, and the Caribbean, as well as in India, slave owners branded their human chattels with marks of ownership (often their initials) in a degrading practice that recalled the identification of livestock. In what amounted to a coalescence of these local and metropolitan/colonial practices, the first colonial regulations on penal tattooing in India (*godna*) were passed by the Bengal government in 1797. But in a departure from extant practice elsewhere, where small letters or figures were marked on the torso or limbs for the purpose of identification on demand, they ordered the tattooing of the name, crime, and date of sentence on the foreheads of convicted felons, thus individuating offenders in an immediately visible way. Recognition of the effect of the permanence and the location of the marks provoked various reforms over the next twenty years, and given the difficulties faced by tattooed ex-offenders in reentering society and the labor market, after 1817 the regulations were only applied to prisoners sentenced to life terms. Derivative regulations were passed in the southern Madras Presidency after 1802, and later on in Burma, but in a display of the unevenness of practices associated with colonial rule, the Bombay Presidency in the west of the subcontinent never adopted them.[59]

There were many problems in the implementation of the regulations. It was difficult to find tattooists (*godnawali*) willing to enact the punishment. Those available were usually low-caste, illiterate women. As a result, into the 1840s practices varied widely. In response to the enquiries of a major prison discipline committee set up in 1836, for example, the resident councilor at the Indian penal settlement in Penang reported the necessity of systematizing the form and language of the inscriptions. Some convicts, he reported, were marked only with the word "murder." *Godna* inscriptions required the tattooing of a relatively large amount of information—name, date of crime, sentence—on a relatively small area of skin, and it seems that penal tattooing developed to incorporate, or in some places was perhaps altogether replaced by, the technically simpler operation of branding. A further layer in the performance of the penal inscription was the opportunity influential prisoners had to bribe tattooists to inscribe light, small, or incomplete tattoos, and/or to efface tattoos using caustic preparations.[60] Finally, given that they were inscribed in different scripts (Bengali, Burmese, Tamil) and took different forms (they incorporated varying levels of detail) officials could not always read or use *godna* in the way that was intended, and so it became a *collective* mark of conviction rather than the ideal of an *individuating* tattoo. Nevertheless, given the association between decorative tattooing and various socially marginal communities, especially women, the low-caste, the outcast, and tribals, the practice of *godna* was a means of both emasculating and de-casting the native criminal.

In 1830, prisoners convicted of thuggee—a rather vague and heavily orientalized offense of murder said to have ritualistic features—were subjected

FIGURE 7.5: "Convict." From F. S. Marryat, *Borneo and the Indian Archipelago, with Drawings of Costume and Scenery* (London: Brown, Green, and Longmans, 1848): 215.

to *godna* but, given its special status as a criminal act, of a different type to ordinary offenders. Thugs were marked with inscriptions on the cheeks, nose, shoulders, hands, and/or arms. These tattoos were supposed to act as a warning against the peculiarly criminal nature of the offender and serve as an individual mark of identity. Colonial categories, however, were inherently unstable. Though officials in India viewed thugs as the scourge of society, in the penal settlements they were found to be reliable convicts, and quickly rose to the top of the penal hierarchy. By the 1840s, thugs transported to penal settlements in Burma had been elevated to positions of relative authority. The effect of their tattoos was thus quite different to their intent, with thug *godna* transformed from marks of dangerousness to marks of collective empowerment.

Increasingly, however, the government of India described *godna* as a barbarous practice, echoing the concerns of officials in Bombay who were under increasing pressure from Bengal to adopt it. It viewed the marking of prisoners with indelible inscriptions as inconsistent with the "civilizing mission" of colonialism and, much to the consternation of local administrators who testified to its identificatory benefits, in 1849 the government decided to abolish it. A parallel discussion on the branding of British military deserters perhaps elucidates this redrawing of the boundaries of attachment between so-called enlightened and barbaric penal practices. One correspondent wrote to *The Times* in 1851: "I ask whether we have any moral right to brand a fellow creature, and whether this same branding does not place us English on a par with the Asiatic who resorts to chopping off a hand, slicing off a nose, or shaving off an ear."[61] This perhaps explains the moral outrage invoked among Europeans when forty years later a British police superintendent in Burma, Malcolm Chisholm, ordered the tattooing of *Memma Shwin* ("bazaar prostitute") on the forehead of a woman named Ma Gnee. However, because many Burmese women had red dots tattooed on the forehead as love charms, the unwilling tattooist engaged for the job was able to take advantage of European ignorance of Burmese script, subvert the extrajudicial punishment, and transform it into a cultural sign: a series of dots with a quite different local meaning.[62]

In melding together local precolonial and European practices in marking the forehead, *godna* was a visible penal form that was unique in the British Empire. There is an interesting parallel, however, from imperial Russia, which in its establishment of distant penal colonies in Siberia in some ways mirrored British colonial India and its overseas convict settlements. By the beginning of the nineteenth century, the punishment of branding was well established in Moscow and imperial Russia. This was because, as Schrader explains, it was not simply a punishment, but a key measure through which officials marked individuals with their place in the social hierarchy, using ascriptive legal-administrative categories (*sostoianiia*). This prevented vagrants and escaped convicts from disengaging from the status system, which clearly defined exile, by refusing to give personal information on arrest. However, branding had unintended consequences. Together with cultural forms, including literature, song, and language, it became a means through which convicts "constructed their own subjectivity as members of a corporate group." Nonetheless, such actions showed implicitly an acceptance of the importance of the status system, and their tattoos remained a potent symbol of an outsider status.[63]

CONCLUSIONS

The cultural dynamics of tattooing and its representation in the long nineteenth century were many and various, but because of the extent of cross-cultural practice and exchange they are always most fruitfully examined from a global perspective. The tropes of savagery and civilization were invoked commonly in Europe and North America, as they were overseas, to capture the significance of the nature, meaning, and performance of the decorative tattoo. In Europe, tattooing was most commonly considered a marginal practice; only in Britain did it find favor among social elites. In Britain's overseas colonies, administrators sometimes viewed tattoos as a means of social elaboration, most particularly in India. There, marked bodies came under the purview of the colonial state in important ways. Tattoos were also seen as a means of detection and identification, and in some places as signs of atavistic criminality. Uniquely in the British Empire, in early colonial South Asia, officials used tattooing /branding to create immediately visible penal marks that were both individual and collective in nature. But supposedly permanent tattoos and brands were always subject to change and reinterpretation. Moreover, colonized bodies challenged the colonial trajectories that transformed their tattoos into representational categories or inscribed their bodies with penal status. Far from being fixed marks, therefore, tattoos were always both visually and discursively unstable.

CHAPTER EIGHT

Smallpox, Vaccination, and the Marked Body

NADJA DURBACH

Speaking at the formation of the Kettering Anti-Vaccination League in 1884, the suffragist and anti-vaccination lecturer Jessie Craigen denounced vaccination for both physically and symbolically marking the body. In feudal times, she declared, serfs were branded "but freemen were unmarked." But in this day and age, Craigen asserted, the common people had become the "doctor's cattle" and the vaccination mark "his brand."[1] Craigen was part of a lively campaign that opposed the English government's introduction of a program of compulsory infant smallpox vaccination in 1853. Those who supported this initiative insisted that empirical evidence had demonstrated that vaccination could in fact both protect the individual from smallpox and curb outbreaks of the disease. However, as many as 25 percent of parents chose to opt out of the procedure once it became possible to declare oneself a conscientious objector to vaccination in 1907.

The contest over compulsory vaccination that took place in England over the course of the second half of the nineteenth century and the first decade of the twentieth was a political, rather than merely a medical, debate.[2] Walter Hadwen, one of the leaders of the anti-vaccination campaign, asserted that "the very moment you take a medical prescription and you incorporate it in an act of Parliament, it passes beyond the confines of a purely medical question and becomes essentially a social and political one."[3] By mandating a medical procedure for all its newborn—and thus presumably healthy—citizens, the government asserted its right to intervene in matters of an intimate bodily nature and thereby challenged widespread and deeply cherished assumptions about the boundary between public and private. This example of the enforcement of health thus serves to illustrate the ways in which the modern body operates, according to Michel Foucault, as a key site, if not *the* key site, of political contestation.[4]

But if bodies are inherently political objects because they are "the matrix in which are written the rights a society allots to people of various genders, races, and classes,"[5] the experience of—and the meanings invested in—the body is nevertheless always historically contingent. To understand anti-vaccinationism requires unpacking the culturally specific politics and logics of the body at play in England during the nineteenth and early twentieth centuries. At stake in debates over compulsory vaccination was the right to decide whose body would be marked by whom and what this "branding" might mean. To resist the multiple scars of a public vaccination was to oppose being both physically marked by the procedure and socially marked by what many considered a discriminatory practice. As Craigen's critique illustrates, anti-vaccinators expressed these concerns by invoking a range of structured oppositions—free/slave, medieval/modern, pure/polluted, English/foreign, healthy/diseased, among others—that provide insight into their worldview, however uneven or contradictory.

This chapter will locate English anti-vaccinationism as part of a much broader discourse of corporeal marking by analyzing the movement's investment in these binary categories and more particularly in the divide between human and animal, divine and material, and natural and artificial. Anti-vaccinators' commitment to upholding these distinctions reveals profound and pervasive cultural anxieties about the integrity of the body itself and its relationship to the modern state.

SMALLPOX, VACCINATION, AND COMPULSION

Smallpox was periodically epidemic throughout the eighteenth and nineteenth centuries. It was highly contagious, often fatal, and generally both disabling and disfiguring. Although it was not the primary cause of infant mortality in the nineteenth century, and in fact trailed far behind many other contagious diseases as a leading cause of death,[6] it was the only disease that could actually be prevented through a medical technology—namely vaccination—rather than merely through sanitary measures.

Vaccination had become a relatively widespread practice after the 1798 publication of Edward Jenner's *Inquiry into the Causes and Effects of the Variolae Vaccinae*. Jenner, a Gloucestershire physician and surgeon, had observed that milkmaids who regularly came in contact with cows infected with cowpox rarely contracted smallpox. After a series of experiments, he was able to demonstrate that exposure to cowpox protected the human body against an attack of smallpox. Vaccination—the deliberate insertion of cowpox into the human body—was born. But this was not an entirely novel procedure. Vaccination was based on the model of smallpox inoculation. Smallpox inoculation—also called variolation—had been practiced in Asia, Africa, and the Middle East before being imported into England in 1721 by Lady Mary Wortley Montagu, who had encountered it while living in Constantinople. Artificially inducing an attack of smallpox by deliberately inserting a small amount of the disease directly into the body generally resulted in a very mild case of infection, allowed the patient much greater control over the timing and treatment of the symptoms, and left him or her immune to further attacks. Throughout the eighteenth century, smallpox inoculation was enthusiastically adopted by the rich, who checked themselves into specialized institutions for the purpose, and by the poor, who quickly assimilated it into their folk healing practices.[7] Jenner had therefore modified a technique that was already part and parcel of elite and popular medicine in order to make the process safer and more effective. His innovation—inoculating with cowpox instead of smallpox—was advantageous because cowpox provided the individual with a similar immunity to smallpox. As Figure 8.1 suggests, it caused only a mild reaction in humans and, perhaps most significantly, as an animal disease, was not transmissible from person to person.

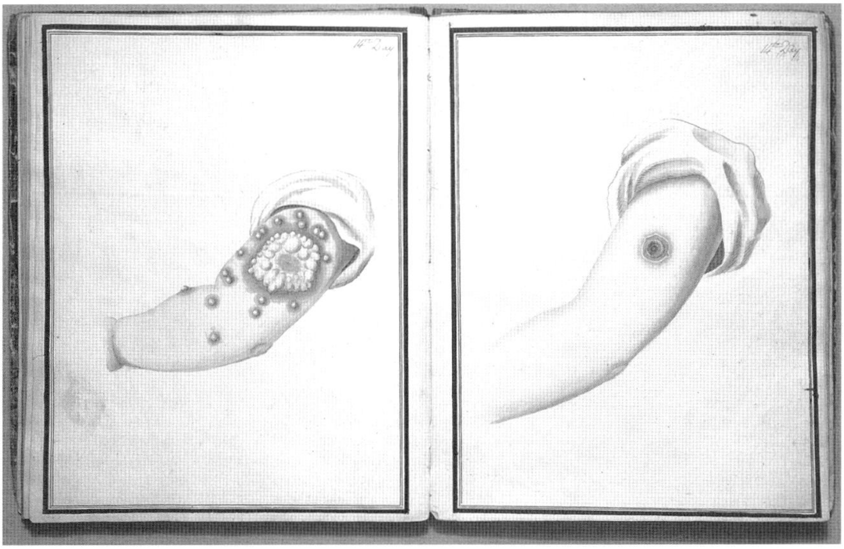

FIGURE 8.1: Watercolor drawings showing smallpox inoculation (variolation) on versos and cowpox inoculation (vaccination) on rectos; *14th Day Smallpox and Cowpox*. Watercolor by G. Kirtland, 1802. Wellcome Library, London (L0015949)

Although vaccination did not immediately replace variolation, and indeed skepticism and outright resistance to the use of cowpox arose almost immediately, Jenner was celebrated by his contemporaries and by the end of his life had been awarded a total of £30,000 in grant money from Parliament (Figure 8.2). Indeed, the English government was enthusiastic about the possibility that vaccination could help to control outbreaks of smallpox, and in 1840 it introduced a free vaccination service aimed at the poor and working classes. Until the late 1840s, Poor Law medical officers only treated paupers—those who had become destitute and were prepared to enter government-run workhouses. The 1840 vaccination service, however, was an exception to this rule. It provided free vaccinations to encourage the lower classes to protect themselves (and thus the rest of the population) against an attack of smallpox, which could spread quickly and easily in densely crowded urban centers. The government targeted the working classes in particular because, until the emergence of germ theory in the last decades of the nineteenth century, the dominant miasmatic theory linked dirt and disease. "The great unwashed," were thus widely deemed responsible for generating and propagating smallpox and other deadly diseases.

Few members of the working poor actually availed themselves of this gratuitous program of public vaccination, however, precisely because the procedure was tainted by its association with the dreaded and heavily stigmatizing New

FIGURE 8.2: *Jenner and his two colleagues seeing off three anti-vaccination opponents, the dead are Littered at their feet.* Colored etching by Isaac Cruikshank, 1808. Wellcome Library, London (M0005397)

Poor Law. Many also preferred to continue to inoculate with smallpox—what they claimed was "the real thing"—rather than to substitute cowpox, which as we shall see provoked a range of concerns. Although it took another decade before the government actually compelled the poor to be vaccinated, it moved immediately to prevent them from continuing to practice smallpox inoculation. In 1841, Parliament outlawed variolation, fearing that it was more likely to provoke than to curb an outbreak of the disease.

The introduction of a government-sponsored vaccination program in 1840 marked the very beginning of a shift in attitudes toward public health that took another ten years to be realized. During the 1830s and 1840s, the Chadwickian sanitary idea had prevailed, which focused on the building of sewers, street cleaning, and other forms of municipal sanitation.[8] By the 1850s an emerging medical profession moved to take public health out of the hands of sanitary bureaucrats and place it squarely in the domain of the medical. As the first medical procedure that could claim to arrest the spread of an infectious disease, vaccination embodied the promise of scientific medicine as the route to public health and was thus key to this process.

To make public health a more scientific pursuit and to buttress their new role as guardians of the nation's well-being, in 1850 a group of influential doctors formed the Epidemiological Society of London, which began to investigate the problem of smallpox. The society argued that smallpox posed a serious danger to the national community, but that vaccination, if performed by a trained medical practitioner, could protect individuals from its ravages. Because every diseased person was "a centre of contagion," and every unvaccinated population a "nidus for the disease to settle in and propagate itself," only compulsory vaccination, its *Report on the State of Small-Pox and Vaccination* argued, could ensure the health (and by implication also the wealth) of the social body.[9] The Epidemiological Society proposed, therefore, a compulsory vaccination act applicable to all infants and carrying stiff penalties for noncompliance.

Given that the voluntary system introduced in 1840 had failed to secure the widespread vaccination of the lower classes—those repeatedly deemed most responsible for the spread of smallpox—and that the medical profession itself was now lobbying heavily for legislation, the government moved to make the procedure compulsory. In 1853 it mandated that every baby born in England or Wales was required to be vaccinated—either by a private physician or by a state-paid public vaccinator—by the age of three months. A machinery was already in place under the 1840 legislation to provide free vaccination via the medical services of the Poor Law, so it was more expedient for the government to continue this arrangement than to devise an entirely new bureaucracy. The administration of compulsory vaccination thus remained under the supervision of the Poor Law guardians.

In 1867 and 1871, Parliament passed new legislation intended to make the vaccination administration more efficient and to tighten loopholes in the system. These acts mandated the employment of public vaccinators in every district, required that Poor Law guardians keep vaccination registers to track compliance and hire vaccination officers for this purpose, introduced a fine of twenty shillings for nonvaccination, and allowed for the repeated prosecution of parents of unvaccinated children. This meant fines could be levied or prison sentences imposed multiple times until the child was either vaccinated or had reached the age of fourteen years.

By the time these stricter laws were introduced an anti-vaccination movement was already in place. What many perceived as more draconian legislation only strengthened the organized campaign, which attacked the procedure itself, condemning it as ineffective and dangerous. But anti-vaccinators also objected to the principle of compulsion, which, they argued, ran roughshod over individual and parental rights. Anti-vaccinators staged demonstrations, published pamphlets, held debates, and wrote letters to national and local newspapers in order to shape public opinion and exert pressure on the government to repeal the compulsory clauses of the vaccination acts. But the anti-vaccination movement as a whole was much broader and more plebeian than the organized campaign with its middle-class leadership might suggest. The parents who consciously chose not to vaccinate their children—and who often suffered economic and physical hardship for what they maintained was their "conscientious objection" to the procedure—were largely drawn from the ranks of manual laborers, artisans, and small shopkeepers. In the context of the contest over compulsory vaccination these resistors identified themselves as members of a persecuted working class.

Anti-vaccinators condemned the compulsory vaccination acts as class legislation that discriminated against the working poor. Although they applied to all babies, the laws were unevenly enforced. Because the working poor could not afford to pay the steep penalties for noncompliance even once, let alone repeatedly as the law allowed, the state regularly seized their property in lieu of monetary payment, or sentenced them to weeks in prison. Members of the middle and wealthy classes who objected to, or were merely wary of, the procedure, however, could easily afford to pay these penalties. But even if they did not, these resistors were rarely pursued by the vaccination officers hired to oversee the enforcement of the acts unless they were active in the repeal movement and thus identified as committed anti-vaccination propagandists. This resistance to prosecuting the middle and upper classes stemmed in part from a widespread belief that this social group was not in fact responsible for the spread of smallpox, which was still associated with dirt and urban squalor. But it was also because the vaccination officers were themselves generally drawn from the working classes and would likely have found it awkward to prosecute, let alone imprison, those they perceived as their social betters.

Working-class anti-vaccinationists attacked the vaccination acts not merely because they were not uniformly applied. They were also unpopular precisely because of the ways in which the vaccination administration linked those who were forced to avail themselves of the free public service (because they could not afford to pay a private physician) to paupers. Paupers were a heavily stigmatized group after the passage of the 1834 New Poor Law, and the English working poor fought hard to distinguish themselves from those who were unable to support their families and had become entirely dependent upon the state. That the vaccination services continued to be overseen by the Poor Law meant that the procedure could never be completely divorced from the stigma of poor relief and the specter of the workhouse.

Compulsory vaccination was thus bound up in larger discourses of autonomy and dependency that played out at the site of the working-class body. Although public vaccination was intended to be nonpauperizing, working-class campaigners argued that in forcing them to vaccinate their infants the state was preventing them from governing themselves and making autonomous decisions about their families. They maintained that compulsory vaccination challenged their claims to be freeborn Englishmen with inalienable rights to their own and their children's bodies, and rendered them little better than paupers. As Figure 8.3 illustrates, anti-vaccinators felt that the long arm of the state, here represented by the police, had reached into the nursery itself. This type of propaganda suggested that mothers were being compelled to submit their babies to be marked by public vaccinators, decried in the text on the flap of this anti-vaccination envelope as "slayers of infants" and represented visually as death itself. The vaccination mark was thus seen by many as a symbol of servitude. In 1856, a working-class anti-vaccinator declared, "they might as well brand us," a critique as we have seen that Jessie Craigen also levied almost thirty years later.[10] By consistently invoking the branding of slaves and cattle these campaigners underscored the stigmatizing nature of the procedure. As Clare Anderson has argued elsewhere in this volume, branding was both a highly charged symbolic practice and one with very real physical and social consequences. By invoking the brand, anti-vaccinators drew attention to vaccination as a painful process that physically and permanently marked the body. But they also suggested that it was a symbolic act that socially marked particular—in this case working-class—bodies as both dangerous to the community and ultimately as property of the State. Debates over compulsory vaccination were thus not merely about the best means of achieving public health, but cultural contests over the meanings of the marked body.

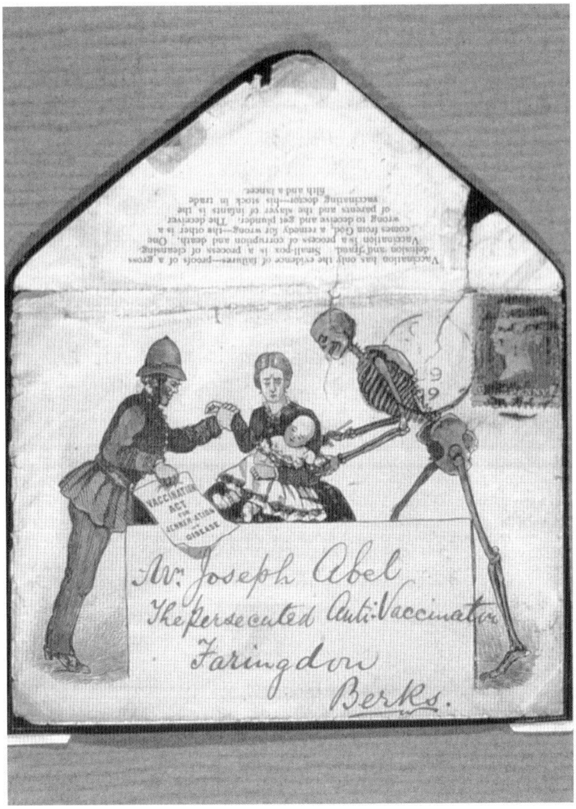

FIGURE 8.3: Anti-Vaccination Envelope, 1879. Collection of Denis Vandervelde.

THE MARKED BODY

Pro-vaccinators repeatedly underscored that smallpox was an extremely disfiguring disease that significantly marked the body, leaving survivors literally scarred for life (Figure 8.4). They frequently cast smallpox as an "insatiable monster" that went "licking round its victims," its "clammy grasp" on the "throat of all society."[11] The unvaccinated, proponents of vaccination warned, fell "victim" to the "devouring monster," and became loathsome creatures themselves.[12] Those who had witnessed epidemics firsthand often stressed the horrific nature of smallpox. In 1891, the former vaccinator-general of Trinidad described the "horror and disgust" that a smallpox patient inspired: the "hideous loathsome aspect of the face, the horrid smell, the frightful pits and scars," he insisted, could only be fully appreciated by those who had attended to the sick during an epidemic.[13] The patient "attacked with this fell disease becomes a mass of living corruption, so hideous," declared the general superintendent of vaccination for Madras, "that the mother sickens at the sight of her child and turns away." If the child escapes with his or her life, he warned in 1865, "it is so frightfully disfigured as scarcely to be recognized."[14] But Jenner, pro-vaccinators triumphantly argued, had tamed this "beast of prey" by extracting "the claws and the teeth."[15] As this monstrous language illustrates, despite frequently attacking anti-vaccinators for their lurid rhetoric and shocking propaganda, pro-vaccinators used equally sensationalist discourses to promote their position. An 1898 illustration in *Punch* repeated the visual trope of the skeletal symbol of death seen in the anti-vaccination envelope in Figure 8.3. This sinister image intimated that relaxing the vaccination laws, or what they termed "de-Jenner-ation," would lead only to a rise in infant mortality (Figure 8.5).

Although smallpox could kill and left telltale pockmarks on its victims, anti-vaccinators stressed that vaccination also marked the body in ways that could be equally transformative. In the nineteenth century, vaccination was performed with a lancet, a surgical instrument that was used to cut lines in the flesh in a scored pattern. (The hypodermic needle, invented in the middle of the nineteenth century, was largely associated with drug addicts and thus was not used in this context until well into the twentieth century.) Vaccine matter, which came either from a preserved supply of calf lymph furnished by the government-run National Vaccine Establishment, or straight from the arm of a previously vaccinated child, was then smeared into the cuts. Public vaccinators were instructed to vaccinate in at least four different places on the arm, which

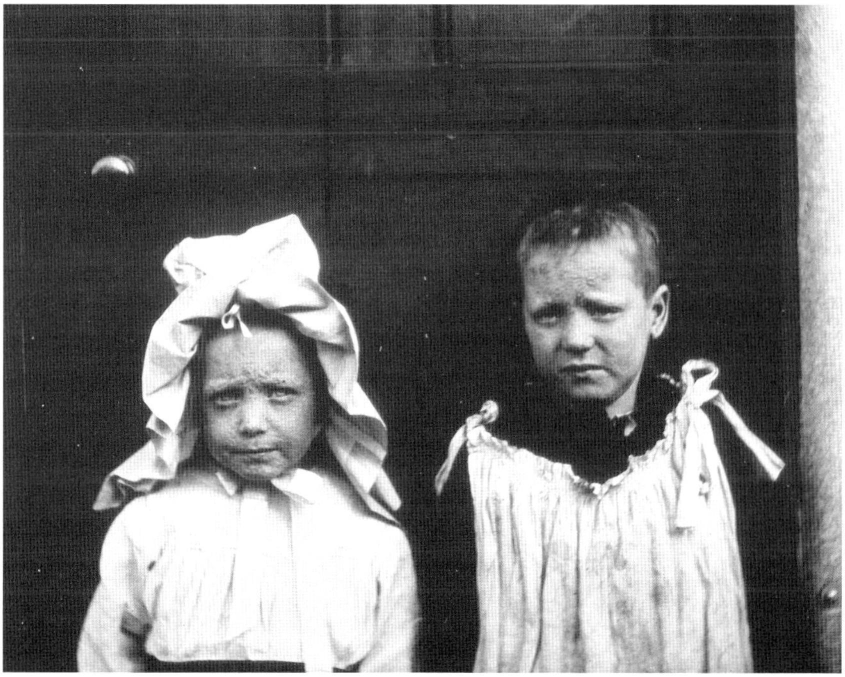

FIGURE 8.4: Gloucester smallpox epidemic, 1876: two convalescent children. Photograph by H. C. F., 1896. Wellcome Library, London (V0031469)

FIGURE 8.5: *Death as a skeletal figure wielding a scythe: representing fears concerning the Act of 1898, which made vaccination against smallpox compulsory.* Wood engraving by Edward Linley Sambourne, 1898. Wellcome Library, London (V0018753)

left large permanent scars that were easily discernible to the naked eye on a part of the body that was not infrequently exposed to the public gaze. These multiple vaccination cicatrices could be so deforming in and of themselves that the middle and upper classes generally preferred to pay for the services of a private vaccinator and could therefore request a single scar in a more inconspicuous bodily location. Young women, loathe to spoil their chances in the marriage market, could, if paying out of pocket, be vaccinated on the foot, in order not to mar the appearance of their delicate limbs. Thus, the nature, placement, and visibility of vaccination scars themselves literally marked bodies as belonging to specific social groups: it was primarily the working classes who used the free public vaccination stations and so bore the physical trace of this encounter throughout their lives.

As Figure 8.6 illustrates, vaccination not only left unsightly scars but the process also left behind open sores that could easily become infected. In working-class households, where access to clean water was limited, these wounds were hard to care for, which meant serious infections (such as gangrene) could follow even a successful vaccination. If gangrene produced anxiety among parents, so did the possibility that diseases could be passed through the blood, marking the body from the inside out. Given that pub-

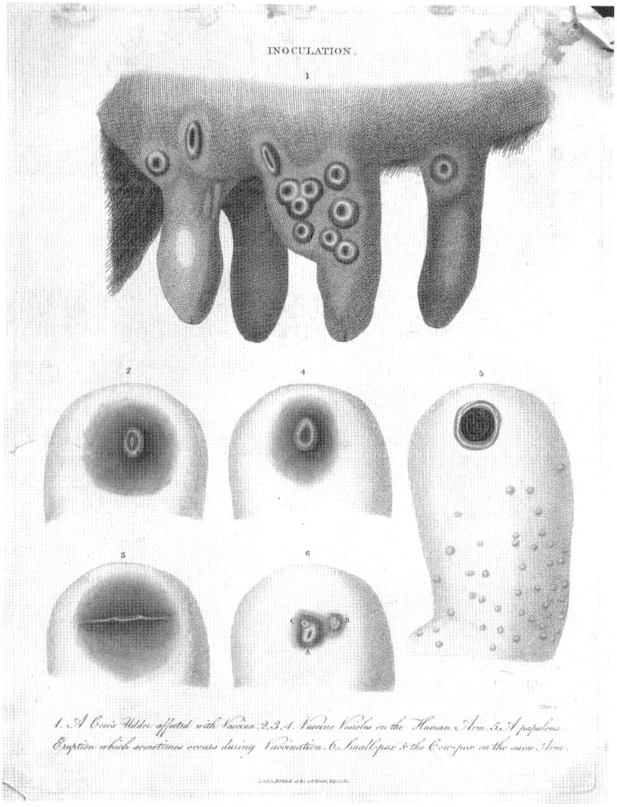

FIGURE 8.6: *A Cow's Udder with Vaccinia Pustules and Human Arms Exhibiting Both Smallpox and Cowpox Pustules.* Colored engraving by J. Pass, 1811. Wellcome Library, London (V0016678)

FIGURE 8.7: *A monster being fed baskets of infants and excreting them with horns; symbolising vaccination and its effects.* Etching by Charles Williams, 1802 (?). Wellcome Library, London (V0011073)

lic vaccination was generally performed by the arm-to-arm method rather than with preserved calf lymph, no matter how healthy a child seemed, anti-vaccinators warned that he or she could harbor hereditary diseases. Because lymph was transferred directly from one child into the body of another, each vaccinated infant, they claimed, was in fact a repository for the diseases of the entire community, some of which could leave the body severely disfigured. Vaccination, its opponents asserted, could be the conduit for scrofula, cancer, leprosy, plague, idiocy, consumption, epilepsy, erysipelas, deafness, blindness, and a variety of other hereditary and contagious conditions. As Figure 8.7 reveals, the fear that vaccination opened up a Pandora's box of diseases that could corrupt the bodies of little children was circulating in popular culture as early as 1802.

If anti-vaccinators were anxious about the spread of leprosy and plague, what they feared most was syphilis. Syphilis left lesions and chancres on the body and eventually destroyed the cartilage so that those suffering often had, for example, only remnants of a nose remaining. The syphilitic's body was thus easily identifiable. This was particularly problematic because, unlike scrofula and erysipelas, syphilis had a moral dimension. It thus not only physically marked the body, but it also carried with it a social stigma. In the nineteenth century it was widely believed that syphilis could be spread not merely by sexual contact but also by heredity.[16] A child marked by syphilitic lesions was thus an advertisement for the sexual impropriety of his or her parents. Vaccinating one's baby, opponents of the procedure implied, thus left one's own and one's family's sexual practices open to scrutiny. If the child contracted syphilis through contamination with "bad blood," anti-vaccinators argued, the moral standing of the parents would be compromised as doctors refused to concede, despite scientific evidence to the contrary, that the procedure was itself responsible for spreading this loathsome disease.[17]

Pro- and anti-vaccinators were thus engaged in a debate about the ways in which both vaccination and smallpox marked the body from the outside in and the inside out. Both groups were concerned about bodily disfigurement and used the trope of the marked body for propaganda purposes. Those who supported vaccination were quick to underscore the transformative nature of smallpox. They argued that smallpox was so monstrous and so easily transmitted that it was far preferable for a person to suffer through what they considered merely a minor procedure that caused but a week of physical discomfort and left, they claimed, only a few bodily scars, than to risk the ravages of the disease itself. But for anti-vaccinators the assessment of comparative risk was really just the tip of the iceberg. Their concerns about the procedure went far beyond whether it was an effective and safe preventive against smallpox. They were equally anxious about the ways in which vaccination threatened to undermine what they considered the essential divide between the human and the animal, the sacredness of the body, and the laws of nature.

THE ANIMAL/HUMAN DIVIDE

Anti-vaccinators consistently argued that even if the vaccine lymph used for vaccinating babies was pure, it was objectionable because it derived from animals. As Figure 8.6 clearly indicates by providing a close-up view of diseased udders, lymph was the product of a cow. By inserting animal matter into the human body, vaccination thus challenged a cherished divide between man and beast. In the early nineteenth century, cartoonists satirized popular fears surrounding cowpox inoculation. In 1802—the year Jenner was awarded £10,000 by Parliament—Charles Williams depicted vaccination as a monstrous cow devouring baskets of babies and then excreting them out as horned and tailed beasts (Figure 8.7). The same year James Gillray produced "The Cow-Pock—the wonderful effects of the new inoculation," in which the vaccinated sprout horns from their heads and fully formed cow carbuncles from their noses, cheeks, and buttocks (Figure 8.8). As these popular caricatures suggest, throughout the nineteenth century the bovine origin of

FIGURE 8.8: *The Cow-Pock—or—the Wonderful Effects of the New Inoculation!* Wellcome Library, London (M0005398)

lymph influenced understandings of its impact upon the human body. Some feared that the vaccinated might in fact assume bestial behaviors: as late as 1896 parents claimed that they were unwilling for "a beast [to] be put into their children" as they believed they might come "to low and to browse in the fields like oxen."[18] These types of anxieties mirrored the discourses of maternal impression—the widespread belief that the form of an unborn child could be altered by something the mother experienced while pregnant. Indeed, many nineteenth-century medical professionals substantiated the popular conviction that if an expectant mother received a fright, longed for a particular food, or had a bad dream, her baby would be physically marked.[19] The belief that vaccination could perform a similar transformation was an extension of ideas about the pliability of the child's body already in circulation during this period.

Although not all of those who resisted vaccination believed the procedure would literally make their bodies more bovine, many feared that calf-lymph could transmit animal diseases. The 1865 cattle plague forced the slaughter of thousands of herd animals, devastated the meat and dairy industry, and resulted in legislation to control the importation of diseased livestock. It therefore suggested to the public that cows were particularly dirty and contagious animals.[20] Parents thus feared that using vaccine matter derived from a cow could communicate foot-and-mouth disease or other disorders that were epidemic among animals. A father summoned to the Woolwich Police Court in 1891 asserted that he had resisted the vaccination of his child because it was "well known that the bulls go mad every seven years, and that the cows make them mad." When these same cows are used for vaccinating children, he reasoned, the children go mad. "The madhouses are full of vaccinated children."[21]

It was not merely cows that provoked concern. Vaccine matter was often manufactured by inserting different animal poxes (such as monkeypox and swinepox) into the cow. One could never be certain, therefore, anti-vaccinators warned, how many different animal diseases were present in the lymph. Indeed, the original vaccine virus was said to be derived from an inflammatory disease that attacked the skin of horses known as the horse-grease. Anti-vaccinators underscored the folly of integrating these epizootic diseases into the healthy human body. They satirized this practice by arguing that repealing compulsory vaccination would still allow those who supported the procedure to "bovinize, equinate, assinate, or assassinate, to their hearts' content with any lymph they please," while still protecting others who did not want their bodies polluted in this promiscuous manner.[22]

This resistance to what anti-vaccinators considered contaminating the body with animal products also emerged in the context of other Victorian social debates. It was, for example, central to the vegetarian movement that began to take shape in England in the 1840s. Vegetarians maintained that all animal products polluted the natural purity of the body and resisted using animal matter in any form.[23] The *Vaccination Inquirer*, the journal of the repeal movement, explicitly linked the anti-vaccination and vegetarian campaigns: it satirically reported that a doctor seeking to vaccinate a child had reassured the mother by insisting that he would use only "the purest calf-lymph." "Then sir," replied the woman, "that settles it, for we are vegetarians!"[24] Vegetarians often campaigned for food reform in general, critiquing the adulteration of food staples such as tea and flour, and condemning the sale of rotten meat to those who could not afford to purchase more expensive and fresher cuts. They drew attention to the relationship between vaccination and food adulteration, interpreting them as parallel forms of bodily contamination. If some animal products were not fit to eat, they declared, then neither were they fit to be inserted into the body: "What is not fit for a child's bread and butter is not fit for its blood," argued the activist W. J. Furnival.[25] In 1890 the *Vaccination Inquirer* reported on cab horses sold for sausage meat and beef tea. Cab horses with "greasy legs," were always rejected, it claimed. But although "the greasy heels are too horrible to be allowed to approach a sausage," they are, it continued sarcastically, "just the thing to exude 'pure lymph' for insertion in little children."[26]

Anti-vaccinationists feared not only that the human body was being contaminated with animal products but also that humans were being treated no better than animals as they were compelled to be cut into and experimented upon. In the last decades of the nineteenth century an animal rights movement emerged that condemned animal experimentation, particularly the practice of vivisection, which involved experimenting on living beings.[27] Animal rights activists were often also anti-vaccinationists who denounced the harvesting of vaccine matter as a cruel practice, for it involved binding the cows to a table, scarifying them, inserting disease into the scars, and then collecting the lymph from the vesicles that appeared on the skin. Vaccination was not only deeply imbricated in the practice of vivisection, they argued, but was itself a form of "infant vivisection" as the operation involved scarifying the arm in several places with a lancet, inserting the lymph, and then harvesting fresh lymph for use in other operations from the vesicles that appeared eight days later. After the 1871 law was passed, parents could in fact be fined up to twenty shillings if they refused to allow their baby to be used at the government-run public vaccination stations as the "stock-baby" from which lymph would be drawn.

But anti-vaccinators and anti-vivisectionists, although involved in related reform movements, were often in conflict with each other. They competed for charitable dollars, for government recognition of the importance of their cause, for the limited attention span of public opinion, and over whose body was the proper object of empathetic identification. Anti-vivisectionists and anti-vaccinators appeared to agree that scientific medicine exploited the most vulnerable members of society and explicitly argued that the bodies of animals, women, children, lunatics, and the working classes were all equally likely to become unwilling subjects in a medical experiment.[28] But exactly whose body had greatest claim to protection, both under the law and in the social conscience of the nation, was sometimes a source of tension. If their support for the anti-vivisection movement and the rights of defenseless animals allowed anti-vaccinators to promote themselves as civilized and humane, they also deployed this position to chide some so-called humanitarians for denying the children of the poor and working classes the basic human right to be protected from harm.[29] They repeatedly argued that animal rights activists, particularly the wealthy patrons of the Royal Society for the Prevention of Cruelty to Animals, were much more concerned with the welfare of their horses, dogs, and cats than they were with working-class babies. Although they contributed large sums of money to protect their pets from torture, anti-vaccinators attacked what they interpreted as the reluctance of many animal rights activists to do the same for working-class children.[30] Anti-vaccinators sought to expose what they claimed was the callousness and inhumanity of the landed gentry in particular who considered their pets and farm animals part of their own domestic units but seemed to deny lower-class infants inclusion in the national, and maybe even human, family.

THE SACRED BODY

Vaccination not only compromised what in the years after the publication of Darwin's *On the Origin of Species* many maintained was a crucial divide between human and animal, but it also challenged firmly held and widely shared beliefs about the sacredness of the body. If vaccination could physically deform the body by marking the skin and spreading disfiguring diseases, it also threatened to pervert the spirit. Although the age of empire was a time of increasing secularization, this period also witnessed the growth of a variety of spiritual and religious movements, many of which were heavily invested in the purity, integrity, and sacredness of the body. Some anti-vaccinators were thus opposed to the procedure primarily because it violated their religious beliefs. William Hume-Rothery—a former Anglican clergyman who emerged as a leader of the movement in the 1870s—insisted that if anti-vaccinationism was a political campaign, it was also a religious movement. Religion and politics, he maintained, were "always united as soul and body."[31] Similarly, "a Christian" argued in *Jenner or Christ?* that vaccination was the "most outrageous blasphemy against God, [and] against Nature."[32]

Those who objected to vaccination on religious grounds proposed that physical health and spiritual health were intimately related. They believed the physical body was a glorious tabernacle that should be a fit dwelling place for a healthy vibrant spirit.[33] These anti-vaccinators often came from dissenting religious communities—Quakers, Baptists, Methodists, Congregationalists, Wesleyans, Unitarians, the Salvation Army, and other chapel-based sects—and interpreted vaccination as a decidedly "unChristian" rite, a type of "devil worship," that could transform an angelic child into an "anti-Christ."[34] A baby's body was particularly sacred to these religious anti-vaccinators precisely because they believed it was fresh from its maker and thus in its purest and most perfect state: it was sacrilegious, they asserted, to tamper with the body of the child "just after God has given it you."[35] These campaigners thus condemned public vaccinators because they saw them as "baby-defilers" who contaminated the body and corrupted the soul of those as yet unsullied by the evils of the world. They interpreted vaccination as a perversion of the Christian sacraments that were supposed to secure the safety of the soul, and by extension the body in which it was housed.

To underscore the spiritual dangers of vaccination these campaigners often evoked a millenarian discourse. They gestured to the Book of Revelation, which presaged that "foul and evil sores came upon the men who bore the mark of the beast." The trope of the mark of the beast (a symbol of the damned and a sign of the apocalypse) appeared frequently in anti-vaccination propaganda. In the 1860s, Thomas Orton, secretary to the Sheffield Anti-Compulsory Vaccination

Society, issued a handbill titled "Compulsory Vaccination the Mark of the Beast"; similarly, in 1894, "A Worcestershire Parent" released a pamphlet entitled "The Parliamentary 'Mark of the Beast.' "[36] For some of those who supported the anti-vaccination cause, the physically marked body represented, therefore, not only ill health but it also signaled the dawn of an apocalyptic age.

If anti-vaccinationism had ties to nonconformist religious sects that figured the body as a sacred vessel for the soul, it also appealed to theosophists, spiritualists, and Swedenborgians. Adherents to these newer heterodox spiritualities were drawn to the campaign because they believed damaging the body could also endanger the soul and because they saw their spiritual alternatives as part of a much wider movement that included medical and religious alternatives to the status quo.[37] These spiritual movements tended to embrace what Joy Dixon calls an "occult body politics" that imagined all bodies as interconnected life forms. This "immanentist theology," Dixon argues, allowed for the continuance of an older model of the body that was not associated with the "possessive individualism" of the modern period.[38]

That many people fervently believed vaccination perverted the spiritual body and undermined religious beliefs became most evident to the British state in the context of the colonial administration of compulsory vaccination. Just as the government had insisted on vaccinating its domestic citizens, so too it attempted to vaccinate its imperial subjects. In India, however, inoculators—who generally came from a specific caste group—had long practiced variolation. Indians incorporated ritual practices into the process, such as special diets, and prayed to a smallpox goddess called Sitala to invoke her protection over the body. When vaccination was introduced into India in the first decade of the nineteenth century, as part of the East India Company's demonstration of their benevolence toward the indigenous population, it was largely ignored. But when it was made compulsory in many provinces in 1880, Indians actively resisted the operation. Unlike variolators, public vaccinators were unknown in the local communities, operated outside the caste system through which social and occupational identities were given meaning, and were seen as agents of the imperial state. Not only did they fail to include any ritual element in their operations, but, perhaps more significantly, they also used vaccine matter rather than smallpox itself. Because the cow was a sacred symbol in Hindu culture, inserting cowpox into the body violated Hindu strictures against eating beef.[39] Enforcing this procedure seemed to many little different from greasing the gun cartridges—that needed to be bitten before use—with animal fat, an event that provoked the Indian Mutiny of 1857. Vaccination thus undermined indigenous religious and healing practices. But to some, vaccination appeared an even more sinister practice. For certain Hindu and Muslim groups the operation appeared to be a way for the colonial state to identify the next prophet or messiah, who many believed had been born to overthrow the Raj and free India from imperial rule. The child would be easily identifiable to vaccinators, for he was said to have milk instead of blood in his veins.[40]

Both at home and in the colonies, then, vaccination disrupted religious and spiritual ideas about the relationship between the body and the soul. Although it was not primarily a religious movement, widespread resistance to vaccination suggests that even during a time of increasing secularization, medical technologies were not separate from but rather imbedded within a much wider set of cultural and religious beliefs and practices.

UNNATURAL PRACTICES

If some anti-vaccinators were concerned that the procedure threatened the humanity and sacredness of the body, many attacked it because they considered the practice to be grossly unnatural. Vaccination, its opponents maintained, was a scientific technology that was not only opposed to, but also interfered with, the healing powers of nature. Although cowpox was a naturally occurring substance, to insert it into the human body, anti-vaccinators asserted, defied nature, for it introduced disease into an otherwise healthy system and thus compromised the constitution. To keep the body healthy, they argued, one should safeguard it from disease rather than incorporating "putrid matter" into the blood stream. Health was a simple issue, they argued, that was achieved through proper diet and exercise, controlling the environmental factors that led to the emergence and spread of disease, and using only natural therapies to prevent and treat illness.[41]

Throughout much of the nineteenth century, orthodox medicine was still based largely on what have been called "heroic therapies": purging, bloodletting, blistering, and other invasive and immoderate ways of dealing with illness. Toxic medicines such as opium or mercury were standard prescriptions, which meant that the side effects of the treatment could be as hazardous as the disease itself. Anti-vaccinators tended to reject this brand of orthodox medicine and allied themselves with naturopathic healers—medical botanists, hydropaths, and hygienists in particular—who used herbs, water, and other organic remedies to prevent and cure illness. Although alternative medical practitioners often used equally harsh purgatives such as lobelia (commonly called pukeweed or vomitroot because it caused severe vomiting) they nevertheless appealed to the healing powers of nature in an effort to contrast seemingly gentle herbs with what they identified as damaging mineral and chemical substances. In addition, as physical puritans, adherents to naturopathic medicine often promoted a vegetarian diet and abstinence from alcohol and tobacco. These health reformers, though followers of distinct sects, shared a belief that disease is not a foreign entity to be suppressed but rather the body's attempt to throw off impurities derived from unhealthy habits, such as intemperance or a flesh-based diet.[42] They maintained that their therapies worked with the body to rid itself of all unnatural elements rather than introducing so-called foreign matter into the body, which could only be debilitating.

Practitioners of alternative medicine argued that diet, exercise, bathing, and the use of natural herbal remedies were important because they helped to purify the blood, the seat of bodily health. Throughout the nineteenth century both professional and popular opinion imagined good blood as the key to a healthy constitution. "The blood is the life," maintained a pamphlet on smallpox, "and *pure* blood is *healthy* life." Each drop of blood, it continued,

> is conveyed over the entire human frame once in less than five minutes, carefully leaving in proper *situ*, like a skillful builder, such concrete, foundation, bricks, mortar, flooring, roof, scantling, plaster, wood, and metalwork, iron, bell-handles, chimney-tops, windows, drainage, kitchen-range, and wash-house, as an erection for modern occupancy requires to pass the survey of the lynx-eyed inspectors of a building society, or fire insurance board.[43]

Blood was thus quite literally imagined to be the foundation of a healthy body.

Vaccination was problematic to many, therefore, because it polluted the blood. Regardless of the nature of vaccine matter, adding something to the blood was adulteration of the body's most important substance, anti-vaccinators insisted. Inoculating the body with a foreign, and what they considered dirty substance was not merely unnecessary, but it actually compromised the constitution, they contended, and thus interfered with the body's natural ability to secure itself from disease. Rather than artificially inducing immunity by manipulating the body's natural defenses, anti-vaccinators proposed that one should follow nature's path and strengthen the constitution. It was not therefore that they rejected the theory of the immune system that began to emerge in the last decades of the nineteenth century. Rather, campaigners against compulsory vaccination proposed that it was better not to influence the body's ability to ward off disease through artificial, as opposed to natural, means.

The route to health, anti-vaccinators argued, was to promote the circulation of good blood, which would keep diseases in check. By the late nineteenth century, those who supported the practice of vaccination explained its success by arguing that the germ of smallpox needed a suitable environment in which to grow and breed. As cowpox and smallpox were almost identical diseases, scientists maintained that they required the same type of sustenance. If one was inoculated with cowpox, the germ would devour all the available food, leaving nothing for any subsequent smallpox microbes to survive upon. An 1884 article in a popular newspaper compared this process to the Australian rabbit infestation. Vaccine destroyed the smallpox germ's habitat just as in Australia "the rabbits, though the smaller animal, are beforehand in devouring the grass, and leave nothing for the sheep to eat."[44] Anti-vaccinators agreed that disease germs required a suitable soil in which to grow, or appropriate nutrients to eat. But since these did not exist in perfectly healthy blood, they argued, to combat germs one must only cultivate good blood. An anonymous pamphleteer maintained that the real safety from smallpox was to "live a pure and healthy life. . . . Germs will not live unless they find appropriate nutriment."[45] Vaccination, its opponents declared, was an unnatural practice because it not only interfered with the production of good blood, but it was actually a form of blood pollution that did not aid, but rather interfered with the body's own ability to fight off the invading forces of disease.

If pure blood was key to health, anti-vaccinators also argued that it was necessary to secure the body from dangerous environmental conditions that might produce disease. In the nineteenth and early twentieth centuries (even during the heyday of the germ theory) it was widely believed in both professional medical and lay circles that the root cause of disease was dirt. Although germ theorists denied that disease could spontaneously generate, they nevertheless conceded that filthy environments could harbor the microbes that cause specific diseases. Thus, germ theory did not do away with older miasmatic conceptions of disease processes, but rather, as both David Barnes and Alison Bashford have argued, provided a new language to articulate the pervasive and abiding problem of dirt.[46] Indeed, anti-vaccinators repeatedly deployed the germ theory of disease precisely to undermine the practice of vaccination. They argued that personal and municipal hygiene could prevent the growth and spread of microbes. Sanitation, anti-vaccinators insisted, could therefore eradicate germs in a more natural manner. This was clearly a widespread popular belief, for Arnold's Soap advertised in the 1890s that disease could be prevented and removed by "washing its germs out of the body."[47]

For anti-vaccinators, vaccination was a dangerous practice not only because it suggested that sanitary measures were unnecessary but also because instead of ridding the body of dirt, it inserted a filthy substance directly into it. To anti-vaccinators, "pure lymph," the substance that vaccinators claimed to use, was an oxymoron. Lymph was, they consistently maintained, an impure substance of questionable origin. How could one have "Clean dirt, and healthy cow-disease!"[48] Anti-vaccinators deemed inoculation with cowpox an unnatural practice, therefore, because instead of safeguarding the body from dirt and disease, the procedure appeared deliberately to contaminate healthy bodies with what they considered filthy matter.

THE SELF, THE SOCIAL, AND THE STATE

Anti-vaccinationism flourished in the mid- to late nineteenth century, the heyday of dissenting reform movements. As such these protesters were often dismissed by their contemporaries as anti-everything cranks. Henry Hyndman, the leader of the Marxist-Socialist Social Democratic Federation, famously quipped that he did not want his own movement to become "a depository of odd cranks," such as "humanitarians, vegetarians, anti-vivisectionists and anti-vaccinationists, arty-crafties and all the rest of them."[49] Many scholars have been equally dismissive, casting campaigners against compulsory vaccination as antiscientific, eccentric, and superstitious for rejecting what has repeatedly been hailed as one of the greatest advances in medical science.[50] But the widespread adoption of vaccination and other forms of immunization as a largely safe and effective means to control the spread of epidemic diseases is due at least in part to scientific advances made in response to this virulent opposition that was launched during the nineteenth century.

The anti-vaccination campaign not only forced further research into making the vaccines and the process through which they were administered safer, but the movement also stimulated public debate around the ethics of enforcing health, which ultimately resulted in the first widespread use of the category of "conscientious objector."[51] That no other immunization was ever made compulsory in England suggests that anti-vaccinationists had in fact convinced the government that mandating medical interventions was an ineffective way to ensure that the public would actually cooperate in disease-prevention programs and thus protect themselves and others from epidemics. Anti-vaccinationism was thus neither a marginal, nor merely a medical movement. The nineteenth-century debate over compulsory vaccination raised complicated questions about the rights of parents, the role of the state, the politics of risk, the boundaries of government intervention in personal life, and ultimately the right to control the body. It was thus central to wider discussions about the appropriate relationships among the self, the social body, and the state that took place at the very intimate site of the marked body.

CHAPTER NINE

Picturing Bodies in the Nineteenth Century

STEPHEN P. RICE

"For so long as we have bodies, it is our *duty* to understand them."[1] So wrote the American physician and editor William Alcott in 1833, near the start of his career as one of the most influential health reformers of his day. Alcott joined a growing number of voices in the first half of the nineteenth century that argued that knowledge of the human body had great social and political consequence, and that the time had come for the study of the body to reach beyond the grim dissecting table and the quiet den of the "natural philosopher." But if Alcott and others came together in their call for a more widespread understanding of the body, they diverged widely on just what they understood, and on the broader purposes to which that understanding could be put. For Alcott, the body was like a machine "fearfully and wonderfully made," whose endless complexity revealed the glory of God's creation. More urgently, and in the metaphor he became best known for, the body was like a house (his popular book *The House I Live In* was published in 1834) whose various systems were governed by laws that needed to be understood for the "house" to be properly used and maintained. This correct use of the body was central to Alcott's concerns, and to the concerns of health reformers more generally. They argued that a great number of social ills originated in ignorance of or inattendance to the "laws of health," laws that were now being discerned with the rigor and certainty of science. Failure to abide by these laws struck a heavy toll not only on the individual but also on the nation and on the "race": hence, the urgency in Alcott's call to "duty" in understanding the body.[2]

Whether they took it as their duty or not, people in nineteenth-century Europe and North America embraced the call to understand the body with much greater avidity, and in a far different context, than was the case the previous century. Medical doctors, social reformers, political theorists, and members of the clergy, along with painters, novelists, photographers, and people simply wanting to make sense of themselves and the world they lived in, found in the body a subject of endless fascination and an instrument of great importance. This chapter focuses on some of the efforts of these people and on how cultural representations of the body—especially visual representations—gave form to different, changing, and contested ideas about the human body. These representations, even those that claimed simply to depict truths about real bodies, also operated prescriptively in one way or another, either as metaphors or as reiterations of particular conceptions of embodiment that could not help but be charged with broader social and political meanings. It is the study of these broader meanings—of how "representation attaches meanings to bodies," in the words of Rosemarie Garland Thomson—and of how these meanings often functioned to promote hierarchies premised in conceptions of bodily difference, that has been the focus of much of the recent scholarship on the body.[3] In this light, descriptions of women's bodies in mid-century novels, or wood-engraved illustrations of Africans appearing in *Harper's Magazine*, or sculptures of idealized Anglo-Saxon men, defined and ordered not only the imagined bodies on display but also society and politics more generally. One feature of this scholarship, and a consideration of this chapter, is to elucidate how representations of the body have worked both to define the normal and to distinguish and subordinate the broadly construed pathological.[4] While it seems true that the normative body—generally perceived as orderly in both its form and its function—dominated the visual field of bodily representation through the nineteenth century, those nonconforming, disorderly bodies also left their mark on the representations under investigation here, sometimes in not so obvious ways.

This chapter is divided into three parts. The first focuses on the end of the eighteenth century and the first third of the nineteenth century and considers how the long-standing analogy between the human body and the social and political body functioned in the context of democratic revolutions and imperial expansion. To consider these questions, the discussion will examine two portrait paintings, one completed near the beginning of the period at hand, and the other near the end. The second part examines how new sciences of interpreting bodily features—especially those of the face and the head—made their way into a variety of cultural forms through the middle decades of the century. The focus here will turn to wood-engraved illustrations, and to the elucidation of a visual order that employed these new sciences to help bolster a social hierarchy of class. Part three considers the last decades of the nineteenth century, when a broad cultural anxiety about degeneration—national, social, and racial—suffused Western thinking and gave rise to a host of measures to produce and represent vital bodies. Here the focus will settle on the work of Eadweard Muybridge, whose animal locomotion photographs in the 1870s and 1880s contributed in novel and contradictory ways to the representation of bodily vitality.

ORDER, DISORDER, AND THE BODY POLITIC

When used as a noun to describe an aggregate, the word "body" always carries with it a sense of relation and order. A body of work, a body of knowledge, a body of evidence—the word gives a collection some kind of organization. So too with the human body, that functional assembly of parts and operations visible and unseen. Lived with its own immediacy and available to others in ways we cannot always control, the human body has long stood not only as an instance of order but also as an image or symbol of that order. The phrase "body politic," which dates at least to the 1500s, does not necessarily evoke the image of the ordered human body any more than the phrase "body of evidence" does, but because the visible aggregate in the "body politic" is a collection of bodies (or at least of functions materially and symbolically performed by bodies), the phrase is never too far distant from that image.

Mapped onto the body politic, the human body helps not only to imagine a political or more broadly social organization but also to express that organization in terms that can be useful to those wishing to preserve it. Thus, just as the human body works because each of its parts performs its assigned function, so the polis works because those who make it up perform their assigned functions. The appeal of this metaphoric relation between the human body and the body politic has been multifarious: it points both to difference in functions (the heart plays one role, the arms another) and to hierarchy of functions (the head governs, the rest obeys), it represents this difference and hierarchy as natural, and it speaks to lived experience. At the same time, the analogy between the human body and the body politic has worked by offering a fundamental distinction between the healthy state of order and the diseased state of disorder (whereby "state" could mean either a general condition or a political arrangement—the word "constitution" contains a similar double meaning), and when attached to particular medical understanding or conception of pathology more generally, it suggests ways to diagnose and treat social and political "disease."[5] In these and other ways, then, the human body has served, in the words of one recent study, as a "*conceptual* principle of an ordered (or disordered) polity."[6]

Always side by side, the separate registers of the human body and the body politic easily converge, as when representations of the human body carry clear social and political meanings. Some have argued that such representations become especially common in times of political instability, when old power arrangements are threatened or new arrangements remain unsettled.[7] The end of the eighteenth century was just such a time, with its political revolutions, expanding and waning empires, colonial intrusions and unrest, and emerging nationalist movements. As real bodies materialized these developments, representations of bodies offered the means to conceive order out of what to many appeared to be an ever-expanding chaos. During the French Revolution, for instance, images of the body circulated endlessly in print, images that operated as metaphors for understanding the events of the day. Indeed, one study finds that "corporeal images were at the very center of the metaphoric language used to describe the revolution in progress." In this account, the metaphor of the body "offer[ed] to politicians and men of letters alike the illusion of an organic ordering of the human community," while at the same time providing a source of continuity to help ease the transition from absolutism to self-rule, from " 'the body of the king' to the great 'body of citizens.' "[8]

But the relation between the human body and the body politic is material as well as figurative, and if the metaphoric body presents a means to conceive order in the aggregate (the "body" of citizens), physical bodies—in all their apparent variety, and in their specific and often competing needs and desires—can just as easily present a source of disorder. This helps to explain why late eighteenth-century political discussions that focused on the rights of "the citizen" or "the individual" or of "man" tended to call forth a version of personhood wholly removed from material embodiment. The difficulty of arraying specific bodies within a scheme of universal claims, or of reconciling professions of public interest with the ineluctably partial view afforded by the body, directed many political thinkers to place a disembodied subject at the center of the body politic. This was true at the founding of the United States, whose leaders could, for instance, imagine human equality only by abstracting the human from any body in particular. Of course they did have certain bodies in mind, and as Karen Sánchez-Eppler and others have argued, this abstracted personhood served both to conceal the range of real bodies in the body politic and to maintain the power of some bodies over others. But the abstracted "person" who was both the guarantor and the beneficiary of self-government rested uneasily on top of real persons whose material bodies undid, it was said, their rightful claims to citizenship. What Sánchez-Eppler and others have found is that in the early decades of the nineteenth century those excluded from political personhood resisted these exclusionary efforts by pressing the body back into the abstraction in a number of ways, effectively materializing or "fleshing out" (in the words of one study) the "body of the nation."[9] Although their efforts confronted great resistance and episodes of terrible violence, by 1870 the Constitution of the United States acknowledged the literal embodiment of citizenship with new language that specified both gender (as a category of exclusion) and race (as one of inclusion) in the national body politic.[10]

Taken together, two American paintings—one begun in 1774 and finished in 1789, and the other painted in 1829—capture both the tension between abstracted personhood and material embodiment in the body politic, and the added pressures that material bodies were placing on those abstractions by the middle third of the nineteenth century. The first is Charles Willson Peale's portrait of the architect William Buckland, who was born in England in 1735 and moved to the North American British colonies in 1755 after learning the building trades (Figure 9.1). There he worked until his death, designing and overseeing the construction of churches, courthouses, and homes of wealthy merchants and planters in Virginia and Maryland. When Buckland sat to have his portrait painted by Peale in Annapolis, Maryland, sometime early in 1774, the British Empire was on the threshold of crisis. Colonists in Massachusetts had, just a few months earlier, displayed their contempt for royal authority by throwing hundreds of chests of tea into Boston harbor, and that spring Parliament responded with a series of punitive acts that many colonists deemed "intolerable" and that would spur the meeting of the first Continental Congress in September. Peale's sympathies had long been with the patriot cause, but Buckland seemed to have been circumspect about his views on the brewing crisis (and he may never have been forced to take a side as he died not long after Peale began work on the portrait).[11]

The portrait shows Buckland seated at a table with a drawing and tools spread before him. Staring directly at the viewer with a slight grin on his face, Buckland leans awkwardly on the table, his right hand raised and holding a drafting pen while the other rests stiffly at his side. Buckland's body lifts into vertical alignment at his head, which perfectly parallels a column directly behind him. Behind that and to the left is a building with classical design elements, evidently still under construction given the presence of scaffolding around it. By showing Buckland's hand pulled up toward but resting below his head, Peale produces a scene of bodily coordination and hierarchy that would have carried its meanings into the social and political realms as well. But it is in the juxtaposition between an ill-composed body in the foreground and a carefully composed and virtually complete building in the background that Peale engages more deeply with ideas about the human body and the body politic. If Buckland appears satisfied with his work on the building behind him, he also appears to have little left in reserve to "design" and "construct" his own body. Not only does he lean at awkward angles, but his vest is open midway down, with a plume of white shirt sticking out. The awkward pose and open vest, along with the simple homespun clothing he wears, all suggest that Buckland's physical appearance is unimportant to him, that his body really is

FIGURE 9.1: Charles Willson Peale, *William Buckland (1734–1774)* (1774; 1789), Yale University Art Gallery, Mabel Brady Garvan Collection.

of no substance. But if the bodily Buckland is unimportant, the edifice under construction—in 1774 still too early to represent a new nation, but evocative nonetheless of classical antiquity and its republican tradition—stands as the focus and the source of fulfillment. In his willingness to place his own interest beneath the common good figured in the building, Buckland models what would have been called "civic virtue," an ideal that was widely evoked in the 1760s and 1770s and that was central to the rhetoric of the patriot cause.[12] In Peale's rendering, then, Buckland is the idealized citizen, removed from those bodily concerns that partialize and localize one's view. Although the portrait represents some body in particular, the painting as a whole points to the abstracted personhood of "the citizen."

At the same time, Buckland's placement in the painting offers a glimpse at the extension of republican values into territories beyond the horizon. If "westward the course of empire takes its way," then reading the top half of the painting from right to left—on a map, from east to west—one sees a trajectory of progress and even conquest in which Buckland plays a central role.

The upper right side of the painting is dark and virtually shapeless at the perimeter, but forms begin to emerge closer to the column. Sweeping across the column—and through Buckland's bright, confident, and commanding head—one comes to the sharp lines of the building, and then to the grid of the scaffolding, which plot outward to the left, like surveyor's lines preparing the way for future claims. That the grid is empty can be understood in two ways, both of which suggest Peale's desire to maintain a level of abstraction and negation in his rendering of the body politic. Read literally as scaffolding, the empty grid voids the presence of the laborers—many of whom in colonial Virginia and Maryland were indentured or enslaved—who materialized Buckland's design. Read figuratively, as surveyor's lines, the empty grid also voids the presence of the indigenous peoples who inhabited those regions to the west that were being prepared for settlement when Peale completed the portrait in 1789. (Indeed, just two years earlier the Confederation Congress passed the Northwest Ordinance, which established a mechanism for settling western territory and incorporating up to five new states into the United States.)

The second painting, also a portrait of an architect and done by the Philadelphia painter John Neagle in 1829, has a much harder time at presenting an abstract ideal and at concealing those bodies that do not conform to that ideal (Figure 9.2).[13] Now more than half a century into its "experiment in self-government," the United States looked far different in 1829 than it did around the time of Peale's portrait. What had been thirteen states arrayed east of the Appalachian mountains and joined in a common cause against Great Britain now numbered twenty-four states extending nearly halfway across the continent. As the nation continued to press westward, it suffered a deepening strain along a divide between North and South over the issue of slavery. Just a decade before Neagle painted his portrait, the crisis over whether Missouri would be admitted to the union as a slave state brought to the fore the possibility of a nation ruptured beyond repair. This sense of a fractured social and political realm greeted antebellum Americans at almost every turn. It was in the spring of 1829 that Andrew Jackson was inaugurated president, marking a new phase in American politics characterized by a broad and boisterous electorate and a newly revived partisanship. In the social realm, Americans in the 1820s were undertaking what was an uneven move from craft and household to industrial production, and by the end of the decade some men and women were calling themselves members of a new working class who were now joining together to resist what they viewed as their exploitation by their employers. And it was in 1829 that David Walker, a free black man in Boston, published his *Appeal to the Coloured Citizens of the World*, with its call for violent resistance to slavery and its apocalyptic vision of America's future.

Neagle's painting is of William Strickland, who designed the building for the Second Bank of the United States (the structure behind him in the painting)

FIGURE 9.2: John Neagle, *William Strickland* (1787–1854) (1829), Yale University Art Gallery, Mabel Brady Garvan Collection.

that Andrew Jackson sought to undo during his first term in office.[14] The portrait shares many similarities with Peale's, including the centered figure in the foreground with the neoclassical building in the background, the glimpse of a distance beyond the building all the way to the left, and the one hand pulled up toward the head and holding a tool. At the same time, there are several differences that suggest a more fractious or atomized body politic than that imagined by Peale. First, Neagle has Strickland looking to the side and away from the viewer, as if disengaged from both the painter and the viewer. Second, Strickland is on the whole much more composed in his person than is Buckland. His posture erect; his coat, vest, and collar purposefully layered and arrayed; and his partially whiskered face all suggest a greater attention to style and physical appearance than what Peale offers. These two qualities of detachment and a devotion to personal interest (qualities that Alexis de Tocqueville would soon locate in a new "individualism" that he found to be characteristic of American democracy) are reiterated in the manner in which Strickland holds a sheet of paper—certainly an architectural drawing—fastened to a board. If, in the portrait of Buckland, the drawing is made public in its placement on the table in the foreground, here Neagle has Strickland pulling his drawing close to his body, concealing it from view. Buckland's public interest and devotion to a common good are thus replaced by a clearer devotion to private interest, as if revealing the drawing might somehow compromise, not so much the making of the building, which appears to be complete, but the self-making of the individual, still and always in progress.

In addition, the arrangement of the raised hand with the more elevated and central head here fails to convey the integration of functions that remained commonplace in the bodily images of the body politic. If "head" and "hand" pull together in the Peale portrait, in Neagle's rendering they are only awkwardly and even improbably joined.[15] The distance from Strickland's shoulder to the point beyond the edge of the canvas where his elbow must rest is too great for anatomical accuracy, and the angle at which his hand comes into view only heightens the sense of disconnection. Of course it is supposed to be Strickland's hand, but Neagle seems unsure of how it connects to the rest of Strickland's body. At a time when "hands" were pulling apart from "heads" in craft shops in the industrializing world (by 1829, Philadelphia was a center of organizing efforts by journeyman craftsman), the portrait raises the uneasy spectacle of a social body undone, as if the hand belongs not to Neagle, but to some almost unseen attendant whose function is simply that of a "hand." Read this way, the rising-up hand—the visible presence of wage-working laborers, who by the late 1820s were not only "rising up" in a nascent labor movement but were being uneasily incorporated into the body politic as states eliminated property restrictions on voting—may not be holding the drawing close to Strickland's chest but pulling it away, dramatizing a class conflict between "hands" and "heads," or wage earners and employers, over property and ownership. Strickland's embodiment is not of some ideal like civic virtue, then. Wrapped up in his own self-interest, and visibly contested in his claims to ownership, he is not the abstracted person that Sánchez-Eppler notes, but closer to a real person, one who embodies countervailing forces that have him both pulling together and coming undone.

A cartoon printed as a lithograph and published in London in 1836 even more graphically represents a human body as a social body disordered and coming undone (Figure 9.3). Titled *The Body Politic on the March of Intellect*, the cartoon shows a sickly man hobbling along with two walking sticks, his body covered with a dozen unhappy heads or faces. His shoulders, then, are two heads; at his knees are two faces exclaiming in unison "we are wretched knee-gociators"; his knotted feet are downward-pressing faces, pinched and crying out "Oh my *Nose* again! I wish you were *Toes* again!"; and his hands have been replaced by two frowning heads, each grasping a walking stick in its wide-open mouth. The image is meant to be grotesque; the man's body is bent and oddly wrapped, he is clearly disabled by his condition, and the heads and faces that cover his body like boils or tumors are grimacing or complaining. "To the tune of 'Let us All Be Unhappy Together'" the print reads at the bottom, in case the misfortune of all those heads and faces were not apparent.

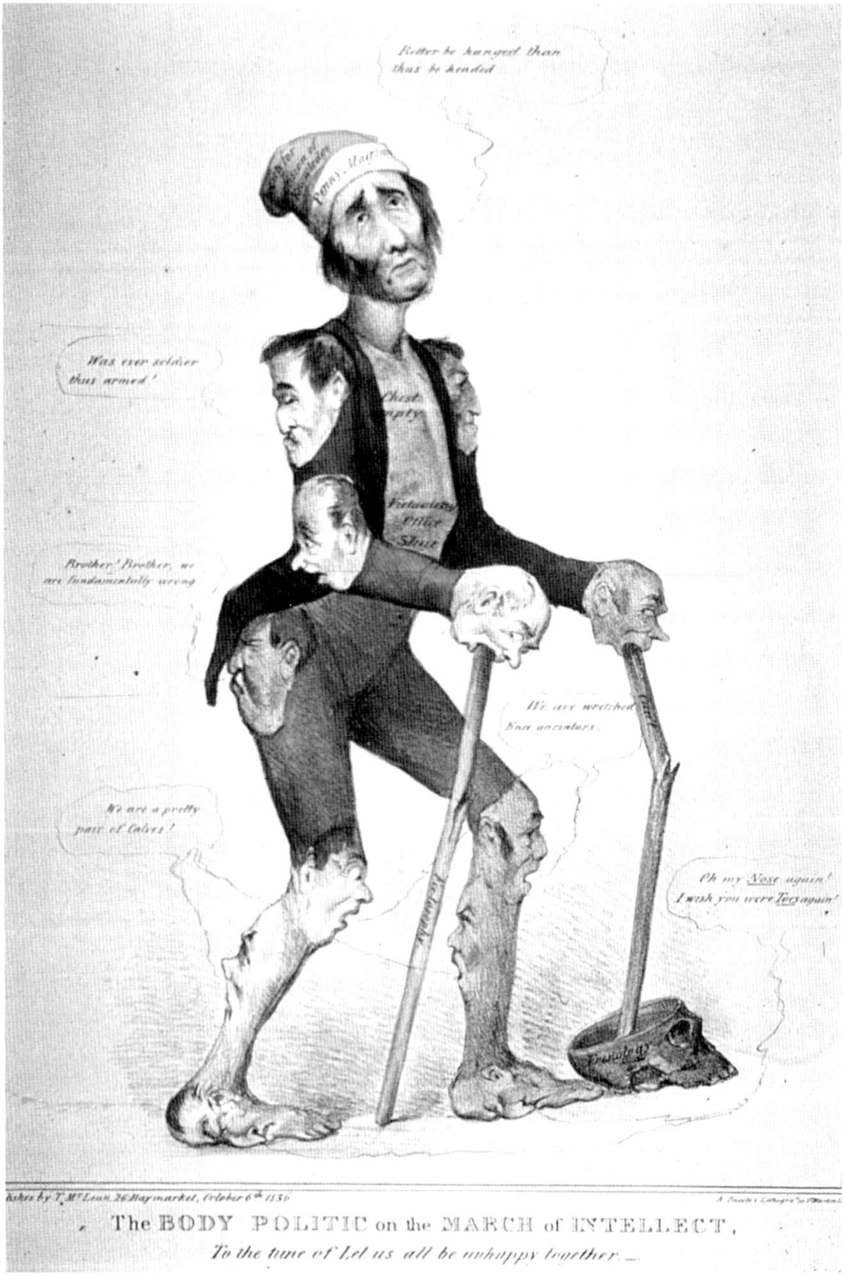

FIGURE 9.3: *The Body Politic on the March of Intellect*, Lithograph. ([London]: T[hos.] McLean, October 6, 1836), National Library of Medicine (A021836)

Layered with meanings, the cartoon imagines the unfortunate state when the difference and hierarchy of ordinary bodily function gives way to some kind of more democratic impulse, where lower parts are raised to the authority of heads. The cartoon alludes to this impulse in the text on the man's hat, which reads "Society for Confusion of Knowledge" in mocking reference to the Society for the Diffusion of Useful Knowledge and the various workingmen's educational organizations that had proliferated through Great Britain over the previous decade. Below that the hat reads "Penny Mag.," referring to the society's magazine. These popular educational societies and useful knowledge magazines were part of a broader movement for democratic reform in Great Britain, and by the time the cartoon appeared, Parliament had already taken a few measured steps toward that reform. The cartoon, though, offers an unfavorable view of the direction the British body politic was taking. "Was ever soldier thus armed?" the head on the man's right shoulder wonders. The answer is no, and the question draws the contrast between the apparent inability of the leveling body politic with the imagined power derived from subordination and command. When "head, shoulders, knees, and toes" becomes "head, heads, heads, and heads," the body—both of the individual and of society—can no longer properly function. The cartoon also places the man on a path of continued decline. As he shuffles forward with his lead stick snapping in two, part of him recalls a more healthful past ("I wish you were toes again"), while another part—the head at top that appears to be in some kind of wretched command—imagines a hopeless future. "Better be hanged," he exclaims in a punning phrase, "than thus be headed."[16]

READING BODIES

To be "headed" one way or another would have carried even further meaning when "The Body Politic on the March of Intellect" was published in 1836. By then, it had become commonplace both in Europe and the United States to view the body—especially the face and head—as a visible sign of some internal character of the individual. In this way, the body materialized character; or, put another way, character could not help but be incorporated in the body. One could thus accurately discern the essential truths about a person—his or her intelligence, or disposition, or moral quality—by studying features of his or her body. To "thus be headed," then, would have meant not only to be moving in a particular direction but also to be visibly equipped in such a way as to reveal some essential truth about oneself. In this way, mid-century viewers of Neagle's portrait of William Strickland would likely have found greater meaning in Strickland's face than in the seeming severance between head and hand. The long nose, thin lips, weak chin, and especially the high, expansive forehead, would have been understood to represent not just what Strickland looked like but what he *was* like as a man. As industrialization, urban concentration, global migrations, and imperial conquest brought together people in new and often disconcerting ways, there was great appeal in the idea that how people looked said a great deal about who they really were.

The belief that bodily traits revealed truths about the individual's character dated back to the ancient world, but by the early nineteenth century this belief had gained the credence and apparatus of science. Two fields of knowledge—physiognomy and phrenology—came to dominate the science of reading bodies. Physiognomy, which assumed a new form and popularity with the publication of multiple editions of Johann Casper Lavater's *Essays on Physiognomy* (originally published in German as *Physiognomische Fragmente*) in the last quarter of the eighteenth century, directed attention to the individual's face, claiming that characteristics such as the placement of one's eyes or the contours of one's brow invariably revealed the depths of one's being. Just as popular was the new science of phrenology, which located dozens of human qualities in different regions of the brain and argued that these regions could be studied—and the character of the individual discerned—by the careful examination of the head.[17] Part of the intended humor in the "Body Politic on the March of Intellect" cartoon resides in its mocking of phrenology, with the unfortunate man now overtaken by heads, and with his walking stick labeled "Can't"—a word meaning inability but sounding like emptiness (or "cant")—snapping as it plunges into a skull labeled "Frenology."

Phrenology and physiognomy arose in the context of a general, even widespread interest in understanding the structure and function of the human body.[18] But unlike popular anatomists and physiologists (such as William Alcott), proponents of phrenology and physiognomy trained people to observe bodily difference and to evaluate personal and collective qualities on the basis of that difference. The chart in Figure 9.4, for instance, published in France toward the end of the nineteenth century, in addition to spelling out the regions of the brain as taught by phrenology, offers distinctions between different brows, eyes, noses, lips, and ears, and it instructs its reader how to look at and evaluate a person's hands. Of course, there were differences between bodily signs over which one had little or no control—the distance between one's eyes—and bodily signs over which one had considerable control. As men and women read other people's bodies, they understood that people were reading their bodies as well, and by attending to their clothing, their carriage, and their overall deportment, they sought to exert whatever control they could over the meaning being made.[19]

Phrenology and physiognomy soon found their way into a range of cultural representations of the body, both verbal and visual. Writers from George Eliot to Honoré de Balzac, Nathaniel Hawthorne, and Mark Twain, crafted the physical traits of characters, and attended to characterization more generally,

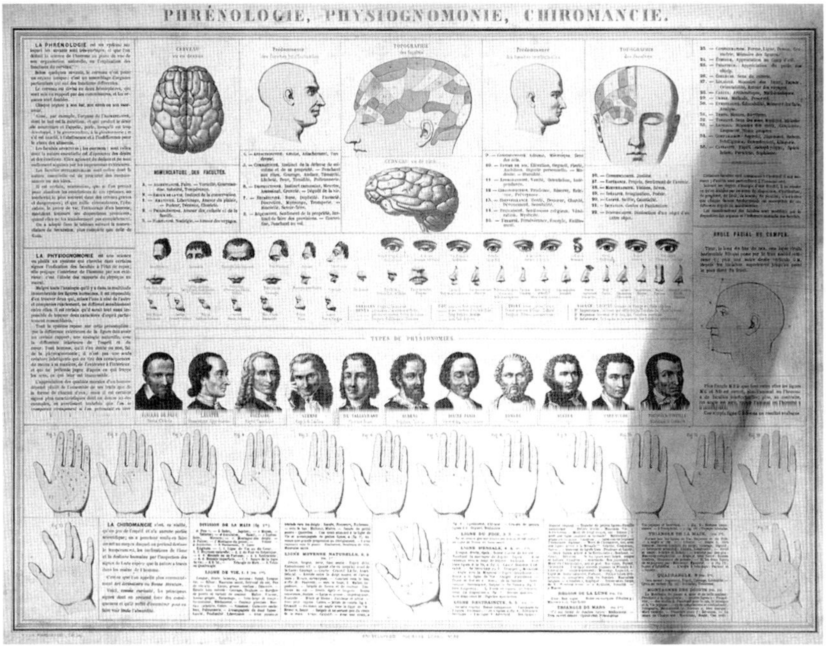

FIGURE 9.4: Chart showing the basic elements of phrenology, physiognomy, and palmistry, with diagrams of heads and hands, and portraits of historical figures. Lithograph (Paris: Maison Bouasse-Lebel, late nineteenth century), Wellcome Library, London (V0009525)

in terms that sometimes explicitly drew on one or both fields of knowledge. From the 1820s onward, one can find in portrait paintings—like Neagle's portrait of Strickland—signs of phrenological or physiognomatic thinking: unusually large or oddly shaped heads, heads turned to one side or another as if to reveal certain regions of the skull, educated men with exceedingly high brows, and so on. Similarly, in their work on portrait busts, American sculptors Hiram Powers, John C. King, and others sometimes modified the facial characteristics of their subjects in terms that were consistent with the science of reading bodies and to accentuate ideal personal qualities.[20]

Through most of the nineteenth century, wood engraving was by far the most popular means for producing and circulating visual representations of the body. By the 1850s, illustration had become commonplace in books and periodicals, and illustrated weeklies such as *The London Illustrated News*, *L'Illustration*, and *Harper's Weekly* were among the most popular periodicals of the day. All were illustrated using wood engravings, which—unlike lithographs and engraving on metal—are printed in relief, like raised type, making it possible to print words and images together with the same press. The wood-engraved illustrations that circulated on both sides of the Atlantic helped to constitute a new mass culture and were perhaps the most widely seen pictures of the nineteenth century.

Not surprisingly, physiognomic and phrenological reasoning routinely informed the representation of the human body in wood-engraved illustrations. But rather than serving to elucidate the qualities of specific individuals, as was usually the case in painted and sculptural portraits, phrenology and physiognomy in wood engraving were more likely to be put to the service of characterizing human types.[21] Thus heads, faces, and bodies, along with more mutable gestures and postures, came to define clear conventions for representing so-called typical Africans, members of the working class, unvirtuous women, respectable men, and so on. These conventions—what we could now call stereotypes—simplified what was an increasingly complex social reality, shifting human multiplicity from the level of the individual to the level of the group, and establishing a visual order that noted not just difference, but hierarchy.[22] This social typing may or may not have served a universal human need to define and diminish the other, but it certainly served a specific historical effort to establish and maintain a social order premised in perceived differences of race, gender, region, and nation.[23] "The practice of typing," notes one recent study of illustration in nineteenth-century America, "helped Americans master social and political difference in increasingly heterogeneous, disorderly, and perplexing nineteenth-century reality marked by commercial competition between regions, emerging industrial conflict, and the chaotic expansion of the body politic."[24]

The social typing in wood-engraved illustrations was in many ways simple and obvious—especially the typing based in what were coming to be understood as the natural differences of gender and race—but sometimes it offered more subtle and complex meanings. Take the illustration titled *Small Power Planer*, which appeared as one of a series illustrations for an article published in Philadelphia's *Graham's Magazine* in 1852 (Figure 9.5). The kind of metal-working machinery depicted was still relatively new in the 1850s, and the article and illustrations sought to explain its operations and prospects to a middle-class audience eager for knowledge of the marvels of machinery. What is striking in the illustration is the absence of stereotypical working-class features in the tender: missing are the burly arms and working-man's cap that had already become conventional for representing male craft or factory workers. Nonetheless, the illustration abides by a set of conventions. Like many illustrations of machines with their white male operatives published in middle-class magazines at this time, the work of machine tending here is represented as hardly a form of manual labor: the operative has one hand tucked into his pocket and the other lightly grasping one small part of the machine, and his lithe physique indicates that little in the way of physical presence is necessary for this work. At the same time, the picture suggests that machine tending is primarily a mental form of labor—the quiet and concentrated gaze of the tender makes this clear, but what especially does so is the high forehead that any viewer familiar with phrenology would have known to read as evidence of superior intellect.

In visually framing the work of machine tending as a form of mental rather than manual labor, *Small Power Planer* addressed contentious questions about the "social body" in the industrial age. By the 1840s it was conventional in popular discussions of work and mechanization to use the words "hands" and "heads" to distinguish between wage workers and their employers. Usually reserved for male workers, the word "hands" highlighted the sense that the labor performed was purely manual, the work of the hand. Likewise, the "heads" in the newly industrial society—the owner of a shop and the men hired to oversee the work of others, but also bankers, retailers, and members of the professions—engaged in what was understood as mental labor. Observers who imagined "hands" and "heads" in this way insisted that the former should submit to the authority of the latter. For them, the human body offered a conservative "language of class" that helped an emergent middle class codify its social authority.[25] If the head governed the hand in the body of the individual, then "heads" should similarly govern "hands" in the body of society. Thus,

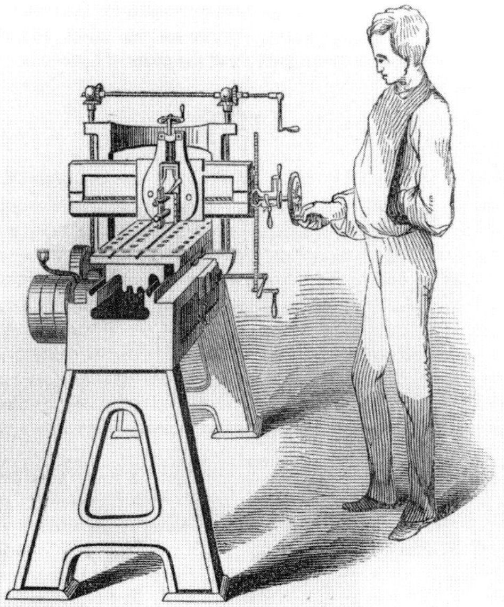

NO. 6.—SMALL POWER PLANER.

FIGURE 9.5: *Small Power Planer*. From "Machinery, for Machine Making," *Graham's Magazine* 41 (November 1852): 475. Courtesy of the American Antiquarian Society.

when Abraham Lincoln—in an address before the Wisconsin State Agricultural Society in 1859 (a year before he was elected president)—imagined that "as the Author of man makes every individual with one head and one pair of hands, it was probably intended that heads and hands should cooperate as friends; and that that particular head, should direct and control that particular pair of hands," he represented not only a personal order, but a social order.[26] *Small Power Planer* joined this discussion by suggesting that mechanization, far from reducing the worker to a mere "hand," instead elevated him to a "head," whose work was now to manage the physical force and productive activity of the machine, much like an employer or overseer would manage the productive activity of workers. The text accompanying this illustration reiterates this understanding. "The head, and not the hand, is to be the chief instrument" in the industrial age, it claimed; "the intellectual and no longer the physical forces are to predominate, even in the merest labor."[27] In this way there was no opposition between workers and managers in the new industrial age, as the work they did was virtually one and the same.

If this was the favorable view of mechanization shared by the author of the article and the artist who drew the illustration, the engraver who transformed the drawing into a wood engraving may have had a different view. As wood engraving became an important printing trade, the craft underwent dramatic changes, like so many crafts during the first half of the nineteenth century. Engravers who learned the trade in the 1820s and 1830s were, by the 1840s, establishing or coming to work for large engraving firms that often produced illustrations by a strict division of labor. The designer who initially composed the illustration typically was not the same person who drew the image onto the block of wood to be engraved. The engraver was yet another person, and by the mid-1850s, when illustrated weekly magazines called upon engravers to produce large illustrations of timely events, blocks of wood were bolted together, drawn on by several different individuals, and then unbolted and divided among multiple engravers. Although the work of wood engraving continued to require considerable skill, these changes undermined the status of the persons who did the actual engraving, making them more like "hands" than creative and self-directing "heads."[28]

Small Power Planer thus not only depicts a newly industrialized work setting, but it was produced when the craft of wood engraving was industrializing. And while the designer represented the worker's body in such a way as to contain the struggle and contestation over this industrial setting, the engraver may have found in the shadow of the worker and the machine a way to inscribe that contestation in the image. The designer of an engraving such as this one would have spent considerable time rendering the details of the machinery, but shading or shadow would have been drawn only suggestively, leaving it to the engraver to decide exactly how it would finally look. Here the shadows cast by the machine and the tender combine into a single form that appears to be a second, more massive body. While the tender seems to be lost in thought, this shadowy body spreads across the floor and rises up a wall behind him, dark to his light, sprawling where he is contained, all body and arms to his transcendent head. It is as if this looming presence—monstrous in its abnormal shape and amalgamation—threatens to overtake the poised scene of production. Reading the body at the center of "Small Power Planer," then, elicits a favorable view of mechanization, while reading the monstrous body in the shadow—where the engraver would have found greater expressive freedom—elicits a much darker view of what happens when humans are joined with machines (and a view that suggests the "spectre haunting Europe" evoked at the beginning of the *Manifesto of the Communist Party*, published just four years earlier). Perhaps in posing this second body, the engraver found a way to challenge both his employer and the new industrial order.

VITAL BODIES

By mid-century, people were imagining that the ingenious mind rather than the powerful body would be the principal means by which nations or humankind would rule the world. But as the head would be the "chief instrument" of the new age, a new set of concerns emerged about the fate of the rest of the body. Through the first half of the nineteenth century, physicians and others writing about health in Europe and the United States argued that prolonged mental exertion could result in a host of physical ailments—including dyspepsia, severe headaches, and unremitting physical fatigue—that they sometimes characterized as the "diseases of a literary life." But it was not until 1869 that physicians established the diagnosis of neurasthenia. Sufferers of neurasthenia were understood to have taxed too heavily the limited supply of "nerve force" that animated their different bodily functions. Their vitality sapped—usually, according to the diagnosis, from excessive mental activity or strain, but sometimes also from overeating, or from undue sexual activity—these neurasthenics had reached a state of debility that left them by the wayside. Through the last three decades of the nineteenth century, physicians developed a host of therapies to treat their neurasthenic patients, from diet therapy, to electroshock therapy, a "rest cure" that enforced cessation of virtually all physical activity (usually prescribed for women), and a "work cure" that enforced just the opposite. The relevant contexts to these concerns are many. The principles of thermodynamics, elucidated and quickly popularized in the 1850s and 1860s, insisted that there was a limit to the available supply of energy at our disposal, and that reserves of energy and of resources more generally, were every day being depleted and needed to be conserved and used efficiently. In the United States and across Europe, industrialization and the growth of new nonmanual occupations that placed fewer physical demands on the body contributed to a sense that members of the ascendant middle class would lose their physical force and, ultimately, their social and political authority.[29]

It was in this context of late-century concerns about bodily fatigue and the attendant national and racial decline that new practices and representations of male bodily fitness proliferated through North America and Europe, most pointing in one way or another to what John Kasson has called the figure of the "Revitalized Man."[30] This is not to say that physical fitness did not emerge as a broad cultural concern before that, but it was in the closing decades of the nineteenth century that scientific and medical theories about energy and fatigue, in conjunction with social and political imperatives to sustain individual and national vitality, gave rise to an unprecedented interest in producing and representing fit male bodies.[31] Signs of this interest were everywhere, from the paintings of Thomas Eakins and George Bellows, to the stories of Jack London (the protagonist of his 1907 novel *The Iron Heel*, a heroic labor revolutionary named Ernest Everhard, is rapturously described by the narrator as a "superman" with "bulging muscles" and a "thick and strong" neck), to the sculpture of Douglas Tilden, to mass-produced wood engravings and photographs of celebrity prizefighters and bodybuilders.[32]

These celebratory visions of male bodily strength were heavily racialized, as in most cases it was not only vitality that was on display, or even male vitality, but white male vitality. This fact that helps to illuminate the lynching photographs taken throughout the United States during this period. Whereas the white male bodybuilders and fitness experts wished to advertise the "revitalized" man, the act of lynching often was a determined effort to render and then advertise a "devitalized" man. Those who committed the crime of lynching had much more in mind than ending a life, as the act frequently involved the public desecration of the victim's body, followed by the display of that desecrated body. Recent studies of lynching in the United States from the close of Reconstruction to the middle decades of the twentieth century have revealed that this public display extended to photographic representations of lynching victims, often shown with those who committed or abetted the violence, which then circulated as postcards. In lynching photographs, the juxtaposition of defiled and lifeless black bodies with engaged and even electrified white bodies—in *The Autobiography of an Ex-Coloured Man*, James Weldon Johnson describes a lynching in which the violent fervor raced through the gathering white crowd "like an electric current"—may have addressed the broader cultural anxiety about degeneration and decline, revealing at the same time the racialization of this anxiety.[33] The power and vitality of the white participants in the lynching, usually standing and sometimes grinning and pointing, was thus heightened in contrast to the abject powerlessness of the victim or victims. And that power extended to the viewers of the photographs as well, who found in them an explicit warning, or a ghastly confirmation of racial authority, or a source of reactions mixed but still deeply felt.[34]

Because of its seeming ability to provide unmediated visual access to objects and events, the camera became an important instrument for collecting and distributing information about the body. Photographs carried a greater sense of empirical truth than any painting, lithograph, or wood engraving could ever do, and by the 1870s, a number of projects were under way that sought to use the camera for the scientific study of the body. One such project that contributed to the representation of bodily vitality in novel ways was the animal locomotion study of Eadweard Muybridge. Using complex arrangements of cameras and tripping devices, Muybridge photographed a variety of animals in motion—most famously horses and humans—in order to reveal the precise bodily positions of various stages of such ordinary activities as walking, running, and throwing. His human motion photographs, taken in 1884 and 1885 at the University of Pennsylvania and published in a volume titled *Animal Locomotion* in 1887, are especially striking. There one finds hundreds of photographic sequences (and thousands of photographs) of men, women, and children—most of whose bodies are unclothed and fully on view—performing dozens of different activities: men running, jumping, and carrying bricks up ladders; women walking up and down inclines, sweeping floors, and scooping up children. There is an aura of science throughout. Each plate includes sequences from two or three different angles photographed simultaneously, demonstrating an empirical impulse that is enhanced even further by the grids against which most of the various activities are carried out. The pair of sequences titled "Descending Stairs" shows a woman stepping down a short set of stairs placed against two grids, one to the side of her and one in front (Figure 9.6). In this series, as with several others, Muybridge has attached to his model a device designed to make clearer the movement of her hips—when viewed from the "rear 90-degree" angle—as she proceeds through her steps.

Muybridge's photographs were in some ways quite conventional as late-century representations of vital bodies. He was, for instance, particularly concerned with representing white male bodily fitness. Most of his male models perform athletic activities or feats of strength, their powerful bodies curved and contoured by their muscles on display, and only one of them—a "mulatto and professional pugilist"—is identified by race, presumably because he deviates from the normative white body.[35] Muybridge's female models, by contrast, generally perform tasks less demonstrative of physical strength, many of which are associated with domestic duties: carrying a pitcher of water, making up a bed.[36] (One study suggests that this division of tasks within Muybridge's photographs—"with the woman confined to a restricted domestic space and the man subjected to rigorous physical activity"—alluded specifically to the discourse of neurasthenia.)[37] At the same time, Muybridge added

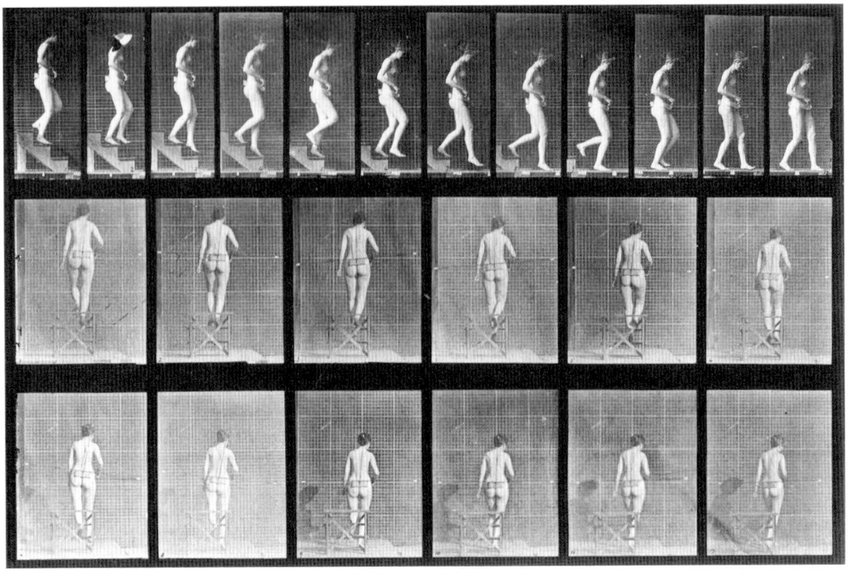

FIGURE 9.6: "Descending Stairs." From Eadweard Muybridge, *Animal Locomotion* (1887), plate 130. Collections of the University of Pennsylvania Archives.

to the project of representing vital bodies in several important ways. Not only does he provide photographic evidence of physical fitness and activity, but he also multiplies that evidence in ways not seen before. Each page of his study is simply filled with photographic images, so that the viewer quickly takes in not just one instance of vitality, but dozens of them. The knowledge that each sequence unfolded over a very short period of time only heightens the sense of a world crowded with vital and active bodies. In addition, Muybridge's bodies are not only "revitalized" in their display of physical fitness—they are vitalized in the sense of being put to motion. If one runs one's eyes from one photograph to the next in sequence, one gains the illusion of motion, of swinging arms and legs lifting and dropping. Although Muybridge's project originated from an impulse to "break down" motion into constituent parts, his photographic sequences thus generated the very motion he sought to arrest. Muybridge understood the appeal of this aspect of his work, and by 1879 he had developed a device (which he called the "zoopraxiscope") for projecting a rapid succession of pictures made from his photographic sequences, so that viewers could see the motion of his subjects restaged before their very eyes.[38]

It is this pull between bodies at rest and bodies in motion that places Muybridge's project at the threshold between not only two different approaches to the body but also two different modes of representing the body. Critics have long noted how Muybridge plied a narrow course between the respectable work of studying bodies as objects of scientific knowledge and the more questionable diversion of looking at bodies as objects of desire.[39] As his project was being conceived, supporters at the University of Pennsylvania expressed concern about this latter possibility, and they wanted to be sure that the undertaking be conducted and perceived as a work of science. The photographs themselves address these concerns with their various furnishings of science, and when they were published, supporters of the study highlighted its scientific nature.[40] Even the publication scheme for *Animal Locomotion* worked to ensure that the photographs would serve their more noble intentions. Rather than being put onto the market in inexpensive volumes containing selected sequences or even individual photographs printed, perhaps, using the recently developed halftone process, the publisher sold the volume by subscription as a series of large-format pages, each of which contained a single plate printed using the much more expensive photogravure process. Interested parties could purchase any 100 plates for $100, or the entire series of 781 plates for $600. As one recent study notes, the volume's "luxurious format, prohibitive prices, and restrictive circulation" effectively limited its audience to members of a particular social set, whose purposes for the photographs seemed more certain to extend beyond whatever "visual pleasure they afforded."[41]

But however much Muybridge sought to restrain that visual pleasure, the sense of scientific control frequently breaks down in his photographs. This happens in various ways: by the harsh sunlight that sometimes intrudes into the even tones of the photographs (the set Muybridge used was outdoors); by disturbances to the grid, as when models brushed up against the threads that produced the long lateral lines; and by the sexual energy that circulates through many of the photographs, energy that derives not so much from the nakedness of many of the striding and bending bodies, but from the activities that Muybridge put his models up to—men flexing their muscles while staring into the camera, women falling onto mattresses or toweling themselves dry.[42] Thus, a plate such as "Descending Stairs," with its effort to reduce bodily motion to poles tipping against a grid, is joined by plates such as "Sitting Down and Placing Feet on Chair," which in a different mode could just as easily have been titled "Leaning Back and Smoking Cigarette" (Figure 9.7). Unlike the trussed-up body in "Descending Stairs," the body here presents itself as an object of desire, caught in a moment of illicit pleasure.

If the detached and reductive approach to the body seen in the former sequences seems at home in a limited edition of scientific still pictures, the inviting and fulsome approach seen in the latter seems much more at home in the mass medium of "motion pictures." The zoopraxiscope—with its serialized images and projected motion—pioneered this new medium, but so too did the conception and construction of the photographic sequences in *Animal Locomotion*. Just as filmmakers would be doing within a decade, Muybridge staged brief stories that he sought to capture on film as they unfolded over time. The photographic representations of these stories that he then offered for

FIGURE 9.7: "Sitting Down, Placing Feet on Chair." From Eadweard Muybridge, *Animal Locomotion* (1887), plate 247. Collections of the University of Pennsylvania Archives.

viewing were not simply the unmediated product of his cameras, but instead were sequences of images that he had cropped and in some cases reordered in ways that served imaginative rather than scientific purposes. These kinds of manipulations would become standard practices among filmmakers. The fact that Muybridge was already conducting his work in this way in the 1880s makes him, according to Marta Braun, "the first man to construct, direct, photograph, and edit the kind of narrative fantasies that are at the core of motion pictures."[43]

Motion pictures would become one of the most important venues for picturing bodies in the twentieth century. It is the nineteenth century, though, that has been the focus of this chapter. In the paintings, prints, and photographs considered here, one sees a variety of means for visually representing the body and a host of purposes those representations might serve. But within this variety, in the images I have taken up, there is a common tension between the body as a symbol or instance of order and the body as an instance or source of disorder. For Charles Willson Peale, then, the civic virtue that would underwrite a new political order demanded not only William Buckland's visible inattention to his own particular body but also the absence of those bodies whose relation to the new body politic was troubling. By the late 1820s, this U.S. body politic seemed in many ways to be even more unruly than it was at the nation's founding, challenging John Neagle's ability to coordinate the hand of William Strickland with the head, or to keep out signs of those laboring bodies who materialized Strickland's designs. Social and political dislocations in Great Britain at about the same time inspired the sickly figure in *The Body Politic on the March of Intellect*, whose conservative appeal for a now-lapsed state of health and good order is unmistakable. *Small Power-Planer*, published at mid-century in the specific context of industrialization, offered a similarly conservative appeal in its depiction of machine tending as a form of mental labor (and of machine tenders as more like members of a middle class than a working class), but the wood engraver who prepared the image for printing—likely to be witnessing the disorganizing effects of industrialization firsthand—appears to have challenged that appeal by presenting a monstrous amalgam of human and machine. Finally, toward the end of the century, Eadweard Muybridge embarked upon a photographic project that pushed bodies up against a grid, even as it entertained a much less orderly vision of the body. William Alcott's call to duty may have echoed in Muybridge's ear, but there was also pleasure to be had, and the body—in its lived experience as well as its representation—could well serve both.

CHAPTER TEN

From Mimetic Machines to Digital Organisms

The Transformation of the Human Motor

ANSON RABINBACH

Along with major shifts in the nature of industrial and postindustrial work at the beginning of the twenty-first century, there has also been a deep crisis of the metaphors mobilized to frame and embody the nature of what we call work. A new image of the symbiosis of body and machine has emerged, which the historian Bruce Mazlish has aptly called the "Fourth Discontinuity"—an allusion to Freud's three great illusory "divides" or discontinuities of man and the cosmos (Galileo), man and animals (Darwin), and man and nature (Freud)—because today the dominion over machines can no longer be taken for granted.[1] One can even go a step further and argue that the metaphor of the human machine or human motor has as much to do with the transformation of work throughout modern history as do machines and industrial processes themselves. The metaphor of the body as machine/motor can be plausibly compartmentalized into three simple and distinct historical types: "mimetic," "transcendental," and "digital." Each of these, in turn, can be represented by a different kind of technology.

The mimetic technology of the eighteenth century is exemplified by a clockwork whose mechanical precision is capable of replicating certain biological processes with remarkable verisimilitude. This form of human or animal machine is exemplified by: the artisanal automata of the great clock makers, for example, the "defecating duck" of Jacques de Vaucanson or the "writing boy" of Pierre Jaquet-Droz. The transcendental materialism of the Industrial Revolution is in turn illustrated by motors (*Kraftmachine*) that convert energy to produce motion: the steam engine, the automobile, and the Taylorist worker. Today's digital metaphor derives its inspiration from computers as the new "human machines" and is best understood in terms of artificial intelligence, "micro-worlds," or "digital organisms." What I am interested in is not simply these changes in the operative metaphor (which is well known) but in how and whether these shifts in the metaphor serve as vehicles through which we negotiate the divide between the artificial and the natural and conjure up very different notions of the biopolitics of work.[2]

The heyday of the automata was the eighteenth century, the era when craftsmen of extraordinary skill endowed their intricate mechanisms with movements of such delicacy that they seemed to mock the boundary between life and technology. Most celebrated among them was Jacques Vaucanson, who became an overnight sensation at the age of twenty-eight. In 1738, he astonished Paris with a life-sized faun capable of playing the flute with such precision that incredulous audiences accused him of concealing a tiny musician inside its body. Vaucanson followed with his most celebrated creation, a mechanical duck that flapped its wings, pecked at its food, drank water, and, finally, evacuated a fetid pellet after passing it through a digestive system. Vaucanson's remarkable fowl (the prototype of Disneyland's mechanical puppets) became a popular attraction, drawing large crowds to the King's Theater in London in 1742 and touring Germany for two years hence. Vaucanson combined his wondrous performing simulacra with a perceptual and pedagogical aim, the illustration of a physiological principle (that digestion occurs by a chemical process rather than by pulverizing the food). Yet however much he had endeavored to make it imitate all the actions of the living animal, Vaucanson stopped short of identifying life with the machine.[3] His reticence demonstrates that he understood that though his duck might simulate physiological function and illustrate a natural principle, it could never attain the "self-moving power" that was life.

These simulacra were not merely models; they strove "not only to mimic the outward manifestations of life, but to follow as closely as possible the mechanism that produced these manifestations."[4] The first automata were both epistemological machines, functioning illustrations of a biomechanical mode of explanation, and at the same time, performative simulacra that seemed to embody the self-moving power and capacity for generation that inevitably eluded them.[5] Locating the automata between the performative and the pedagogical is a sign of their author's own ambivalence vis-à-vis these wondrous machines.

The automata were also a combination of inscription and simulation, technologies that could write or inscribe actual writing (as in the writing boy) or sound (as in the flute player) in a machine that reproduced the effects of our sen-

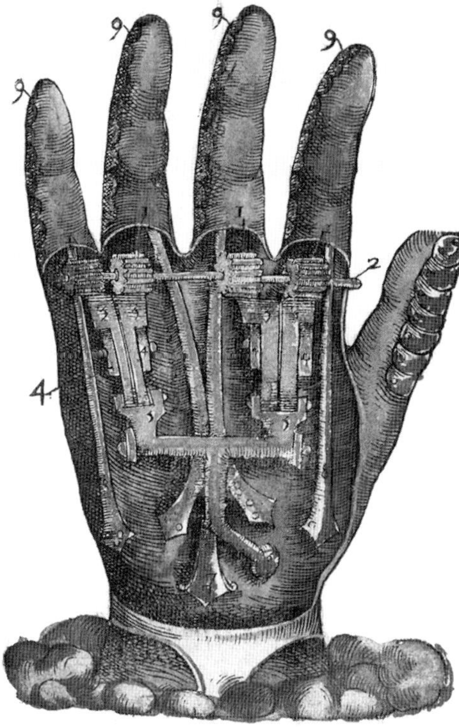

FIGURE 10.1: *Illustration of mechanical hand*. From Ambroise Paré, *Instrumenta Chyrurgiae et Icones Anathomicae* (1564). Wellcome Library, London (L0043496)

sory apparatus while assuming its external form. Mimesis, as Walter Benjamin reminds us, is not merely the imitation of nature by artifice, but it is a faculty that is the "weak remnant of a powerful compulsion to be and act similarly."[6] The unmistakable presence of the occult in Benjamin's theory of technical mimesis points to why two central aspects of the eighteenth century automata were so often combined: as scientific creations and performance pieces, their powerful attraction rested on their unfulfilled ontological promise—the reproduction of nature and ultimately of life. They were designed to appear to embody the capacity to generate their own motion.

Subsequently, in the late nineteenth century, the automata divided into two distinct types, those that were representations of specific physiological processes and those that just replicated their mechanical effects. These second order of simulacra were no longer replicas but prostheses, extensions of human sense organs. In his excellent book on sound technologies, James Lastra has drawn the consequences of this distinction. If the classical automata mimetically replicated or duplicated organic or corporal movement, nineteenth-century devices like the phonograph and the telephone abandoned the quest for imitation but were capable of more effectively copying nature—perhaps even more perfectly than nature itself.[7] This process, which might be called the demystification or *entzauberung* of the automata, greatly diminished the phantasmagoric character of simulation, while perfecting inscriptions of sound or, more precisely, the material form of sound. As simulacra they no longer had any need to imitate the attributes of living beings, and, ontologically speaking, such simulacra are indifferent to the question of what constituted the "life" of the beings that were being created in the laboratory.

The problem, classically posed by Descartes, that no matter how superbly produced, the automata lack self-moving power—a soul, emotions, language, spontaneity—and thus remain merely *mimetic* beings, mere *correspondences to life,* was rendered moot. Even earlier, during the eighteenth century, we can already observe certain gestures toward the *entzauberung* of the automata even in staunch materialists. Julien Offray de La Mettrie, the famous author of *L'Homme-machine* admitted that living machines are more than machines.[8] In the end, the classical automata of Vaucanson, Droz, and Menzel could do no more than *illustrate* the principle of self-moving power that they so patently failed to embody.

The eighteenth-century automata were not "working" machines in the strict sense, though from a scientific point of view they did perform "work." Their energy (or force) was provided by their creators while they simultaneously served as entertainment and illustrations of the principles of physiology. By contrast, the productivism of the Industrial Revolution was governed by a very different conception of force, one that entirely rejected self-moving power as a phantasm and by the realization that human society and nature are linked by the primacy and ultimate interchangeability (convertibility) of all productive activity, whether of the body, technology, or nature. As the principles of thermodynamics came to be understood, a new social imaginary emerged that presupposed an entirely different metaphor of the motor.[9] As Hermann von Helmholtz, its most passionate popularizer liked to point out, the discovery of the laws of thermodynamics proved that the authors of the automata were hopelessly implicated in what he called their "mimetic error"—the belief that beasts and human bodies corresponded to apparatus that "moved themselves energetically and incessantly . . . and were never wound up."[10] "Self-moving power" was a chimera: all beings and machines are moved by energy converted into motion.

The machines of the preindustrial age differed from mid-nineteenth century productivism in their lack of a unifying, transcendental metaphor. In the Newtonian universe diverse forces (gravity, wind, water, horse) pushed, pulled, or turned machines, generating motion. In the Helmholtzian universe, force or *Kraft* is converted into work by motors (natural, human, and technological). Unlike the metaphor of the machine, the metaphor of the motor is productivist: it refers not simply to the mechanical generation of movement but to the industrial model of a calculable and natural channeling of energy *converted* from nature to society and back again. To borrow a line from Henry Adams, the nineteenth-century energeticist metaphor of production was framed by "incessant transference and conversion" whereas eighteenth-century labor operated in the framework of "creation."[11] The productivism of the Industrial Revolution was governed by the realization that human society and nature are linked by the primacy and ultimate interchangeability (convertability) of all productive activity, whether of the body, technology, or nature.

The metaphor of the motor first appeared during the first quarter of the nineteenth century.[12] After Nicolas Sadi Carnot's discovery of "the motive power of heat" in 1824, it became clear that all the forces of nature are essentially different varieties of a single, universal energy or *Kraft*. The discovery of thermodynamics revealed the mimetic machine to be an epistemological dead-end, because energy is always universally present in all nature and technology.

If the eighteenth-century machine was a refraction of the Newtonian universe with its multiplicity of forces, disparate sources of motion, and reversible mechanisms, the nineteenth-century metaphor of the machine was the thermodynamic engine, the servant of a powerful nature conceived as a reservoir of undiminished motivating power. The machine is capable of work only when powered by some discrete external source; the motor, by contrast, is regulated by internal, dynamic principles, converting calories into heat, and heat into mechanical work.[13] The body, the steam engine, and the cosmos were thus connected by a single and unbroken chain, by an indestructible energy, omnipresent in the universe and capable of infinite mutation, yet immutable and invariant.

The law of energy conservation accorded the concept of energy or *Kraft* undisputed primacy in the explanation of the natural world. Physics became the supreme science: the discovery of the laws of *Kraft* raised the concept of work to the dignity of a universal principle of nature, irrespective of the so-called moral perfection of servants, or of any other workers. Helmholtz frequently mentions the implications of his new understanding of nature for the meaning of work. The shift from mechanical forces to the language of a universal *Kraft* eliminated the need for a spiritual understanding of labor; the work ethic was eclipsed by the quantitative economy of energy. "Hence, in a mechanical sense," he noted, the idea of work has become "identical with that of the expenditure of force."[14] The new technology of the industrial age thus produced a new image of the body whose "origins lie in labor-power," and which is not simply analogous to, but essentially identical with, a thermodynamic machine: "the animal body therefore does not differ from the steam-engine as regards the manner in which it obtains heat and force, but does differ from it in the purpose for, and manner in which the force gained is employed."[15] Helmholtz did not simply demote the living creature to a machine, but he transposed the character of an energy converting machine—an automata—to the body, the industrial dynamo, and indeed, to the universe itself.

FIGURE 10.2: *Patent Gymnasticon, or Machine for Exercising the Joints and Muscles of the Human Body*. National Library of Medicine (A024623)

Before the motor came into its own, the metaphor that defined labor was that of a generative activity derived from eighteenth-century theorists of labor: John Locke, Adam Smith, the Abbe Sieyès, and of course, Karl Marx in his famous 1844 *Economic and Philosophical Manuscripts*. There Marx famously saw labor as the anthropological universal that is both a meaning-bestowing and a historically constituting activity, a "metabolic exchange" between history and nature. The emancipation of labor is the emancipation of society and the individual, of production and expression. After 1859, however, Marx increasingly regarded the distinction between concrete and abstract labor in the language of labor power, as an act of *conversion* rather than *generation*. Marx thus shifted his focus from the emancipation of humankind *through* labor, to emancipation *from* productive labor by an even greater productivity. Marx became a "productivist" when he no longer considered labor to be an anthropologically "paradigmatic" mode of activity, and when, in harmony with the new physics, he saw labor power as an abstract magnitude (a measure of labor time) and a natural force (a specific set of energy equivalents located in the body). For his previous generative view of labor, emancipation occurs *within* labor itself, whereas from the point of view of *conversion*, it occurs only *apart* from labor, in the form of shorter hours or reduced physical and mental exertion. Consistent with the energeticist view, freedom, in Marx's famous formulation, begins where the realm of necessity ends, and not, in a utopian realm where the two are fused. That crucial distinction was elaborated by a generation of German Social Democratic theorists, most notably Karl Kautsky, who significantly rejected both the Prometheanism of Bolshevism and its authoritarianism. Work did not necessarily limit creativity, Kautsky noted, but truly productive labor in art, science, and human well-being could be achieved only as a consequence of reducing the working day.[16] It is no coincidence that the eight-hour day was the most lasting contribution of the Socialist International.

Helmholtz, too, was aware of the social vision implicit in the idea of self-moving power: redemption from painful labor, a society of perpetual idleness. "Perpetual motion was to produce labor-power (*Arbeitskraft*) inexhaustibly without corresponding consumption, that is to say, out of nothing."[17] The inventors of the automata envisaged a body without fatigue, without discontent, and without aversion to work. But they also revealed their ignorance of how motors convert the supply of nourishment into heat, and heat into force. Perpetual motion could never be replicated, because no novel source of energy was ever produced in nature. Conversion of force did not merely solve the problem of mimesis; it superseded it by reducing the mimetic machine to the illusion of a body that "creates energy out of itself."[18] The "transcendental materialism" of the nineteenth century produced a powerful metaphor of how nature, technology, and the human body all operate under the same dynamic laws of force—a homogeneity that is much more than the reduction of life process to the model of industrial technology. For those, like Helmholtz, who grasped the secrets of the cosmos in terms of the industrial dynamo, life was equivalent to force, and force was the physiochemical principle that governed the universe. "Life" was redefined to mean not mechanical motion, nor anything having to do with living being, but rather the particular form taken by the universal force of motion that propels all of nature. For Helmholtz, there is no essential distinction between the work of the industrial dynamo, the hefty blows of the preindustrial blacksmith, the delicate movements of the lacemaker, or the precise fingering of the concert violinist. Nature became a vast cistern of protean energy awaiting its conversion to work.

By the time he wrote *Das Kapital*, Marx regarded the distinction between concrete and abstract labor in the language of labor power, as an act of *conversion* rather than of *generation*. Labor power is the motor of history, the natural force that makes possible its own supercession by the machine. "The existing productivity of labor, from which it proceeds as its basis, is a gift, not of nature, but of a history embracing thousands of centuries."[19] Insofar as the physical aspect of labor becomes first naturalized, and ultimately, mechanized, the distinction between the emancipation of labor as a creative and organic reality, and the metamorphosis of the material process of production is blurred.[20] The realm of necessity is characterized by the regulation of the expenditure of energy and freedom by the liberation of human energy from external constraint. "But, it [control of the production process] nonetheless still remains a realm of necessity, beyond it, begins that development of human energy which is an end in itself, the true realm of freedom, which, however, can blossom forth only with this realm of necessity as its basis. The shortening of the working day is its prerequisite."[21]

According to Engels, Marx considered the discovery of labor power his most important achievement. His use of the term credits the early nineteenth-century French engineers who analyzed steam engines and the discoveries of Lord Kelvin and Helmholtz, who had discovered the concept of *Arbeitskraft* as the principle by which labor power becomes convertible, quantifiable, and equivalent to all other forms of labor power (in nature, in the universe, or in machines). In *Das Kapital* labor power appears as a force equivalent to other forces of production. The concept of labor power is also a quantitative measure of the expenditure of human energy in production. Marx stresses this point in *Das Kapital*: "Tailoring and weaving, although they are qualitatively different productive activities, are both a productive expenditure of human brains, muscles, nerves, hands etc. . . . merely two different forms of the expenditure of human labor power."[22] Or even more strikingly: "On the one hand, all labor is an expenditure of human labor-power in the physiological sense, and it is in this quality of being equal, or abstract, human labor that it forms the value of commodities. On the other hand, labor is an expenditure of human labor-power in a particular form and with a definite aim."[23] The energetic basis of labor-power is the link between labor-power as a magnitude and labor-power as a natural force. Marx conceives of labor-power as both a measure of force, or socially necessary labor-time, and as a specific set of energy equivalents located in the body.

In the capitalist West, it was not Marxism but the American system of Frederick Winslow Taylor that put the Helmholtzian revolution in synch with the industrial workplace, inaugurating what Charles Maier has called "the heralded utopian change from power of men to the administration of things."[24] Taylorism, it should also be noted, was but one among many scientific approaches to labor that shared a fetishism of corporal rationalization and promised the end of class conflict through the greater productivity achieved by conserving the workers' energy. On the continent a variety of approaches claimed to be capable of producing a more rational and a more scientifically governed laboring process—*science du travail, Arbeitswissenschaft, human science of* work, ergonomics, and so on.

The Galileo of modern fatigue research, the Turin physiologist Angelo Mosso, whose *La Fatica* (1891) became the standard work for generations of scientists, and who developed the technology of fatigue measurement, identified the chief obstacle to work not in the weakness of moral character but in the depletion of physical and mental force.[25] Fatigue, which could be inscribed and measured by newly invented devices (the ergograph), was, unlike purely "subjective" tiredness, measurable, quantifiable, and always represented the objective limit or "optimal" point of exertion, the outermost boundary of the human motor. The new physics and the physiological theories that emanated from it dissolved the old moral categories by absorbing them into the fragile scaffolding of a universe built out of labor power. Fatigue thus emerged at the threshold of the body's economy of energies; it was the corporal horizon of a mechanical universe, with its own internal laws of energy and motion.

The pioneering ergonomist and fatigue expert, Jules Amar, argued in *Le Moteur Humane* (*The Human Motor* 1913) that a worker's movements could be carefully observed and charted "with a view towards eliminating all that is unnecessary."[26] Amar was firmly convinced that all tasks utilizing the same tool required a similar output of energy, and that when a muscular action was exercised in the same fashion, it always resulted in the same tracing. Any variation could be explained "principally because the worker is lacking in skill."[27] Comparing the results obtained by a "good workman, skillful and well trained" with those of a less practiced apprentice, he found that the accomplished journeymen normally adopted an efficient economy of motion that starkly contrasted with those of an apprentice, whose "chief defects are irregular and spasmodic action leading to unduly rapid fatigue."[28] Amar called his most important insight, that "a muscle returns more speedily to its condition of repose in proportion to its rapid performance of the work," Amar's law.[29] This discovery, he claimed, had practical consequences: by correcting the trifling defects of position, "the worker's fatigue is diminished without injury to his daily output."[30]

Amar studied the mechanics of the human body in microscopic detail, under experimental conditions, and with strictly mathematical and scientific rigor. He was the first to apply the different techniques of measurement developed since Étienne-Jules Marey's pioneering tracings to different types of work, which he reproduced in his laboratory: the work of the hands, the legs, and of different tools.[31] Amar also investigated the transformation of chemical and caloric energy into work and the nature of respiration, diet, clothing, and hygiene. He took measurements in the field, conducting experiments sanctioned by the colonial authorities in Algerian prisons and in small industrial workshops. He developed techniques of measurement to a high degree of technical sophistication, inventing almost all of the basic techniques of modern ergonometric measurement—including the first exercise bike (ergocycle). He pioneered the training of apprentices and the study of their aptitudes for different types of work—which were in part based on "morphological" comparisons of different body types.[32] With Amar's systematic research the search for the dynamic laws of fatigue were superseded by the search for the dynamic regularities of work. Amar was motivated by a single principle: for the performance of a maximum amount of work with a minimum amount of fatigue there was an optimum speed and position that could be scientifically predetermined. He applied these researches to different national and ethnic groups, creating a comparative anthropology of labor power. Indefatigable in his pursuit of the laws of economic expenditure of energy, Amar applied his theory to a wide variety of activities

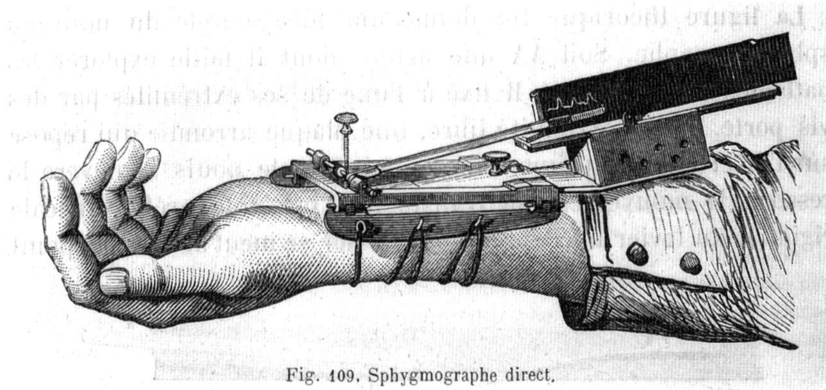
Fig. 109. Sphygmographe direct.

FIGURE 10.3: *Direct Sphygmograph*. From Étienne-Jules Marey, *La Circulation du Sang* (Paris: Masson, 1881). Wellcome Library, London (L0012232)

from writing, to musical instruments, "to those of the athlete, the sportsman and the soldier."[33]

In 1919, Amar reviewed his first decade of experimentation: "tried first, and many times upon myself, this method has for ten years given proof of its simplicity and reliability. During that time it has been applied to about a thousand persons—Parisian working-men, soldiers, and natives of North Africa, it is therefore of universal applicability, and for that reason eminently scientific."[34] Amar was convinced of the intimate connection between social efficiency and social harmony. "On the average one third of the available energy of man is wasted," he argued, while "methodical organization could increase the industrial output in the same proportion."[35]

Before World War I, the hope that the state would perform the task of ensuring a thorough rationalization of the workplace in the interests of capital and labor remained utopian. However, the arrival of the American Taylor system—as scientific management—in Europe brought a competing model of rationalizing labor power into sharp conflict with proponents of the science of labor. Both the European science of work and Taylorism claimed validity as a scientific approach; each claimed that precise analysis of the worker's minute movements could produce efficiency, each was preoccupied with economizing motion, each adapted the body to technology, and each claimed to stand above class interests and ideology. Yet in contrast to the science of work, which relied on laboratory research far from the workplace, the Taylor system relied on engineers who, in close cooperation with plant management, worked at the level of the individual firm to reduce costs and maximize profits.

The state played no role in Taylor's initial conception of the workplace. Scientific management was directed at owners of industry, offering them a method of rationalizing production while inhibiting socially disruptive activities. French proponents of the Taylor system, like Henri Le Chatelier, by contrast, argued that the American system was socially neutral, promising greater production and higher wages. Its staunch opponents, like Amar and the labor physiologist Jean-Marie Lahy, argued that the Taylor system ignored the physiology, and above all the health, of the worker. It was, as they saw it, unregulated "superproduction."[36]

World War I did not so much further the invention of new techniques of testing aptitude, combating fatigue, diagnosing and treating psychological illness, improving efficiency and boosting output, as much as it eased the extensive employment of these techniques for the first time. The war provided those trained in psycho-technical methods with a vast laboratory to demonstrate the utility of their knowledge. By 1915, a new generation of unskilled workers, composed largely of women and adolescents, were at work in industrial and munitions plants with little or no training, at lower wages, and with longer hours. The war gave legitimacy to the aptitude test, the neurasthenia (and shock)

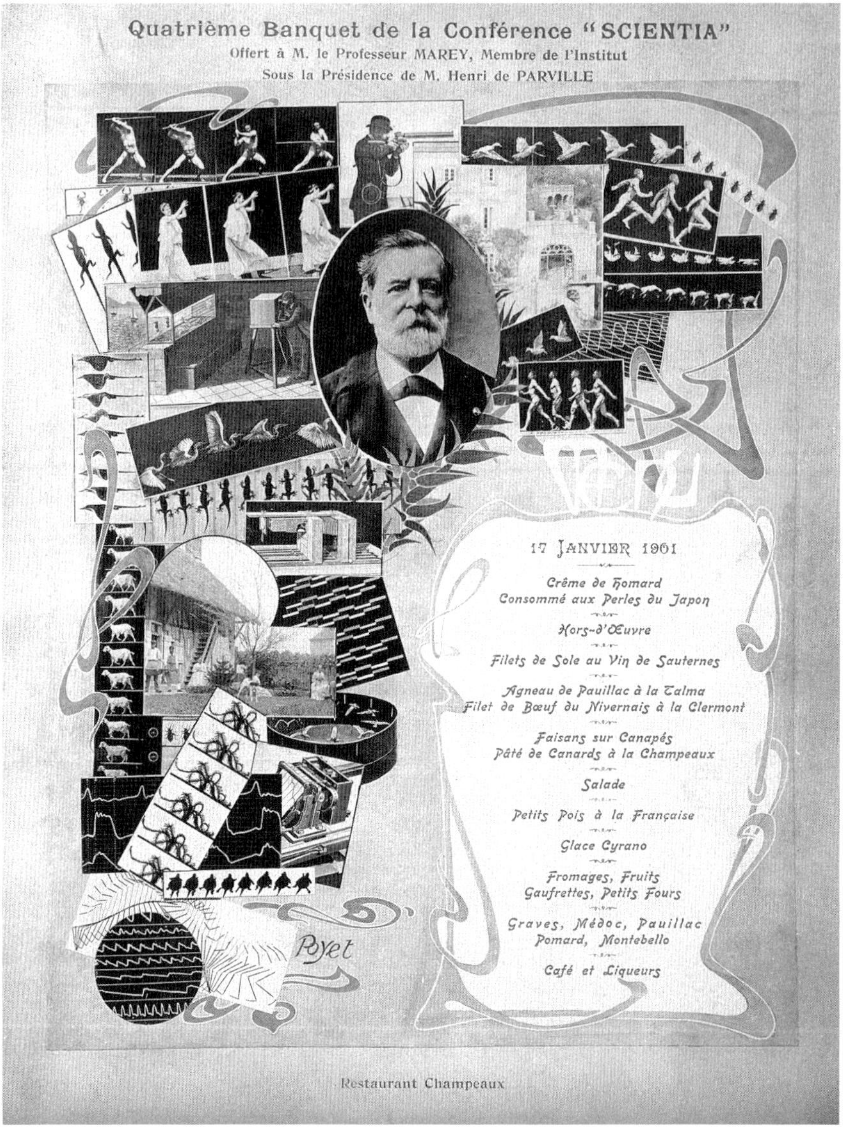

FIGURE 10.4: *Étienne-Jules Marey*. Collotype [?] collage by Poyet, 1901. Wellcome Library, London (V0003855)

diagnosis, and the industrial reeducation and rehabilitation of the maimed and wounded—especially through the science of prosthetics.[37] In Germany, too, as early as 1915 aptitude testing developed by the "psycho-technicians" Walther Moede and Curt Piorkowski were used to select aviators, drivers, and radio operators among others.[38] The war also furthered the scientific study of what was termed "human economy"—the point of intersection between production techniques, mobilization of national resources, and the politics of demogra-

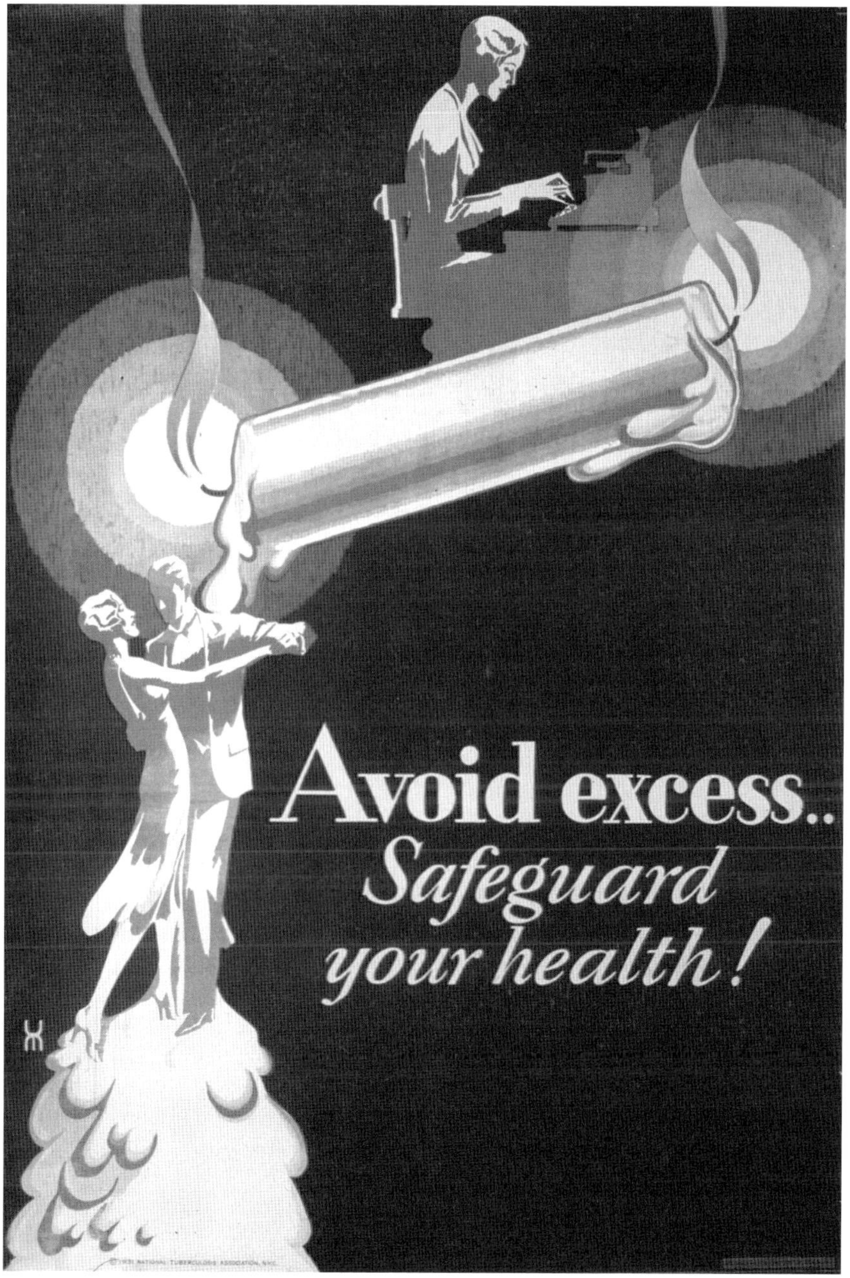

FIGURE 10.5: "Avoid Excess—Safeguard Your Health!" (New York: National Tuberculosis Association, ca. 1931). National Library of Medicine (C01752)

phy. As Fritz Giese, a leading psycho-technical expert in Weimar Germany remarked of his own wartime service "the war not only presented psychology with new knowledge, it also created new subject matter, which, without that sad occasion would surely have remained estranged from it."[39]

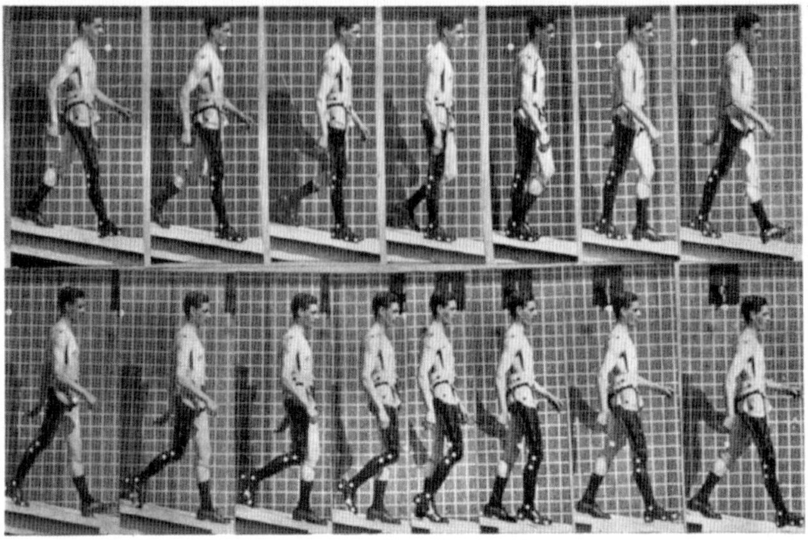

FIG. 177. [To face p. 174.

FIGURE 10.6: "Man with Artificial Leg Shown in the Course of Walking." From Ernest Muirhead Little, *Artificial Limbs and Amputation Stumps* (London: H. K. Lewis, 1922). Wellcome Library, London (L0042698)

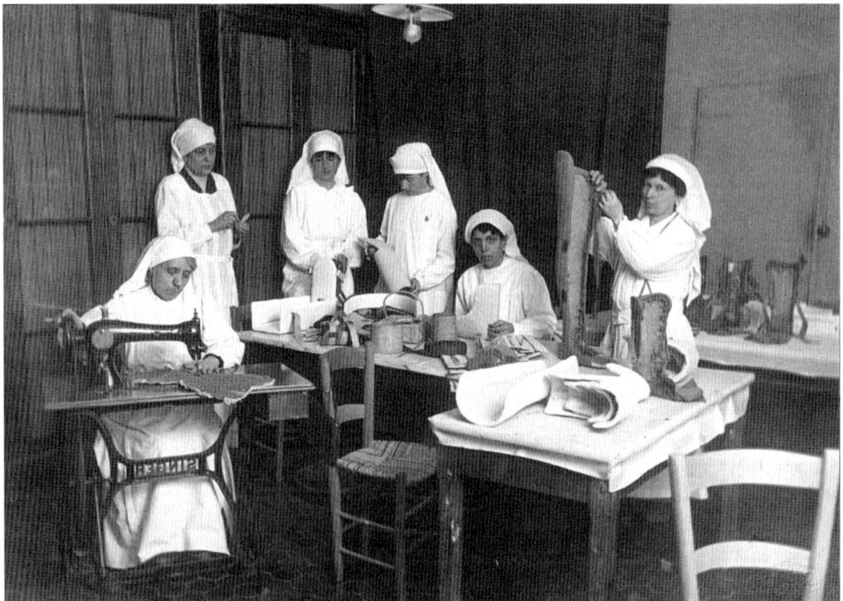

FIGURE 10.7: "Artificial Limb Factory in Rome: Six Women Working at Benches, One Using a Sewing Machine and One Stitching the Back of a Full-length Leg." Photograph, 1914/1918. Wellcome Library, London (V0029754)

After the war, professional academic or quasi-academic subdisciplines with applied or "practical" interests in labor management (industrial physiology, industrial medicine; industrial psychology; industrial sociology) emerged in almost all European countries with the expressed purpose of forestalling or circumventing a potentially cataclysmic breakdown. In the interwar era not merely the social relations of the workplace but the working body became the arena of the psychosocial contestation over labor power. By extension, the pervasive fear of the body's total breakdown represented a threat to industrial civilization. The preoccupation with corporeal fatigue as the universal form of resistance to work represented both the limits of the excessive demands of industry and, somewhat paradoxically, the last line of defense against the superexploitation of the industrial Moloch. This was especially true in Germany where industrial psychology first achieved professional legitimacy. The chair in applied industrial psycho-technics established at the Institut für Industrielle Psychotechnik at the Technische Hochschule Charlottenburg shortly after the war, and occupied by Georg Schlesinger, became a showplace for government-sponsored efforts to improve the efficiency of industrial labor during the early 1920s. By mid-decade, engineering students in German universities were required to complete a course in scientific management and scientific factory organization, including the latest developments in applied psychology.[40]

During the 1920s, a number of forward-looking physiologists employed several strategies of "psycho-technics," a term invented in 1903 by the psychologist William Stern to describe the uses of applied psychology for investigating the external conditions affecting human work in the school and in the occupations.

Popularized by Hugo Münsterberg in his 1914 book *Psychotechnik*, psycho-technics sought to combine quasi-Taylorist methods to improve "psycho-physical capacity for performance" with reduction of fatigue and its concomitant accidents. The movement embraced the vision of the body as a thermodynamic system, traversed by flows of energy and fatigue, while at the same time, expressing fears that a new urban order, what Andreas Killen has called an "Electropolis," a highly rational, yet crisis-ridden product of industrial civilization that threatened the integrity of the human motor. If the body conjured up an image of order, the possibility of disruption, confusion, social chaos, and systemic breakdown was so omnipresent that by 1926 it was no longer compensated as an illness by the German welfare state.[41]

German psycho-technicians were a heterogeneous community made up of academic psychologists, engineers, and physiologists who shared a commitment to rationalization and a belief that their approach to optimizing labor was superior to the allegedly purely economic rationality of Taylorism. Beyond that, their approaches diverged, sometimes significantly. In the early Weimar era at least three distinct strands of psycho-technics emerged. Professionally, the movement was composed of both engineers and academics. In reality, the two were hard to distinguish, as evident in the case of Schlesinger, who was called the "German Taylor" for his enthusiasm for scientific management, envisioning of workers as "working machines, perpetually exhausting and regenerating themselves."[42] Schlesinger and Walter Moede (his successor at the Institut für Industrielle Psychotechnik in Berlin), conceived of "psycho-technics" as a practical *Heilslehre* that could compensate for some of the "irrationalities" of capitalist production methods, by rationalizing work methods, evaluating behavior in the firm, and preselecting individuals for tasks from "youth to old age."[43] The academics regarded the Taylor system with profound suspicion, and in 1927, in an effort to undermine Taylorist assumptions, Edgar Atzler and a collaborator, Günther Lehmann, attempted to calculate the expenditure of energy and the food required by Taylor's famous human "ox," the Dutchman Schmidt, whose legendary brute strength and dumb obsequiousness were recounted in the most famous passage of Taylor's *Principles of Scientific Management*. Schmidt, despite his superhuman productivity, ate more than he produced. Their calculation was that Schmidt expended the equivalent of more than 5,515 calories in ten hours, making him less than cost efficient. The conclusion they drew was that "the process of work had to be adapted to the special nature of the human motor," extracting optimal not maximal output.[44]

A second direction was represented by the tragic figure of Otto Lipmann, who was an academic psychologist and a socialist. Lipmann sought to fashion a version of industrial psychology more consistent with a politically sensitive, rational, and humane worldview. By siding with the needs of industry to "sort out" the most unproductive workers and maintain those who were both reliable and effective, Lipmann rightly saw that psycho-technics had simply become one tool among other, harsher, management practices. Lipmann believed the scientific credibility of industrial psychology would be entirely sacrificed if, as Moede proposed, industrial profitability was to be the only criteria of success, and if the new science did not also "meet the class interests of the working class." Still, the purpose of industrial psycho-technics was not to intervene in class conflict as he believed it could only remain neutral and demonstrate what was the optimal energy expenditure; labor power was a larger, cultural problem, one that engaged the worker's entire being, including his or her satisfaction, health, mental well-being, and general moral sensibility, as well as the worker's relation to the larger social body.[45]

A third direction was represented by the nationalists, most prominently, Fritz Giese and Karl Arnhold—technocrats and "reactionary modernists"—both of whom in the 1930s embraced the Nazi cause. In post–World War I Germany, maximizing labor power, Giese felt, held the key to national regeneration; and psycho-technics the key to maximizing labor power. But, as Killen points out, though indebted to Münsterberg's work, Giese sought to go beyond what he perceived as the limitations of the "human-motor" model favored by his American counterpart. Labor power was in his eyes not reducible to mere calculations of energy expenditure.

By 1922, 170 psychological testing stations had been created across Germany, under the auspices of the Ministry of Labor. A veritable psycho-technical craze was born and Giese—using the mechanical devices he improved upon—was contracted to provide measurements on the employees, mostly women, at the telephone exchanges of the Imperial Post Office (RPM). Between 1925 and 1933, Giese's political views underwent a significant shift. In his pro-American study, *Girlkultur*, he expressly favored the more sporting American body culture over Prussian militarism.[46] In his writings of the early 1930s, however, with the American model tarnished by the world economic crisis, Giese embraced the new body culture valorized by the Nazis, the physiognomy of the new human type of the National Socialist future. After 1933, he exempted attributes like courage and nobility—the masculine virtues—from psycho-technical analysis, placing them beyond experimental determination.[47] Though already committed to such a view early, the implications of this gendering of the psycho-technical subject would only become clear with Giese's embrace of the National Socialist body ideal of the "sovereign" male warrior in a hyper-gendered social order.[48]

Arnhold's DINTA (Deutsche Institut für technische Arbeitsschulung) was founded in the fall of 1925, receiving considerable support from Ruhr Industrialists and from the conservative DNVP. Arnhold conceived of his mission to put psycho-technics on anti-Socialist and nationalist course, to create "a new Industrial species [the] carrier of a German ethos from the old Germanic epoch."[49] Aggressively anti-American and antirepublican, the DINTA was absorbed into the German Labor Front and became the representative organization of industrial psychology in Nazi Germany.

In February 1933, the editors of the *Psychotechnischen Zeitschrift*, proclaimed their allegiance to National Socialism. In June, Moede's *Industrielle Psychotechnik* announced its support of all those "practitioners of applied psychology and psycho-technics who affirm the new state." In fact, the psycho-technical experts were quick to embrace the new order.[50] Apart from those Jews and Socialists who were officially proscribed from performing their functions, most academic psychologists followed suit, proclaiming in many different and creative ways, their ideological allegiance and the usefulness of psychology for the Nazi state. Tragically Lipmann, who had never held a university post, was hounded from his library and institute and committed suicide in October 1933.

Though worlds apart ideologically, Helmholtz's characterization of the universe as *Arbeitskraft*, Marx's theory of the relentless transformation of labor power into the powerful engine of capital, and Frederick Winslow Taylor's utopia of the worker's body subordinated to the rational intelligence of the engineer, were all variations on the theme of the metaphor of the human motor, of the working body as a medium for converting energy into work. In the totalitarian models of the work-society (in both Soviet Russia and Nazi Germany) "disciplinarity" reigned supreme.

The Bolshevik founding fathers, Lenin and Trotsky, readily grasped the significance of the Taylorist revolution, differing only in the intensity of their utopian impulses and the locales of their respective "hypermodernist" models. Lenin's utopia in *State and Revolution* centered on the German organized *Finanzkapital* described by Rudolf Hilferding, whereas Trotsky's vision was of the Americanized Bolshevism that will crush and conquer imperialist Americanism. Though Lenin modulated his utopia with a long disquisition on the virtues of "co-operation," Trotsky gave Soviet productivism a more Faustian demeanor, culminating in his grand design for the "militarisation of labor" eventually realized by his murderer, Stalin.[51] Stalin's Five Year Plans required far more than a massive militarization of Soviet society in peacetime. As Sheila Fitzpatrick points out, the massive Soviet campaigns to mobilize society for rapid heavy industrial growth during the 1930s were accompanied by a military rhetoric couched almost entirely in the vocabulary of "battles," "fronts," "shock-brigades," and "sabotage."[52] Stalinist productivism criminalized resistance to work with his famous dictum that "People who talk about the necessity of *reducing* the rate of development of our industry are enemies of Socialism, class enemies."[53] Still, the "showpieces" of Stalin's industrialization program—Magnitogorsk and the Stalingrad Tractor Plant—could not overcome pervasive chaos, worker resistance, and low productivity.[54]

Though the Fordist model of disciplinarity dominated the post–World War II era of prosperity (1945–1973), in the last decades of the twentieth century a broad consensus gained ground among many observers of the global economy that the classical Fordist constellation of growth, prosperity, and social equity had reached its limit, forcing capital to adapt new and more flexible consumer-oriented strategies of production and distribution. Instead of standardized mass production, innovative firms stressed flexible batch production, smaller inventories, and niche markets. Instead of highly centralized labor systems, decentralization is highly preferable; even in the public sector, the only growth sector for trade unionism in the United States, outsourcing became the watchword of the new economy. In Europe and the United States, rigidly time-bound factory labor began to suffer an irreversible decline, and even in that sector, flexible work schedules and home work rapidly became the norm. Instead of top-down control over unskilled/deskilled routinized industrial labor, management experts, corporate intellectuals, and neoliberal politicians proclaimed the bright future of more desirable skilled, educated workers, capable of working in tandem with superiors and subordinates. "Democracy" and "communication," not hierarchy and authority, were the watchwords of the new workplace. Does this presage the end of "disciplinarity"? The new workplace has introduced a new form of "discipline": the discipline of permanent, unrelenting, insecurity, and fear.

The disappearance of the older Fordist disciplinary ideal may be one of the most prominent features of the new post-Fordist productivism. Even a cursory glance at the industrial management textbooks lets us get a glimpse of what the new culture of work looks like. Instead of top-down industrial discipline, democratic and "communicative" communities of fewer but more elite workers capable of "intersubjectivity" and "synergy" are in greater demand. Post–Taylorist/Fordist industrial relations specialists reject the obsolete ideal of a homogeneous workforce composed of a dull, uneducated worker subjected to hierarchical control and prescribed tasks. In the early 1990s, it became a mantra for economists like Lester Thurow, Robert Reich, and Charles Heckscher to reinforce the notion that the chief liability of the American workplace was its double Taylorist legacy: intense labor capital conflict and strict rule-bound production, both of which are dysfunctional in a digital environment. The "new economy" favors a flexible strategy toward work and work time; union norms and adversarialism must be sacrificed to "knowledge based" cooperation, profit-sharing, work-sharing, flexible work schedules, incentive pay for innovation, and so on.[55] If the managerial manuals are any indication, the disciplined rule-bound worker has given way to the ideal of "flexible, internally motivated, continuously learning work force," obsolete bureaucratic structure will be reborn as "a strong internal culture to support information sharing and participation in problem solving," commands will give way to "balancing dialogue and discussion," hierarchy will be redefined as the "delegation or shared responsibility in recognition that dispersed activity requires local action and flexibility."[56]

We may conclude that the "work-centered society" of the nineteenth and early twentieth centuries was to a large degree a phenomenon that was both dependent on and necessitated metaphor of the human motor. To the extent that this metaphor has begun to wane, the emphasis on disciplinarity has also begun to disappear. But this does not mean that we are no less a work-centered society; in fact, today's worker may be more disciplined and overworked than his or her predecessors, as Juliet Schor argued in her book, *The Overworked American*.[57] None of this suggests that control of the workforce will diminish; on the contrary, discipline is far more intense and internalized, coerced by the threat of total insecurity, as the dot.com collapse made abundantly evident just one decade later. What has changed is that the body no longer occupies the same space in the metaphoric economy of work. With the gradual eclipse of the metaphor of the human motor, the "work-centered model of society" was emptied of its most compelling metaphor.

EPILOGUE

What metaphors now occupy the space of the "mimetic" machine or the "transcendental" motor? If labor power is no longer the site of the central metaphor of productivism, what does this mean for the new configuration of labor and the model of labor based on information processing rather than the generation of things or the conversion of force? It has sometimes been suggested that computers are in fact a new kind of automata, presaging a "workless" world in which human beings do not so much control the robot as become symbiotically merged with it. As one might have predicted, the phantasm of a body without fatigue and labor—the ghost of the classical automata—has also re-emerged. Historian Bruce Mazlish, for example, envisioned a new "species"—computers implanted in robots (he calls them "combots") that will do far more than perform mundane household tasks. The potential for humans to evolve into a new species in tandem with the post-Cartesian robot, a kind of "thinking machine" which he calls "homo-comboticus," is, he tells us, "open and realizable."[58]

Whether this prophecy will be fulfilled remains to be seen, but there are some readily observable consequences of how digital machines are already altering the way we think about work. The first and most dramatic is the computerized simulacra of the work process itself. Shoshanna Zuboff has studied how paper production follows a digitally simulated computer pattern that occurs virtually just moments before it occurs in a human-free factory environment. This allows the observer of the digital process to preemptively shut down the actual process, anticipating problems. As she explains, "when the textualizing consequences of an informating technology become more comprehensive, the body's traditional role in the production process (as a source of effort and/or skill in the service of *acting-on*) is already transformed."[59] With the eclipse of the great utopias of labor, the body reduced to an element in the *conversion* of force is being replaced by the digital model of work as the computer driven *simulacra* of combined technological, physical, and mental labor. Like Taylor's system, information technology continues to remove skill and knowledge from the "sentient" body, but it now also demands an interactive relationship to its object, a precise interaction with digital "events" that occur before, and hence anticipate, real-time events. These technologies require a profound "reskilling" and an adaptive change in the relations of authority in the workplace. Instead of reducing the "realm of necessity" (labor time) to a few dull hours a week, the new work demands its own realm of freedom in cohabitation with the machine. Consequently, work loses its materiality, corporeality, and spatiality. The manipulation of vast quantities of data also demands not discipline and the subordination of the body to external norms, but flexibility, judgment, engagement, and commitment. This sort of work, as the commercials for laptops on the beach tell us, occurs, not merely in the workplace, but in any place that the worker may be.

Secondly, these developments are not merely a consequence of the technology itself but of how the computer has provided management experts with a new image of organization that apparently does away with cumbersome top-down authoritarian forms of management. As I suggested, an intellectually vigorous new discourse of "antidisciplinarity" has found a niche in the boardrooms of corporations and on the editorial pages of influential newspapers and periodicals. Take for example the *Wall Street Journal*, which in the 1990s campaigned against the lingering consequences of the Taylorist–Fordist workplace, for example, firms sticking to an outdated model in which management distrusts the autonomy of workers, prescribes dull routinized tasks, curbs creativity, and creates a workplace ill suited to "literate, independent-minded workers."[60] Similar visions of neoromantic anarchocapitalism can be found in an influential manifesto by Kevin Kelly, technoguru of the influential periodical *Wired*, under the apt title, *Out of Control: The New Biology of Machines*. According to Kelly, digital machines offer a different biophysical metaphor for the operations of nature than did thermodynamics. Like beehives, computer "swarms" are adaptable, *resilient, evolving, and constantly generating novelty*. According to Kelly, socialism, deep ecology, and the early experiments in artificial intelligence were *all* flawed by an obsolete, top-down, engineering approach.[61] Unlike the old "clockware" that required intricate, and inefficient central planning, he argues, bottom-up, evolutionary, swarm logic allows new forms of life and society to *emerge* spontaneously. Kelly's "swarm consciousness" might have only appealed to the T-shirt and rollerblade capitalists of the dot-com bubble, but it is ultimately indebted to far more serious thinkers who for several decades have contended with the forms of "artificial life" produced by digital machines.

According to the theorists of "a-life," "aliveness" is a property of form and logic, not of carbon molecules. In the words of its foremost spokesman, Christopher Langton, a-life is "the attempt to abstract the logic of life in different material forms."[62] Cellular Automata (and here I rely on Langton's own description) do not merely directly code the "organism" as in creating a pattern or a blueprint, but allows an open-ended set of local rules to "emerge." From the perspective of artificial life, biology is only one variety of a much larger conception of life that reaches beyond our own carbon-based world, perhaps even beyond electronic digital life, to what Kelly calls "hyperlife" a kind of quantum biology that maps the space of all possible life. "Deep evolution" as some of its adherents call it, is an amalgam of environmentalism, computer simulations, sociobiology, and old-fashioned market economics. Like most organicist thinking it betrays a deeply conservative impulse, the desire to project back onto a benign nature the *summum bonum* of all experience. There is a new "ontology" at work here, one which interestingly recalls the way in which thermodynamics regarded the classical "automata" as a kind of "error" based on a misunderstanding of what constitutes "life." This time, however, the error lies in the reductionist idea of life as simply the law-bound conversion of energy. Computers can simulate the flocking behavior of birds, grow vast botanical gardens of flora and fauna, and evolve insect-like creatures that behave in ways remarkably like those in the natural world. Many of the properties that biologists would consider intrinsic to life: evolution, self-organization, emergence, parallel processing, occur in these computer simulations. Such computer-generated discourses are symptomatic of the belief that new organizational systems, like this redefined "life" itself, are "emergent" rather than structured by management. At bottom, the image of a world that is "out of control" promises not chaos, but hypercontrol without central authority. Like Mandeville's earlier *Fable of the Bees*, the lesson of the swarm is that private vices bring public benefits.

A-life's critics point out that the idea of life as "emergence" is itself a phantasm: life eludes computation because "it stands in a permanent non-deterministic and pragmatic relation to its environment." This "slimy" quality, has nothing to do with "carbon chauvinism," argues the biologist Claus Emmeche, but has much more to do with the question of whether biological processes can be *entirely* grasped in syntactical, computational terms. In short, as Emmeche puts it, despite the "ontological" claim that computers have already produced an alternative life form, they have not demonstrated that "we are in fact speaking of a *realization* of new forms of life (in another medium) and not simply a *simulation* of life.[63] Despite the claims of the "a-life" theorists, the capacity of computers to simulate life has not transcended the bounds of eighteenth-century automata—they have not become "life." Like the old automata, they are merely simulacra conceived as proofs, not only that life can be created "in silico," but as Emmeche put it, as "a broad analogy, a class of models of complex calculated systems that share ecological and evolutionary conditions with many of the real organisms found in nature."

Emmeche's conclusion points to the weaknesses of Kelly's overarching hypersynthesis of biology and technology. But, the readiness with which such computer-driven visions of a nonauthoritarian, bottom-up, open-ended, flexible, and "emerging" workplace have been accepted are significant on their own terms. The discourse of emergence and the fact that work is now imagined in digital rather than disciplinary terms, does not do away with exploitation, overwork, or "sweating" in vast computer barns. Just as Taylorism and Fordism relegated the worker to a function, so the workstation is just an interactive site in the Internet universe. At the same time, the replacement of "organization" by "emergence" signifies a fallacy similar to the one that governed assumptions about Fordism, namely that such processes are automatic, autonomous, and independent of the vicissitudes of the market. In the United States, and even more strongly in Europe, from the end of World War II to the late 1960s, the Fordist social contract guaranteed high wages, high purchasing power, a comfortable old age, and above all, the relative security that each generation can expect to achieve a higher standard of living than the previous one. Today, none of these assumptions can be taken for granted.

With this transformation we may also have reached the "end of the work-centered society," but not the abolition of labor, discipline, or the reduction of work time to a socially necessary minimum. Reforms that encourage the reduction of work time, more flexible work situations and arrangements, as well as reductions in unemployment demand a far more radical rethinking of the relationship between politics and market forces. But with the eclipse of the great productivist utopias of the twentieth century—Fordism, fascism, communism—work has ceased to be the defining activity of modernity, and, regrettably, no longer offers the vision of a more just or a more dynamic future.

NOTES

Introduction

Thanks to Eva Åhrén, Clare Anderson, Doug Atkins, Julie K. Brown, Joanna Ebenstein, Elizabeth Fee, Nick Hopwood, Erika Naginski, Myriam Peignist, Paul Theerman, Stephen Rice, the Clark Art Institute, the History of Medicine Division of the National Library of Medicine, and the Wellcome Library, for help in writing this chapter.

The opinions expressed here are those of the author and do not necessarily reflect the positions of the National Library of Medicine, the National Institutes of Health, or the U.S. Department of Health and Human Services.

1. James L. Kernan, invitation to a performance by Marinelli the Anatomical Puzzle, Washington, D.C., November 11,1886. National Library of Medicine. Ouroboros, the circular snake that eats its own tail, was an Ancient Greek symbol.
2. This volume construes "age of empire" broadly. The period 1800–1920 was an age of empire, but it was also an age of industrialization, improvement, class formation, nationalism, revolution, discipline formation, and so on. In terms of the traditional notion of empire, much of the action in this volume takes place in Britain and the British Empire, America and its nonimperial imperium, France proper, Germany, and Austria-Hungary. Readers should look elsewhere for the growing literature on the body in other colonial empires (formal and informal) and the Ottoman, Russian, and Manchu empires. For a sampling of recent scholarship on the body in empire in the nineteenth and twentieth centuries, see Ann Laura Stoler, *Carnal Knowledge and Imperial Power: Race and the Intimate in Colonial Rule* (Berkeley: University of California Press, 2002) and Stoler, ed., *Haunted by Empire: Geographies of Intimacy in North American History* (Durham, NC: Duke University Press, 2006); David Arnold, *Colonizing the Body: State Medicine and Epidemic Disease in 19th-century India* (Berkeley: University of California Press, 1993); Steven Pierce and Anupama Rao, eds., *Discipline and the Other Body: Correction, Corporeality, Colonialism* (Durham, NC: Duke University Press, 2006); Tony Ballantyne and Antoinette Burton, eds., *Bodies in contact: Rethinking Colonial Encounters in World History* (Durham, NC: Duke University Press, 2005).

3. A shelf of significant works would include Michel Foucault, *The Birth of the Clinic: An Archaeology of Medical Perception*, trans. A. M. Sheridan-Smith (New York: Pantheon, 1973); *Discipline and Punish: The Birth of the Prison*, trans. Alan Sheridan (New York: Pantheon, 1978); *The History of Sexuality*, vol. 1, *An Introduction*, trans. Robert Hurley (New York: Pantheon, 1978); Norbert Elias, *The Civilizing Process*, vol. 1, *The History of Manners* (1939; repr. Oxford: Blackwell, 1969); Stephen Kern, *Anatomy and Destiny: A Cultural History of the Human Body* (Indianapolis: University of Indiana Press, 1975); Alain Corbin, *The Foul and the Fragrant: Odor and the French Social Imagination* (Cambridge, MA: Harvard University Press, 1986); Mary Poovey, *Making a Social Body: British Cultural Formation, 1830–1864* (Chicago: University of Chicago Press, 1995); Caroline Walker Bynum, *Fragmentation and Redemption: Essays on Gender and the Human Body in Medieval Religion* (New York: Zone Books, 1991); Roy Porter and W. F. Bynum, eds., *Medicine and the Five Senses* (Cambridge: Cambridge University Press, 1993); Roy Porter, *Bodies Politic: Disease, Death and Doctors in Britain, 1650–1900* (London: Reaktion, 2001); Thomas Laqueur and Catherine Gallagher, eds., *The Making of the Modern Body: Sexuality and Society in the 19th Century* (Berkeley: University of California Press, 1987); Ruth Richardson, *Death, Dissection and the Destitute* (London: Routledge, 1987); Ludmilla Jordanova, *Sexual Visions: Images of Gender in Science and Medicine between the 18th and 20th Centuries* (Madison: University of Wisconsin Press, 1989); Judith Butler, *Bodies that Matter: On the Discursive Limits of Sex* (New York: Routledge, 1997); Sander Gilman, *Difference and Pathology: Stereotypes of Sexuality, Race, and Madness* (Ithaca, NY: Cornell University Press, 1985); Michel Feher with Ramona Naddaff and Nadia Tazi, eds. *Fragments for a History of the Human Body* [*Zone* 3–5] (New York: Zone Books, 1989); Jonathan Crary and Sanford Kwinter, eds. *Incorporations* [*Zone* 6] (New York: Zone Books, 1992); Christopher E. Forth and Ivan Crozier, eds., *Body Parts: Critical Explorations in Corporeality* (Lanham, MD: Lexington Books, 2005).
4. Rosemarie Garland Thomson, *Extraordinary Bodies: Figuring Disability in American Culture and Literature* (New York: Columbia University, 1997); Rosemarie Garland Thomson, ed., *Freakery: Cultural Spectacles of the Extraordinary Body* (New York: NYU Press, 1996); Robert Bogdan, *Freak Show: Presenting Human Oddities for Amusement and Profit* (Chicago: University of Chicago Press, 1990); Andrea Dennett, *Weird and Wonderful: The Dime Museum in America* (New York: NYU Press, 1997); Richard Altick, *The Shows of London* (Cambridge, MA: Belknap Press, 1978).
5. See "Purchases Two Theatres; James L. Kernan Buys Houses in Washington and Baltimore, May Be in Stair-Hanlin Syndicate," *New York Times*, February 12, 1903: 9; "A New Theatre Circuit," *New York Times*, May 28, 1906: 9; "Burlesque Trust Shaky," *Washington Post*, February 18, 1905: 9; "Sells His Theaters; James L. Kernan Relinquishes Burlesque Holdings . . ." *Washington Post*, May 12, 1906: 12; "Kernan Endows Hospital; Theatrical Man Gives $100,000 for Benefit of Crippled Children," *Washington Post*, November 29, 1910: 3.
6. Michel Foucault, *Birth of the Clinic:* Andrew Cunningham and Perry Williams, eds., *The Laboratory Revolution in Medicine* (Cambridge: Cambridge University Press, 1992); John Harley Warner, *Against the Spirit of System: The French Impulse in 19th-century American Medicine* (Princeton, NJ: Princeton University Press, 1993).

7. Ian F. McNeeley, *"Medicine on a Grand Scale": Rudolf Virchow, Liberalism, and the Public Health* (London: Wellcome, 2002); Laura Otis, *Müller's Lab* (New York: Oxford University Press, 2007).
8. According to an anonymous journalist, at the height of his career Marinelli earned more than 100,000 francs per year. See "Hommes serpents et femmes couleuvres," *Lecture pour tous* (February 1908). Thanks to Myriam Peignist for directing my attention to this and providing me with a photocopy.
9. Allison Muri, *The Enlightenment Cyborg: A History of Communications and Control in the Human Machine, 1660–1830* (Toronto: University of Toronto Press, 2006); Anson Rabinbach, *The Human Motor: Energy, Fatigue and Modernity* (Berkeley: University of California Press, 1992); Michael Sappol, *Dream Anatomy* (Washington, D.C.: Government Printing Office, 2006); Mark Seltzer, *Bodies and Machines* (New York: Routledge, 1992).
10. Michael Sappol, *A Traffic of Dead Bodies: Death and Embodied Social Identity in 19th-century America* (Princeton, NJ: Princeton University Press, 2000); Richard Bushman, *The Refinement of America: Persons, Houses, Cities* (New York: Vintage, 1993); John F. Kasson, *Rudeness and Civility: Manners in 19th-century Urban America* (New York: Hill & Wang, 1990).
11. "Hommes serpents et femmes couleuvres." See also Guyot-Daubès, *Curiosités physiologiques: Les homme-phénomès: Force—agilité—adresse* (Paris: G. Masson, éditeur; Libraire de l'Académie de Médecine, 1885), 219–32 (chap. 27–28).
12. See Richard Cope, *Ploughshare of War: The Origins of the Anglo-Zulu War of 1879* (Pietermaritzburg, South Africa: University of Natal Press, 1999); Jeff Guy, *The Destruction of the Zulu Kingdom: The Civil War in Zululand 1879–1884* (London: Longman, 1979).
13. Benedict Anderson, *Imagined Communities: Reflections on the Origin and Spread of Nationalism*, rev. ed. (London: Verso, 1991). For a related discussion of images of the Santal rebellion (India, 1855) published in the *Illustrated London News*, see Daniel Rycroft, *Representing Rebellion: Visual Aspects of Counter-insurgency in Colonial India* (New Delhi, India: Oxford University Press, 2006).
14. For a cultural critique that connects domesticity, patriarchy, capitalism, racism, and imperialism, see Anne McClintock, *Imperial Leather: Race, Gender and Sexuality in the Colonial Contest* (New York: Routledge, 1995): "Imperialism suffused the Victorian cult of domesticity and the historic separation of the private and public" (36).
15. After another bloody war, this time between the Boers (the descendants of an earlier Dutch colonial venture) and the British.
16. See Robert Ross, *Status and Respectability in the Cape Colony, 1750–1870: A Tragedy of Manners* (Cambridge: Cambridge University Press, 1999).
17. The term "evolution" itself did not come into common usage until the 1870s. The word does not appear in *On the Origin of Species*, but Darwin used it liberally in *The Descent of Man*.
18. Charles Darwin, *The Descent of Man and Selection in Relation to Sex* (London, 1871), 405.
19. Even before the publication of *On the Origin of Species,* Darwin implicitly made the argument in his works on living and fossil barnacles. See Darwin, *Living Cirripedia, A monograph on the sub-class Cirripedia, with figures of all the species,* vols. 1–2 (London: The Ray Society, 1851; 1854); *A monograph on the fossil Lepadidae, or, pedunculated cirripedes of Great Britain* (London: Palæontographical

Society, 1851); and *A monograph on the Fossil Balanidæ and Verrucidæ of Great Britain* (London: Palæontographical Society, 1854). Darwin popularizers like H. G. Wells explicitly argued the point; see Wells, "Zoological Retrogression," *Gentleman's Magazine* 271 (1891): 245–53.

20. The literature on Darwin and evolution is vast. See for example, Adrian Desmond, *The Politics of Evolution: Morphology, Medicine and Reform in Radical London* (Chicago: University of Chicago Press, 1989) and *Darwin: The Life of a Tormented Evolutionist* (New York: Norton, 1994); Peter J. Bowler, *Evolution: The History of an Idea* (Los Angeles: University of California Press, 2003); Janet Browne, *Darwin's* Origin of Species: *A Biography* (London: Allen & Unwin, 2006); Jonathan Smith, *Charles Darwin and Victorian Visual Culture* (Cambridge: Cambridge University Press, 2006); and the numerous essays and books of Stephen Jay Gould.
21. See, for example, Stephen Jay Gould, *The Mismeasure of Man* (New York: Norton, 1981); Jan Goldstein, *Console and Classify: The French Psychiatric Profession in the 19th Century* (Chicago: University of Chicago Press, 1987); Robert A. Nye, *Crime, Madness, & Politics in Modern France: The Medical Concept of National Decline* (Princeton, NJ: Princeton University Press, 1984); Daniel Pick, *Faces of Degeneration: A European Disorder, c.1848–c.1918* (Cambridge: Cambridge University Press, 1989).
22. William Seton, "The Century's Progress in Science," *Catholic World* (May 1889), 165: "Lamarck's explanation of the origin of species is not really opposed to Darwin and [Alfred] Wallace's view. It is rather complementary to it." Darwin himself was not entirely consistent. In some of his writings, he slipped, or perhaps tactically conceded to his critics, and endorsed the notion of progress, though not gestational adaptation (1859, chap. 10 and 14).
23. For Haeckel, see Robert J. Richards, *The Tragic Sense of Life: Ernst Haeckel and the Struggle over Evolutionary Thought* (Chicago: University of Chicago Press, 2008); Paul Weindling, "Ernst Haeckel, Darwinismus and the Secularization of Nature," in *History, Humanity and Evolution: Essays for John C. Greene*, ed. James R. Moore (Cambridge: Cambridge University Press, 1989), 311–27; Michael K. Richardson and Gerhard Keuck, "Haeckel's ABC of Evolution and Development," *Biological Reviews* 77 (2002): 495–528.
24. See, for example, Haeckel, *Anthropogenie*, plate 12.
25. Quote is from Haeckel, *Anthropogenie*, 82–83. For monism, see for example, Haeckel, *Der Monismus als Band zwischen Religion und Wissenschaft* (Bonn, Germany: Strauss, 1892).
26. Haeckel, *The History of Creation: Or the Development of the Earth and Its Inhabitants by the Action of Natural Causes* (New York: Appleton, 1876), 434, and *The Wonders of Life* (New York: Harper, 1905), 390; George J. Stein, "Biological Science and the Roots of Nazism," *American Scientist* 76 (1988): 50–58, argues that Hitler's *Mein Kampf* relied heavily on Haeckel's doctrines and was crucial in the development of National Socialism.
27. The literature of eugenics is vast. See, for example, Daniel Kevles, *In the Name of Eugenics: Genetics and the Uses of Human Heredity* (New York: Knopf, 1985); Nancy Leys Stepan, *The Hour of Eugenics: Race, Gender, and Nation in Latin America* (Ithaca, NY: Cornell University Press, 1991).
28. See, for example, Gillian Beer, *Darwin's Plots: Evolutionary Narrative in Darwin, George Eliot and 19th-Century Fiction* (Cambridge: Cambridge University Press,

2000); Peter Morton, *The Vital Science: Biology and the Literary Imagination 1860–1900* (London: Allen & Unwin, 1984); Robert Young, *Darwin's Metaphor: Nature's Place in Victorian Culture* (Cambridge: Cambridge University Press, 1985); John Glendening, *The Evolutionary Imagination in Late-Victorian Novels: An Entangled Bank* (Burlington, VT: Ashgate, 2008).

29. In late nineteenth- and twentieth-century Japan, the evolutionary discourse of hairlessness was mobilized to support Japanese expansionism and imperialism against the European colonial powers and the indigenous, light-skinned "hairy Ainu."
30. For the cultural meaning of mid-nineteenth-century beards in Europe and North America, see Charles Colbert, *A Measure of Perfection: Phrenology and the Fine Arts in America* (Chapel Hill: University of North Carolina Press, 1997), 123–47.
31. Charles Darwin to Charles Lyell, January 10, 1860; letter 2647, Darwin Correspondence Project, http://www.darwinproject.ac.uk/darwinletters/calendar/entry-2647.html.
32. For Darwin on the "marine Ascidian," see *Descent of Man* (London, 1871), chap. 21.
33. The phrase comes from Alfred Tennyson's poem "In Memoriam" (1849), which itself was influenced by the (pre-Darwinian) evolutionary arguments of Robert Chambers, *Vestiges of the Natural History of Creation* (1844).
34. Thanks to Clare Anderson for suggesting the connection with Baartman, who was a notable figure in nineteenth-century race and gender discourse. For Baartman, see Clifton Crais and Pamela Scully, *Sara Baartman and the Hottentot Venus: A Ghost Story and a Biography* (Princeton, NJ: Princeton University, 2008); Sander Gilman, "Black Bodies, White Bodies: Toward an Iconography of Female Sexuality in Late Nineteenth-Century Art, Medicine, and Literature," in *Race, Writing and Difference*, ed. Henry Gates (Chicago: University of Chicago Press, 1985), 223–61; Stephen Jay Gould, "The Hottentot Venus," in *The Flamingo's Smile* (New York: Norton, 1985), 291–305; Rachel Holmes, *The Hottentot Venus* (New York: Random House, 2006); Z. S. Strother, "Display of the Body Hottentot," in *Africans on Stage: Studies in Ethnological Show Business*, ed. B. Lindfors (Bloomington: Indiana University Press, 1999), 1–55; Sadiah Qureshi, "Displaying Sara Baartman, the 'Hottentot Venus'," *History of Science* 42 (2004): 233–57.
35. Darwin, *Descent of Man, 2:398.*
36. Barbara Welter, *Dimity Convictions: The American Woman in the 19th Century* (Athens: Ohio University Press, 1977); Gillian Brown, *Domestic Individualism: Imagining Self in 19th-Century America* (Berkeley: University of California Press, 1992).
37. An evolutionary solution to the problem of hairlessness was to argue that sexual dimorphism in primates was an evolutionary trend: sexual dimorphism is more marked in human beings than in the great apes. The sexual dimorphism argument was then available to be coded in class and racial terms: it was argued that cultural and biological sexual dimorphism was much more highly developed in upper-class European society than among so-called primitive peoples and the European peasantry and working class, where both men and women did hard physical labor, drank alcohol, lacked separate architectural spaces, etc. But the evolutionary perspective on dimorphism raised other issues for cultural critics to chew on: the feminization of upper-class men (who don't do hard physical labor), the sometimes elaborate cultural divisions between man's and woman's

spheres in subaltern groups, and the tendency of sexes to grow more alike with old age.

38. Christine Stansell, *City of Women: Sex and Class in New York, 1789–1860* (Chicago: University of Illinois Press, 1987); Susan Porter Benson, *Counter-cultures: Saleswomen, Managers and Customers in American Department Stores, 1890–1940* (Chicago: University of Illinois Press, 1986); Faye Dudden, *Serving Women: Household Service in 19th-century America* (Middletown, CT: Wesleyan University Press, 1983); Carroll Smith-Rosenberg, *Disorderly Conduct: Visions of Gender in Victorian America* (New York: Oxford University Press, 1986); Ann Douglas, *The Feminization of American Culture* (New York: Knopf, 1977).
39. Judith Walkowitz, *Prostitution and Victorian Society: Women, Class, and the State* (Cambridge: Cambridge University Press, 1980); Timothy J. Gilfoyle, *City of Eros: New York City, Prostitution, and the Commercialization of Sex, 1790–1920* (Norton: New York, 1992); Patricia Cline Cohen, *The Murder of Helen Jewett* (New York: Knopf, 1998).
40. McClintock, *Imperial Leather*, 5.
41. See, for example, Shirley Samuels, ed., *The Culture of Sentiment: Race, Gender and Sentimentality in 19th-Century America* (New York: Oxford University Press, 1992).
42. For homosociality, see Eve Kosofsky Sedgwick, *Between Men: English Literature and Male Homosocial Desire* (New York: Columbia University Press, 1985). For masculinity studies, see Rachel Adams and David Savran, eds., *The Masculinity Studies Reader* (London: Blackwell, 2002).
43. John Spargo, *The Bitter Cry of the Children* (New York: Macmillan, 1906).
44. For the sacralization of children and debates over child labor, see Viviana A. Rotman Zelizer, *Pricing the Priceless Child: The Changing Social Value of Children* (Princeton, NJ: Princeton University Press, 1994).
45. For the history of childhood, see Carolyn Steedman, *Strange Dislocations: Childhood and the Idea of Human Interiority, 1780–1930* (Cambridge, MA: Harvard University Press, 1995); Viviana A. Rotman Zelizer, *Pricing the Priceless Child*; Phillippe Ariès, *Centuries of Childhood*, trans. Robert Baldick (London: Jonathan Cape, 1962); Steven Mintz, *Huck's Raft: A History of American Childhood* (Cambridge, MA: Harvard University Press, 2004); David Nasaw, *Children of the City: At Work and at Play* (New York: Oxford University Press, 1986); Paula Fass and Mary Mason, eds., *Childhood in America* (New York: NYU Press, 2000); Hugh Cunningham, *Children and Childhood in Western Society Since 1500*, 2nd ed. (London: Longman, 2005).
46. Notwithstanding photographic hoaxes and manipulations, which deceived and delighted the Victorian public, there was a widespread belief in, even enchantment with, the facticity of photography, especially scientific, forensic, and social scientific photography. See, for example, George Rockwood, "Photography as Legal Evidence," *The Manufacturer and Builder* 21, no. 4 (1889): 81: "It is an old saying that 'photography will not lie,' and in many ways its veracity is unimpeachable./. . ./ The photographic process will often make legible figures, alterations, amendments, etc., in manuscripts which are invisible to the naked eye. /. . ./A good photograph is sometimes far more reliable than 'expert' testimony, and will bear scrutiny, if not a severe 'cross' examination." For the epistemological status of photographs, see

Jennifer Tucker, *Nature Exposed: Photography as Eyewitness in Victorian Science* (Baltimore, MD: Johns Hopkins University Press, 2005).

47. For photojournalism and muckraking, see George Dimock, "Child of the Mills: Re-reading Lewis Hine's Child-labour Photographs," *Oxford Art Journal* 16, no. 2 (1993): 37–54; Maren Stange, *Symbols of Ideal Life: Social Documentary Photography in America, 1890–1950* (Cambridge: Cambridge University Press, 1989), and "Jacob Riis and Urban Visual Culture: The Lantern Slide Exhibition as Entertainment and Ideology," *Journal of Urban History* 15 (1989): 274–303; Peter Bacon Hales, *Silver Cities: The Photography of American Urbanization, 1839–1915* (Philadelphia, PA: Temple University Press, 1983); Alan Trachtenberg, *Reading American Photographs: Images as History, Matthew Brady to Walker Evans* (New York: Hill & Wang, 1990); Cecelia Tichi, *Exposés and Excess. Muckraking in America,1900–2000* (Philadelphia: University of Pennsylvania Press, 2004), and *Shifting Gears: Technology, Literature, Culture in Modernist America* (Chapel Hill: University of North Carolina Press, 1987); Susan D. Moeller, "The Cultural Construction of Urban Poverty: Images of Poverty in New York City, 1890–1917," *Journal of American Culture* 18 (Winter 1995): 1–16; and Julie K. Brown, *Health and Medicine on Display: Sites and Exhibits at International Expositions in the United States, 1876–1904* (Cambridge, MA: MIT Press, 2009).
48. See, for example, Tom Lutz, *American Nervousness 1903: An Anecdotal History* (Ithaca, NY: Cornell University Press, 1993).
49. Carolyn Steedman, *Strange Dislocations;* Charles W. Taylor, *Sources of the Self: The Making of Modern Identity* (Cambridge, MA: Harvard University Press, 1989); Robert Miles, *Gothic Writing, 1750–1820: A Genealogy* (London: Routledge, 1993).
50. Steedman, 4.
51. Alexander Francis Chamberlain, *The Child: A Study in the Evolution of Man* (London: W. Scott, 1900).
52. Steedman, 4.
53. Mikhail Bakhtin, " 'Forms of Time and of the Chronotope in the Novel' (1937–38)," in *The Dialogic Imagination*, ed. Michael Holquist; trans. Caryl Emerson and Michael Holquist (Austin: University of Texas Press, 1981).
54. U.S. Census Bureau, "Historical Estimates of World Population," http://www.census.gov/ipc/www/worldhis.html.
55. Christopher Lasch, *The Culture of Narcissism: American Life in an Age of Diminishing Expectations* (New York: Norton, 1979).
56. Archaeology and genealogy are two of Foucault's key concepts. Michel Foucault, *The Order of Things: An Archaeology of the Human Sciences* (New York: Pantheon, 1970); *The Archaeology of Knowledge and the Discourse on Language*, trans. A. M. Sheridan Smith (New York: Pantheon, 1972); and "Nietzsche, Genealogy, History," in *Language, Counter-memory, Practice: Selected Essays and Interviews*, trans. Donald F. Bouchard and Sherry Simon (Ithaca, NY: Cornell University Press, 1977).
57. Michel Foucault, *The Birth of the Clinic: Discipline and Punish*. See also Foucault, *The History of Sexuality*, vol. 1.
58. See Raymond Williams, *Keywords: A Vocabulary of Culture and Society*, rev. ed. (London: Fontana, 1983), 87–93. According to Williams, culture is "one of the . . . most complicated words in the English language . . ." (87).

59. David S. Shields, "The Science of Lying," in *Messy Beginnings: Post-coloniality and Early American Studies*, ed. M.J. Schuller and Edward Watts (New Brunswick, NJ: Rutgers University Press, 2003), 230.
60. I say this advisedly, even though E.P. Thompson, in *The Poverty of Theory and Other Essays* (London: Merlin Press, 1978), denounced structural Marxism as an "orrery" that refused the possibility of human agency. From my perspective, the issue of whether language constructs people or people construct language has degenerated into an almost theological disputation. Scholarship needs to account for both sides of the equation: people (bodies) make history (praxis), but history (praxis) makes people (bodies).
61. E. P. Thompson, *The Making of the English Working Class* (London: Victor Gollancz, 1963) and "Time, Work-discipline and Industrial Capital," *Past & Present* 38, no. 12 (1967): 56–97. Thompson, of course, was not the only one moving social history in a cultural direction; there was also the immensely influential Birmingham Centre for Contemporary Cultural Studies.
62. The de-emphasis on class, while understandable, has gone way too far. My own position accords with the formulation of Anne McClintock, discussed earlier in this introduction—that social and cultural categories come into existence in and through each other.
63. See, for example, Bruno Latour, *Science in Action: How to Follow Scientists and Engineers through Society* (Cambridge, MA: Harvard University Press, 1987); and *The Pasteurization of France* (Cambridge, MA: Harvard University Press, 1987).

Chapter 1

1. See, for these figures, Massimo Livi-Bacci, *The Population of Europe: A History of Europe* (Oxford: Blackwell, 2000), 134–35 and 126–63 more generally.
2. Ibid., 136.
3. François Jacob, *The Logic of Life: A History of Heredity* (New York: Random House, 1973), 87; *OED*, 2nd ed. (Oxford: Oxford, 1989).
4. Jacques Donzelot, *The Policing of Families* (New York: Pantheon, 1979); Joshua Cole, *The Power of Large Numbers: Population, Politics, and Gender in Nineteenth-Century France* (Ithaca, NY: Cornell University Press, 2000).
5. William Wordsworth, "We Are Seven," in *Lyrical Ballads, with a Few Other Poems* (Bristol, UK: T. N. Longman, 1798), 110–14.
6. Mrs. [Sarah Stickney] Ellis, *The Mothers of England. Their Influence and Responsibility* (New York: Edward Walker, 1850), 12.
7. Georgiana Bruce Kirby, December 14, 1854, *Diary*, typescript copy at the California Historical Society Library, San Francisco, cited in Erna Olafson Hellerstein, Leslie Parker Hume, and Karen M. Offen, eds., *Victorian Women: A Documentary Account of Women's Lives in Nineteenth-Century England, France, and the United States* (Stanford, CA: Stanford University Press, 1981), 211.
8. Georgiana Bruce Kirby, December 14, 1854, *Diary*, typescript copy at the California Historical Society Library, San Francisco, cited in Hellerstein et al., *Victorian Women*, 211.
9. Queen Victoria to Victoria, Princess Royal, June 15, 1858, cited in Hellerstein et al., *Victorian Women*, 208.

10. Queen Victoria to Victoria, Princess Royal, May 2, 1859, cited in Roger Fulford, *Dearest Child: Letters between Queen Victoria and the Princess Royal, 1858–1861* (New York: Holt, Rinehart, and Winston, 1964), 191.
11. Ellis, *The Mothers of England*, 5; for the dominant view among women's historians in the 1970s and 1980s that Westerners were obsessed with (and invented) innate maternal instinct, see Elisabeth Badinter, *Mother Love: Myth and Reality: Motherhood in Modern History* (New York: Macmillan, 1981).
12. Judith Walzer Leavitt, *Brought to Bed: Childbearing in America, 1750–1950* (New York: Oxford University Press, 1986).
13. P. B. Saur, *Maternity: A Book for Every Wife and Mother* (Chicago: L.P. Miller, 1887), 223–28.
14. Lisa Forman Cody, *Birthing the Nation: Sex, Science, and the Conception of Eighteenth-Century Britons* (Oxford: Oxford University Press, 2005); Richard W. Wertz and Dorothy C. Wertz, *Lying-In: A History of Childbirth in America*, exp. ed. (New Haven, CT: Yale University Press, 1989); Ornella Moscucci, *The Science of Woman: Gynaecology and Gender in England, 1800–1929* (Cambridge: Cambridge University Press, 1990); Barbara Ehrenreich and Deirdre English, *For Her Own Good: Two Centuries of the Experts' Advice to Women*, rev. ed. (New York: Random House, 2005).
15. Donald Caton, *What a Blessing She Had Chloroform: The Medical and Social Response to the Pain of Childbirth from 1800 to the Present* (New Haven, CT: Yale University Press, 1999); and Tina Cassidy, *Birth: The Surprising History of How We Are Born* (New York: Atlantic Monthly Press, 2006), 77–90.
16. Rachel G. Fuchs, *Poor and Pregnant in Paris: Strategies for Survival in the Nineteenth Century* (New Brunswick, NJ: Rutgers University Press, 1992), 23.
17. Margaret DeLacy, "Puerperal Fever in Eighteenth-Century Britain," *Bulletin of the History of Medicine* 63 (1989): 538.
18. Wertz and Wertz, *Lying-in*, 126.
19. Judith Flanders, *Inside the Victorian Home: A Portrait of Domestic Life in Victorian England* (New York: W.W. Norton, 2003), 51–56.
20. In Freud's case, see the new English translation of Sigmund Freud and Joseph Breuer, *Studies in Hysteria*, trans. Nicola Luckhurst (Harmondsworth, UK: Penguin, 2004) in which Luckhurst retains patients' subtle metaphorical references to pregnancy, which had been lost in earlier translations, such as *Studien über Hysterie* (Leipzig Franz Deuticke, 1895). Michael Mason, *The Making of Victorian Sexuality* (New York: Oxford University Press, 1994); historians and critics who in turn have questioned the repressive hypothesis include, of course, Michel Foucault, *The History of Sexuality*, vol. 1: *An Introduction* (1977, repr. New York: Viking, 1990) and Peter Gay, *The Bourgeois Experience*, vol. 1: *The Education of the Senses* (New York: Oxford University Press, 1984).
21. Matt Ridley, *Genome: The Autobiography of a Species in 23 Chapters* (New York: Harper, 1999), 48.
22. Lisa Forman Cody, "The Politics of Illegitimacy," *Journal of Women's History* 11 (2000): 131–56; Felix Driver, *Power and Pauperism: The Workhouse System, 1834–1884* (Cambridge: Cambridge University Press, 1993).
23. Michelle Perrot, "Roles and Characters," in *From the Fires of Revolution to the Great War*, ed. Michelle Perrot, 167–259, vol. 4: *A History of Private Life*, ed. Philippe Ariès and Georges Duby (Cambridge, MA: Harvard University Press, 1990), 198.

24. For early examples, see John Howard, *The State of the Prisons in England and Wales* (Warrington, UK: William Eyres, 1777) and Charles White, *A Treatise on the Management of Pregnant and Lying-in Women* (London: W. Johnston, 1772); John Snow's *On the Mode of Communication of Cholera*, 2nd ed. (London: John Churchill, 1855) is the classic example of the work of this new science.
25. For a good recent account of this history in two countries see Libby Schweber, *Disciplining Statistics: Demography and Vital Statistics in France and England, 1830–1885* (Durham, NC: Duke University Press, 2006), esp. 93–135.
26. William Coleman, *Death is a Social Disease: Public Health and Political Economy in Early Industrial France* (Madison: University of Wisconsin Press, 1982), 149–180 and esp. 161–162 for these statistics.
27. Richard Evans, *Death in Hamburg: Society and Politics in the Cholera Years 1830–1910* (New York: Oxford University Press, 1987).
28. Sir Lyon Playfair, *Subjects of Social Welfare*, 2nd ed. (London: Cassell, 1889).
29. Laman Blanchard, "A Visit to the General Cemetery at Kensal Green," *Ainsworth's Magazine* 2 (August 1842): 177.
30. David Charles Sloane, *The Last Great Necessity: Cemeteries in American History* (Baltimore: Johns Hopkins University Press, 1995); João José Reis, *Death Is a Festival: Funeral Rites and Rebellion in Nineteenth-Century Brazil* (Chapel Hill: University of North Carolina Press, 2003); David L. Pike, *Subterranean Cities: The World Beneath Paris and London, 1800–1945* (Ithaca, NY: Cornell University Press, 2005), 101–89.
31. Lucy Riall, *Garibaldi: Invention of a Hero* (New Haven, CT: Yale University Press, 2007), 358.
32. Rev. P. Charles Augustine, *A Commentary on the New Code of Canon Law* (London: B. Herder, 1921), 101.
33. Edmund Parkes, *Manual of Practical Hygiene: Prepared Especially for the Use of Medical Service on the Army* (London: John Churchill and Sons, 1864), xvii; Parkes's later editions of 1866, 1869, and 1873 were quoted approvingly by early advocates of cremation. Parkes had argued in the 1857 report of the Royal Commission investigating the public health debacle of the Crimean War that doctors were called upon not simply to care for the sick but also to preserve the health of the living.
34. Sir Henry Thompson, *Cremation: The Treatment of the Body after Death* (London: Henry S. King & Co., 1874), 21.
35. Sir Lyon Playfair's calculations for the 1850 Board of Health's investigation into intramural internment were cited in *Pharmaceutical Journal and Transactions* 9 (1850): 446.
36. Thompson, *Cremation*; Henry Thompson, Edmund Parkes, and others got these figures for gas from the calculations of Playfair, who, when he first produced them had no interest in cremation but was aligned then with burial reformers; W. H. Lyttelton, *Scripture Revelation of the Life of Man After Death* (London: Daldy, Isbisther & Co., 1875), xxii–xxiii.
37. William Eassie, *Healthy Houses: A Handbook to the History, Defects, and Remedies of Drainage, Ventilation, Warming, and Kindred Subjects* (London: Simpkins, Marshall, 1872), i–iv; Eassie, *Sanitary Arrangements for Dwellings: Intended for the Use of Officers of Health, Architects, Builders and Householders* (London: Smith, Elder and Co., 1874), 2, 12; see pp. 85–86 for the mundane but important

details of the project on the evils of cesspits; p. 60 for the perfect arrangement of the seat and trap of the water closet.
38. Bertrand Russell, *Marriage and Morals* (London: Allen Unwin, 1929), 214.
39. Hague W. Grant, *The Eugenic Marriage: A Personal Guide to the New Science of Better Living and Healthy Babies* (New York: The Review of Review Companies, 1916).

Chapter 2

1. [George H. Lewes], "Training in Relation to Health," *Cornhill Magazine* 9 (1864): 219–31.
2. Anne Hardy, *Health and Medicine in Britain since 1860* (Basingstoke, UK: Macmillan, 2000).
3. R. Porter, *The Greatest Benefit to Mankind* (London: HarperCollins, 1997), 400–405.
4. J. C. Riley, *Rising Life Expectancy: A Global History* (Cambridge: Cambridge University Press, 2001).
5. J. C. Riley, *Sick, Not Dead: The Health of British Workingmen during the Mortality Decline* (Baltimore: Johns Hopkins University Press, 1997).
6. J. Harley Warner, *The Therapeutic Perspective: Medical Practice, Knowledge, and Identity in America, 1820–1885* (Princeton, NJ: Princeton University Press, 1984).
7. F. Condrau and M. Worboys, "Epidemics and Infections in Nineteenth-Century Britain," *Social History of Medicine* 20 (2007): 80–91 and reply G. Mooney, "Infectious Diseases and Epidemiologic Transition in Victorian Britain? Definitely," *Social History of Medicine* 20 (2007): 595–606.
8. A. Hardy, *The Epidemic Streets: Infectious Disease and the Rise of Preventive Medicine, 1856–1900* (Oxford: Clarendon Press, 1993).
9. Office of National Statistics, *The Health of Adult Britain, 1841–1994*, vol. 1 (London: Stationery Office, 1997), 30–57.
10. F. B. Smith, *The Retreat of Tuberculosis* (London: Croom Helm, 1988).
11. "Registrar General of Births, Deaths and Marriages in England, Forty-second Annual Report," British Parliamentary Papers, 1881 [C.2907], 27, 79.
12. L. S. Feuer, "The Friendship of Edwin Ray Lankester and Karl Marx: The Last Episode in Marx's Intellectual Evolution," *Journal of the History of Ideas* 40 (1979): 633–48.
13. M. J. Cullen, *The Statistical Movement in Early Victorian Britain: The Foundations of Empirical Social Research* (Hassocks, UK: Harvester Press, 1975).
14. V. L. Hilts, "Epidemiology and the Statistical Movement," *Times, Places, and Persons: Aspects of the History of Epidemiology, Bulletin of the History of Medicine Supplements* NS 4 (1980): 43–60.
15. D. Porter, *Health, Civilization, and the State: A History of Public Health from Ancient to Modern Times* (London: Routledge, 1999).
16. C. Hamlin, *Public Health and Social Justice in the Age of Chadwick: Britain, 1800–1854* (Cambridge: Cambridge University Press, 1998).
17. Ibid., 46–48, 135–50.
18. C. Lawrence, *Medicine in the Making of Modern Britain, 1700–1920* (London: Routledge, 1994).

19. C. E. Rosenberg, *Explaining Epidemics and Other Studies in the History of Medicine* (Cambridge: Cambridge University Press, 1992).
20. J. C. Whorton, *Nature Cures: The History of Alternative Medicine in America* (Oxford: Oxford University Press, 2002).
21. C. E. Rosenberg, "Pathologies of Progress: The Idea of Civilization as Risk," *Bulletin of the History of Medicine* 72 (1998): 714–30.
22. C. Lawlor, *Consumption and Literature: The Making of the Romantic Disease* (Basingstoke, UK: Palgrave, 2007).
23. N. Jewson, "Disappearance of the *Sickman* from Medical Cosmologies 1770–1870," *Sociology* 10 (1976): 225–44.
24. J. Henry Bennet, "The Treatment of Pulmonary Consumption," *British Medical Journal* ii (1867): 137.
25. Hughes Bennett, "Phthisis pulmonalis," in *A System of Medicine*, vol. 3, ed. J. Reynolds (London: Macmillan, 1871), 571.
26. T. H. Green, *An Introduction to Pathology and Morbid Anatomy* (London: Henry Renshaw, 1871), 145.
27. J. Henry Bennet, "The Treatment of Pulmonary Consumption," *British Medical Journal* ii (1867), 137.
28. J. Pemble, *The Mediterranean Passion: Victorians and Edwardians in the South* (Oxford: Clarendon Press, 1987).
29. C. R. Drysdale, *Alpine Heights and Change of Climate in the Prevention and Treatment of Pulmonary Consumption* (London: Odell & Ives, 1869). Reduced air pressure at high altitude would also lead to expansion of the lungs, and ozone might act as an antiseptic.
30. "Tuberculosis as a Contagious Disease," *British Medical Journal* i (1880): 704. Cf. H. Cullimore, *Consumption as a Contagious Disease* (A translation of Professor Cohnheim's pamphlet "Die Tuberklose vom Standpunkte der Infections-Lehre") (London: Baillière, Tindall and Cox, 1880).
31. "The Discussion at Glasgow on the Pathology of Phthisis Pulmonalis," *British Medical Journal* i (1881): 818.
32. R. Robertson, "Family History in Relation to Contagion in Phthisis Pulmonalis," *British Medical Journal* ii (1882): 624.
33. M. Worboys, *Spreading Germs: Disease Theories and Medical Practice in Britain, 1865–1900* (Cambridge: Cambridge University Press, 2000), 193–234.
34. A. Ransome, "The Consumption Scare," *Medical Chronicle* 2 (1895): 241–49.
35. R. Shingleton Smith, "Phthisis," *Medical Annual* (1892), 366–84; and J. E. Squire, *The Hygienic Prevention of Consumption* (London: C. Griffin, 1893).
36. C. Gradmann, "Robert Koch and the Pressures of Scientific Research: Tuberculosis and Tuberculin," *Medical History* 45 (2001): 1–32.
37. "Dr Koch's Investigations upon the Treatment of Tuberculosis," *Lancet* ii (1890), 933.
38. K. Ott, *Fevered Lives: Tuberculosis in American Culture Since 1870* (Cambridge, MA: Harvard University Press, 1996).
39. J. F. Coats, "The Spontaneous Healing of Tuberculosis," *BMJ* ii (1891): 933–38; J. K. Fowler, *Arrested Pulmonary Tuberculosis* (London: Churchill, 1892).
40. W. Calwell, "The Hygienic Treatment of Consumption Independently of Sanatoria," *BMJ* ii (1898): 947.
41. P. Baldwin, *Contagion and the State in Europe, 1830–1930* (Cambridge: Cambridge University Press, 1999), 37–243.

42. C.E. Rosenberg, *The Cholera Years: The United States in 1832, 1849, and 1866* (Chicago: University of Chicago Press, 1962).
43. R. Rawlinson, "Old Lessons in Sanitary Science Revived, and New Lessons Considered," *Transactions of the Sanitary Institute* 2 (1880–1881): 127–28.
44. "Review IV, Papers on Sanitary Science," *British and Foreign Medico-Chirurgical Review* 40 (1867): 65.
45. C. Hamlin, *A Science of Impurity: Water Analysis in Nineteenth Century England* (Bristol, UK: Adam Hilger, 1990), 99–116, 220–24.
46. Rosenberg, *The Cholera Years*, 65–81, 151–74.
47. M. Sigsworth, "Cholera in the East and West Ridings of Yorkshire, 1848–1893," (PhD thesis, Sheffield Hallam University, 1991).
48. Hamlin, *Public Health*, 245–74.
49. M. Pelling, *Cholera, Fever and English Medicine, 1825–1860* (Oxford: Oxford University Press, 1978), 203–49.
50. J. Simon, "Proceedings against Cholera under the Disease Prevention Acts and Otherwise," *Ninth Annual Report of the Medical Officer of the Privy Council for 1866*, British Parliamentary Papers, 1867, [3949], xxxvii, 1.
51. W. Luckin, "The Final Catastrophe: Cholera in London, 1866," *Medical History* 21 (1977): 32–42.
52. Simon, "Proceedings," postscript, 33.
53. N. Howard-Jones, *The Scientific Background to the International Sanitary Conferences, 1851–1938* (Geneva, Switzerland: World Health Organization, 1975).
54. M. Harrison, *Public Health in British India: Anglo-Indian Preventive Medicine, 1859–1914* (Cambridge: Cambridge University Press, 1994); J.D. Isaacs, "D.D. Cunningham and the Aetiology of Cholera in British India, 1869–1897," *Medical History* 42 (1998): 279–305.
55. W. Coleman, "Koch's Comma Bacillus: The First Year," *Bulletin for the History of Medicine* 61 (1987): 315–42.
56. C. Gradmann, *Krankheit im Labor: Robert Koch und die medizinische Bakteriologie* (Göttingen, Germany: Wallstein, 2005).
57. R. Evans, *Death in Hamburg: Society and Politics in the Cholera Years, 1830–1910* (Oxford: Oxford University Press, 1987).
58. Worboys, *Spreading Germs*, 252–53.
59. S. Bhattacharya, M. Harrison, and M. Worboys, *Fractured States: Small Pox, Public Health and Vaccination Policy in British India, 1800–1947* (Hyderabad, India: Orient Longman, 2005).
60. G.H. Bornside, "Waldemar Haffkine's Cholera Vaccines and the Ferran-Haffkine Priority Dispute," *Journal of the History of Medicine and Allied Sciences* 37 (1982): 399–422.
61. A. Cunningham, "Transforming Plague: The Laboratory and the Identity of the Infectious Disease," in *The Laboratory Revolution in Medicine*, ed. A. Cunningham and P. Williams (Cambridge: Cambridge University Press, 1993), 209–44.
62. A. Scull, *The Most Solitary of Afflictions: Madness and Society in Britain, 1700–1900* (New Haven, CT: Yale University Press, 1993).
63. R. Porter, *Madness: A Brief History* (Oxford: Oxford University Press, 2002).
64. C.E. Rosenberg, "Body and Mind in Nineteenth-Century Medicine: Some Clinical Origins of the Neurosis Concept," in *Explaining Epidemics and Other*

Studies in the History of Medicine (Cambridge: Cambridge University Press, 1992), 77.

65. N. Pemberton and M. Worboys, *Mad Dogs and Englishmen; Rabies in Britain, 1830–2000* (Basingstoke, UK: Palgrave, 2007), 11–19.
66. R. Porter, "Madness and its Institutions", in A. Wear, ed., *Medicine in Society* (Cambridge University Press, 1992), 277–301.
67. A. Digby, *Madness, Morality and Medicine: A Study of the York Retreat, 1796–1914* (Cambridge: Cambridge University Press, 1985).
68. A. Scull, "The Discovery of the Asylum Revisited: Lunacy Reform in the New American Republic," in *Madhouses, Mad-doctors, and Madmen: The Social History of Psychiatry in the Victorian era*, ed. A. Scull (London: Athlone Press, 1981), 144–65.
69. A. Digby, "Moral Treatment at the Retreat, 1796–1846," in *The Anatomy of Madness: Essays in the History of Psychiatry*, vol. 2, ed. W. Bynum, R. Porter, and Michael Shepherd (London: Tavistock, 1985), 52–72.
70. C. MacKenzie, *Psychiatry for the Rich: A History of Ticehurst Private Asylum, 1792–1917* (London: Routledge, 1992).
71. C. Bronte, *Jane Eyre* (1847, repr. London: Penguin, 2006), 249.
72. Scull, *The Most Solitary of Afflictions*, 175–232.
73. R. Cooter, *The Cultural Meaning of Popular Science: Phrenology and the Organization of Consent in Nineteenth-Century Britain* (Cambridge: Cambridge University Press, 1984); J. Van Wyhe, *Phrenology and the Origins of Victorian Scientific Naturalism* (Aldershot, UK: Ashgate, 2004).
74. A.D. Wood, "'The Fashionable Diseases': Women's Complaints and Their Treatment in Nineteenth-Century America," *Journal of Interdisciplinary History*, 4 (1973): 25–52.
75. C.E. Rosenberg, "The Place of George M. Beard in Nineteenth-Century Psychiatry," *Bulletin of the History of Medicine* 36 (1962): 245–59; Rosenberg, "Contested Boundaries: Psychiatry, Disease, and Diagnosis," *Perspectives in Biology and Medicine* 49 (2006): 407–24.
76. M. Gijswijt-Hofstra and R. Porter, *Cultures of Neurasthenia from Beard to the First World War* (Amsterdam: Rodopi, 2001).
77. Charlotte Perkins Gilman, *The Yellow Wallpaper* (London: Virago, 1981).
78. D. Pick, *Faces of Degeneration: A European Disorder, c.1848–c.1918* (Cambridge: Cambridge University Press, 1989).
79. B.A. Morel, *Traité des dégénérescences physiques, intellectuelles et morales de l'espèce humaine et des causes qui produisent ces variétés maladives* (London: H. Baillière, 1857).
80. P. J. Bowler, *The Non-Darwinian Revolution: Reinterpreting an Historical Myth* (Baltimore: Johns Hopkins University Press, 1988).
81. H-P. Söder, "Disease and Health as Contexts of Modernity: Max Nordau as a Critic of Fin-de-Siècle Modernism," *German Studies Review* 14 (1991): 473–87.
82. B. Haley, *The Healthy Body and Victorian Culture* (Cambridge, MA: Harvard University Press, 1978).
83. M. Worboys, "British Medicine and its Past at Queen Victoria's Jubilees and the 1900 Centennial," *Medical History* 45 (2001): 461–82.
84. W. F. Bynum, "An Experiment that Failed: Malaria Control at Mian Mir," *Parassitologia* 36 (1994): 107–120.

Chapter 3

1. Subscribers to this influential point of view include Jeffrey Weeks in *Sexuality and Its Discontents* (London: Routledge & Kegan Paul, 1985), 92; Joseph Bristow's *Sexuality* (London: Routledge, 1997); Michel Foucault's *The History of Sexuality*, vol. 1 (Random House, 1978); Arnold Davidson, *The Emergence of Sexuality* (Cambridge, MA: Harvard University Press, 2001); and Roy Porter, "Perversion in the Past," *Communicationes de Historia Artis Medicinae* 44, no. 1–4 (1999): 17–25. In *How to Do the History of Homosexuality* (Chicago: University of Chicago Press, 2002), David M. Halperin argues that Foucault never intended a rigid separation of acts from identities. Although Karl Maria Benkert coined the term "homosexual," it was Magnus Hirschfeld's *Homosexuality of Men and Women*, trans. Michael Lombardi-Nash (1913; repr. New York: Prometheus Books, 2000) that gave it legitimacy.
2. Davidson, *The Emergence of Sexuality*, 35.
3. Ibid., 57. I find Michel Foucault's more elastic sense of perversity as a discourse that "makes it possible to stitch together" categories of intent and malice along with medical, psychiatric, psychopathological, and psychological discourse more helpful. See his *Abnormal* (New York: Picador, 2004), 33.
4. I here take issue with Judith Butler's sense that identity is necessarily exclusionary because it is always normative. See her essay in *Feminist Contentions: A Philosophical Exchange* (New York: Routledge, 1995), 50. As Amanda Anderson tartly responds, Butler's subversive performances counter her assumption that the norm necessarily normalizes; cf. *The Way We Argue Now* (Princeton, NJ: Princeton University Press, 2006), 33.
5. On the structuring role of the closet, see Eve Sedgwick, *The Epistemology of the Closet* (Berkeley: University of California Press, 1990). Space considerations mandate the neglect of other perversions important to the nineteenth century: flagellation, sadism, masochism, and fetishism.
6. On the rise of function in the biological sciences and its implications for a perverted identity, see my "Romanticism and the Sciences of Perversion," *Wordsworth Circle* 36, no. 2 (Spring 2005): 43–48.
7. Deidre Coleman, *Romantic Colonization and British Anti-Slavery* (Cambridge: Cambridge University Press, 2005), 23.
8. My thinking about perversity is indebted to Sander Gilman's dialectical structuring of disease and to Hegel's *Phenomenology of the Spirit*, trans. A. V. Miller (Oxford: Oxford University Press, 1977), especially the section on "lordship and bondage," 111–19. See Gilman's *Disease and Representation: Images of Illness from Madness to AIDS* (Ithaca, NY: Cornell University Press, 1988), 1–2.
9. Harry Oosterhuis, *Stepchildren of Nature. Krafft-Ebing, Psychiatry, and the Making of Sexual Identity* (Chicago: University of Chicago Press, 2000), 13.
10. "Inversion" is a way of thinking about homosexuality that explains same-sex desire through the inversion of gender. For example, because a lesbian has a masculine psyche, she desires another female. As such, inversion makes desire intrinsically heterosexual.
11. Diana Fuss argues that "heterosexuality can never fully ignore the close psychical proximity of its terrifying (homo) sexual other any more than homosexuality can escape the equally insistent pressures of (hetero) sexual conformity." See her

introduction to *Inside/Out Lesbian Theories, Gay Theories*, ed. Diana Fuss (New York: Routledge, 1991), 3.

12. I am reminded here of Diane Ackerman's charge that we are often condescending to matter. One symptom of this is our tendency to see matter and essence as fixed. See her *An Alchemy of Mind* (New York: Scribner, 2004), 24.
13. In 1995, Leonore Tiefer argued that the real challenge to sexology was the need to better understand "the social construction of sexuality with regard to race and class as well as gender." Only in this way could it be proven that sex has never been a natural phenomenon. See her *Sex Is Not a Natural Act and Other Essays* (Boulder, CO: Westview Press, 1995), 205.
14. Recent studies have suggested that "middle class" and "bourgeoisie" were ideologies rather than material realities. Dror Wahrman has argued that for Britain, the middle class was a language, an idiom, and a belief. See his *Imagining the Middle Class* (Cambridge: Cambridge University Press, 1995). Sarah Maza argues similarly that the French bourgeoisie was a myth in *The Myth of the French Bourgeoisie* (Cambridge, MA: Harvard University Press, 2003). Her key point is that no one identified themselves as bourgeois because that was always someone else.
15. Malthus first published this in 1798 (ed. Philip Appelman [New York: Norton, 1976]); subsequent editions in 1803 and 1807 became increasingly punitive toward the poor. Malthus insisted that sexual desire was a human imperative.
16. See Wahrman, *Imagining the Middle Class*, 46. Wahrman argues that until 1833, middle class meant masculinity. After 1833, middle class became synonymous with middle-class virtues of hearth and home (403–407).
17. Thomas Laqueur and Londa Schiebinger have argued that around 1800, a one-sex model whereby women are inferior men, is contested by a two-sex model, whereby the sexes become incommensurate. See Thomas Laqueur's *Making Sex: Body and Gender from the Greeks to Freud* (Cambridge, MA: Harvard University Press, 1990) and L. Schiebinger, *Nature's Body* (Cambridge, MA: Harvard University Press, 1990).
18. Felicity Nussbaum, *Torrid Zones: Maternity, Sexuality, and Empire in Eighteenth-Century English Narratives* (Baltimore: Johns Hopkins University Press, 1995), 7–9.
19. On flirtation's powerful if underexamined role in Victorian culture, see Richard Kaye's *The Flirt's Tragedy* (Charlottesville: University of Virginia Press, 2002).
20. Rudi Bleys, *The Geography of Perversion, 1750–1918* (London: Cassell, 1996), 74.
21. Ibid., 73–74.
22. Gautam Chakravarty, *The Indian Mutiny and the British Imagination* (Cambridge: Cambridge University Press, 2005), 39. Chakravarty suggests that "at very least there was insufficient evidence of rape, and that the sipahis may not have violated captive women because of fears of caste defilement" (39–40).
23. Walter Kendrick, *The Secret Museum Pornography in Modern Culture* (New York: Viking, 1987), chap. one.
24. My phrasing here recalls Eve Sedgwick's claim that the discovery of Ancient Greece remade the human body as a site of utopian speculation. See her *Epistemology of the Closet*, 136.
25. Schiebinger, *Nature's Body*, 159.
26. Mungo Park, *Travels in the Interior Districts of Africa* (London: W. Bulmer and Co, 1799).

27. I borrow this point from Joanna de Groot, "'Sex' and 'Race': The Construction of Language and Image in the Nineteenth Century," in *Cultures of Empire*, ed. Catherine Hall (London: Routledge, 2000), 41.
28. Ann Laura Stoler, "Sexual Affronts and Racial Frontiers," in *Tensions of Empire: Colonial Cultures in a Bourgeois World*, ed. Frederick Cooper and Ann Laura Stoler (Berkeley: University of California Press; 1998), 205.
29. Lynn Hunt, "Pornography and the French Revolution," in *The Invention of Pornography*, ed. Lynn Hunt (New York: Zone Books, 1996), 308. On pornography's vexed role in the late nineteenth century in France, see Carolyn Dean, *The Frail Social Body* (Berkeley: University of California Press, 2000).
30. Dorinda Outram, *The Body and the French Revolution* (New Haven, CT: Yale University Press, 1989), 45.
31. Lawrence Birken, *Consuming Desire* (Ithaca, NY: Cornell University Press, 1988), 88–89.
32. Steven Angelides, *A History of Bisexuality* (Chicago: University of Chicago Press, 2001), 35.
33. Siobhan Somerville, *Queering the Color Line: Race and the Invention of Homosexuality in American Culture* (Durham, NC: Duke University Press, 2000), 29.
34. Frank Sulloway, *Freud, Biologist of the Mind: Beyond the Psychoanalytic Legend* (New York: Basic Books, 1979), 297
35. See Gary Dyer, "Thieves, Boxers, Sodomites, Poets: Being Flash to Byron's *Don Juan*," *PMLA* 116, no. 3 (May 2001): 562–78; and Louis Crompton, *Byron and Greek Love* (Berkeley: University of California Press, 1985).
36. Angelides, *A History of Bisexuality*, 47.
37. Randolph Trumbach, *Sex and the Gender Revolution: Heterosexuality and the Third Gender in Enlightenment London* (Chicago: University of Chicago Press, 1998), 8.
38. On the coding of lesbian relationships in the eighteenth century, see Emma Donoghue, *Passions Between Women: British Lesbian Culture 1668–1801* (London: Scarlet Press, 1993). On male friendship as an aperture into a gay male subjectivity beyond the acceptable gay emotions of grief and mourning, see Katherine O'Donnell and Michael O'Rourke, eds., *Love, Sex, Intimacy, and Friendship Between Men, 1550–1800* (Houndmills, UK: Palgrave, 2003).
39. Valerie Traub, "The Psychomorphology of the Clitoris," *Gay and Lesbian Quarterly* 2 (1995): 90–92.
40. Ann Laura Stoler, "Cultivating Bourgeois Bodies and Racial Selves," in *Cultures of Empire*, ed. Catherine Hall (New York: Routledge, 2000), 100.
41. Ibid., 99.
42. See Alan Bewell, *Romanticism and Colonial Disease* (Baltimore: Johns Hopkins University Press, 1999), 18.
43. Sander Gilman, *Difference and Pathology: Stereotypes of Sexuality, Race, and Madness* (Ithaca, NY: Cornell University Press, 1985), 200.
44. Theodric Romeyn Beck, *Elements of Medical Jurisprudence*, vol. 1 (Albany, NY: Webster and Skinners, 1823), 102–103.
45. William Acton, *The Contagious Diseases Act* (London: John Churchill & Sons, 1870), 30.
46. Stoler, "Cultivating Bourgeois Bodies and Racial Selves," 89.
47. *The Phoenix of Sodom, or the Vere Street Coterie* (London: Sold by J. Cook, 1813), 13 [BL Private Case Cup 364 p13]. Ivan Bloch cites this text in his *Sexual Life in England*, trans. M. Paul Eden (New York: Allied Book Company), 336–42.

48. Earl of Haddington, *Select Poems on Several Occasions by the Right Honorable The Earl of H*****N.* (London, 1824), 136 [British Library Private case PC 20a4].
49. The notion of lesbianism as a ghostly apparition is from Terry Castle, *The Apparitional Lesbian* (New York: Columbia University Press).
50. Sander Gilman, *Difference and Pathology*, 213.
51. Ibid., 191. On the complexities of the subfield of colonial science, see Londa Schiebinger, "Colonial Science," *Isis* (March 2005): 52–88.
52. Schiebinger, *Nature's Body*, 168–69 and 190.
53. See Londa Schiebinger, "Women and Gender in Science," in *The Cambridge History of Science*, vol. 4, ed. Roy Porter (Cambridge: Cambridge University Press, 2003), 200.
54. Graham Richards, *Mental Machinery: The Origins and Consequences of Psychological Ideas* (Baltimore: Johns Hopkins University Press), 211–12. Siobhan Somerville shows how African Bushwomen and their intermediate sexuality was used to racialize blackness in terms of peculiarity in *Queering the Color Line*, 26–27.
55. On the sexing, racializing, gendering, and classing of the body and the body's role in colonialism, see the essays in *Bodies in Contact Rethinking Colonial Encounters in World History*, ed. Tony Ballantyne and Antoinette Burton (Duke University Press, 1995).
56. On how the colonial was depicted as a child needing to be fathered, see Lisa Lowe, *Critical Terrains: French and British Orientalisms* (Ithaca, NY: Cornell University Press, 1991).
57. George L. Mosse, *Nationalism and Sexuality Respectability and Abnormal Sexuality in Modern Europe* (New York: Howard Fertig, 1985), 133, 11.
58. See Catherine Hall, *Civilizing Subjects Metropole and Colony in the English Imagination 1830–1867* (Chicago: University of Chicago Press, 2002), 178.
59. David Arnold, "Crisis and Contradiction in India's Public Health," in *The History of Public Health and the Modern State*, ed. Dorothy Porter (London: Wellcome Institute Series in the History of Medicine, 1994), 346.
60. *The Kama Sutra of Vatsyayana* was also translated by F. F. Arbuthnot (Cosmopoli: Kama Shastra Society of London and Benares, 1883). For more on this work, see Lisa Sigel *Governing Pleasures: Pornography and Social Change in England, 1815–1914* (New Brunswick, NJ: Rutgers University Press, 2002), chap. 2.
61. See, for example, Edward Moor's *Hindu Infanticide. An Account of the Measures adopted for suppressing the practice of the systematic murder by their parents of female infants* (London: J. Johnson, 1811).
62. Anna Clark, *Scandal: The Sexual Politics of the British Constitution* (Princeton, NJ: Princeton University Press, 2005), 88.
63. Ibid., 90.
64. Mrinalini Sinha, *Colonial Masculinity* (Manchester, UK: Manchester University Press, 1995), 18–19
65. Philippa Levine, "Venereal Disease, Prostitution, and the Politics of Empire: The Case of British India," *Journal of the History of Sexuality* 4 (1993–1994): 602.
66. Mosse, *Nationalism and Sexuality*, 84–85.
67. R. Hyam, *Empire and Sexuality: The British Experience* (Manchester, UK: Manchester University Press, 1990), 141; Mosse, *Nationalism and Sexuality*, 84.

68. "Homosocial" describes social bonds between persons of the same sex. Eve Sedgwick argues that because homosociality is maintained through homophobia, "homosocial" is part of an unbroken continuum between homosocial and homosexual; cf. *Between Men: English Literature and Male Homosocial Desire* (New York: Columbia University Press, 1985).
69. On this amendment and its significance to patterns of homoerotic writing, see Joseph Bristow, *Effeminate England* (New York: Columbia University Press, 1995). Richard Dellamora argues that a suppressed male panic induced Labouchère to add this last minute amendment; like Wilde, Labouchère frequented the West End and thus could not be distinguished from other bohemian middle-class men. See his *Masculine Desire; The Sexual Politics of Victorian Aestheticism* (Chapel Hill: University of North Carolina Press, 1990), 203.
70. Bewell, *Romanticism and Colonial Disease,* 25.
71. Roy Porter and Lesley Hall, *The Facts of Life: The Creation of Sexual Knowledge in Britain, 1650–1950* (New Haven, CT: Yale University Press, 1995), 133.
72. See Seth Koven, *Slumming: Sexual and Social Politics in Victorian London* (Princeton, NJ: Princeton University Press, 2004), chap. 5.
73. Helena Michie, *The Flesh Made Word: Female Figures and Women's Bodies* (New York: Oxford University Press, 1987), 71.
74. Sander Gilman, *Sexuality: An Illustrated History, Representing the Sexual in Medicine and Culture from the Middle Ages to the Age of AIDS* (New York: John Wiley & Sons, 1989), 298–303.
75. Judith Walkowitz, *Prostitution and Victorian Society* (Cambridge: Cambridge University Press, 1980), 45–46.
76. Stephen Garton, *Histories of Sexuality* (London: Equinox, 2004), 162.
77. Hyam, *Empire and Sexuality*, 3.
78. Cited in Iwan Bloch's *Anthropological Studies in the Strange Sexual Practices of All Races in All Ages* (New York: Anthropological Press, n.d.), 202. Originally published in German in 1902 under the title *Betrage zur Aetiologie der Psychopathia Sexualis, Erster Teil* [vol. 1] (Dresden, Germany: H. R. Dorn, 1902).
79. For an intriguing account of how Roman Catholicism could function as a closet for alternative sexualities, see Ellis Hanson, *Decadence and Catholicism* (Cambridge, MA: Harvard University Press, 1997). He cites this quotation from Hall on page 25.
80. On the Eastern origins of Aryan, see Tony Ballantyne, *Orientalism and Race Aryanism in the British Empire* (Houndmills, UK: Palgrave, 2002), 4–7, 44–46.
81. Ballantyne notes that "nigger" was in general use in the newspapers after the revolt. Ibid., 45.
82. Hyam, *Empire and Sexuality*, 157, 115.
83. Ibid., 115.
84. Ibid., 116, 123.
85. Ibid., 116.
86. Ann Laura Stoler "Sexual Affronts and Racial Frontiers," in *Tensions of Empire Colonial Cultures in a Bourgeois World,* ed. Frederick Cooper and Ann Laura Stoler (Berkeley: University of California Press, 1997), 199.
87. Hyam, *Empire and Sexuality*, 118–19.
88. Judith Walkowitz, *Prostitution and Victorian Society*, 49.
89. Mary Spongberg, *Feminizing Venereal Disease The Body of the Prostitute in Nineteenth-Century Medical Discourse* (New York: New York University Press, 1997), chap. 2.

90. Vern Bullough, *Science in the Bedroom* (New York: Basic Books, 1994), 4. For a good and brief overview of the historiography of sexology, see Chris Waters, "Sexology," in *Palgrave Advances in the Modern History of Sexuality,* ed. H. G. Cocks and Matt Houlbrook (Houndmills, UK: Palgrave Macmillan, 2006), 41–63. On national differences in sexological studies, see Robert A. Nye, "National Sexological Traditions," *Science in Context* 4, no. 2 (Autumn 1991): 387–406. Because the Napoleonic Code did not penalize sodomitical or homosexual acts, French sexologists had no need to beg for legal forbearance (392–93). Low birthrates led to French insistence upon reproductive fertility and familialist ethics; they measured perversions as departures from a procreative sexual norm (398).
91. See Siobhan Somerville's brilliant analysis of how homosexuality and race intersect in Ellis, in *Queering the Color Line*, chap. one. Ellis was named after a distant relative, Sir Henry Havelock, a British general during the Indian Mutiny. I cite the entry in the *Dictionary of National Biography* (Oxford: Clarendon Press, 2004).
92. Havelock Ellis and John Addington Symonds, *Sexual Inversion* (1897; repr. New York: Arno Press, 1975), x. The case history of his own wife, who had lesbian tendencies, is listed under "Miss H." See Bullough, *Science in the Bedroom,* 79. Lesley Hall reminds us that at the time of the publication and suppression of *Sexual Inversion,* British medical curriculum omitted any study of venereal disease, a major problem of the time. See her "British Fin de Siècle Sexology," in *New Sexual Agendas*, ed. Lynne Segal (New York: New York University Press, 1997), 9.
93. Ellis and Symonds, *Sexual Inversion*, 4.
94. Ibid., 4–9.
95. Ibid., 9–12.
96. Cited in Garton, *Histories of Sexuality*, 176.
97. Ellis and Symonds, *Sexual Inversion*, 104.
98. Oosterhuis, *Stepchildren of Nature*, 71. Oosterhuis argues that individuals who found their conditions described in *Psychopathia Sexualis* were not passive victims of forces beyond their control.
99. Ellis and Symonds, *Sexual Inversion*, 158.
100. Mrs. Elizabeth Osgood Goodrich Willard, *Sexology as the Philosophy of Life: Implying the Social Organization and Government* (Chicago, 1867), 16, 462. The *OED* (Clarendon Press, 1989) wrongly claims that the term "sexology" was first used in 1902.
101. Willard, *Sexology as the Philosophy of Life*, 308, 482.
102. Magnus Hirschfeld, *The Homosexuality of Men and Women,* 451. Hirschfeld innovated the use of a questionnaire in sexology. His very first question demands the age, sex, and race of the individual (290). In "Sexology," Chris Waters notes that Hirschfeld cast himself as a hero of reform (53). Eve Sedgwick claims in *The Epistemology of the Closet* that Hirschfeld equated cross-gender behavior with homosexuality (88–89); his claim that we are all bisexual complicates that claim considerably.
103. Magnus Hirschfeld, *Les Homosexuels de Berlin* (Paris: Librarie Medicale, 1908), 13, 14, 36, and throughout. The French *connaitre* implies familiarity with people. *Savoir* implies knowledge of things.
104. Jeffrey Weeks, *Sexuality and Its Discontents,* 68. Weeks is especially good on how the work of the sexologists could be put to competing and often surprising uses. Their definitions were not just passively accepted.
105. Oosterhuis, *Stepchildren of Nature*, 13.

106. Mrs. Elizabeth Osgood Goodrich Willard, *Sexology as the Philosophy of Life*, 7; Ellis and Symonds, *Sexual Inversion*, 159.
107. Ellis and Symonds, *Sexual Inversion*, vi.
108. Ibid., 4. In Britain, early sexology had marginal influence; *Sexual Inversion* was prosecuted as an obscene book.
109. Havelock Ellis, *Studies in the Psychology of Sex*, vol. 1 (Philadelphia: F. A. Davis Company, 1900), 1.
110. Iwan Bloch, *Sexual Life in England*, trans. William Forstern (1958; repr. London: Corgi Books, 1965), 26.
111. Ibid., 21.
112. Ibid., 25–26, 29.
113. Iwan Bloch, *The Sexual Life of Our Time*, trans. M. Eden Paul (1908; repr. New York: Allied Book Company, 1928), ix.
114. On the vexed role of biology in Freud's psychology, see Frank Sulloway, *Freud, Biologist of the Mind Beyond the Psychoanalytic Legend* (New York: Basic Books, 1979). The quotation from Freud is from the *Standard Edition of the Works of Sigmund Freud*, 24 vols., trans. James Strachey (London: Hogarth Press and the Institute of Psychoanalysis, 1953–74), 18:36.
115. Somerville, *Queering the Color Line*, 28–29.
116. Far from allowing sexuality to become coherent (pace Arnold Davidson), the instinct is based on incoherence.
117. Hirschfeld, *Homosexuality of Men and Women*, 454.
118. See Oosterhuis, *Stepchildren of Nature*, 59.
119. Ibid., 105–6.
120. Bloch, *Sexual Life of Our Time*, 546.
121. Richard von Krafft-Ebing, *Psychopathia Sexualis with Especial Reference to Contrary Sexual Instinct, a Medico-legal Study*, trans. Charles Gilbert Chaddock (Philadelphia: F. A. Davis, 1893), 370.
122. Oosterhuis, *Stepchildren of Nature*, 166–67. On the gaps between Krafft-Ebing's case histories and his typology of the metamorphosis from perversity to perversion, see Merl Storr, "Transformations: Subjects, Categories and Cures in Krafft-Ebing's Sexology" in *Sexology in Culture*, ed. Lucy Bland and Laura Doan (Chicago: University of Chicago Press, 1998), 15–17. Storr argues that these examples undermine Krafft-Ebing's sense of the exclusiveness of same sex desire that is needed for perversity to become perversion; I emphasize that these gaps undermine the very difference between perversity and perversion.
123. Krafft-Ebing, *Psychopathia Sexualis*, 255. On changes from edition to edition, see Oosterhuis, *Stepchildren of Nature*, 47–51. Bullough reminds us that he began to describe sex acts in Latin, and adopted more technical language when he discovered that he was being read by the general public (42).
124. Krafft-Ebing, *Psychopathia Sexualis*, 185.
125. Ibid., 187. Oosterhuis argues that these expressions "evaded a clear monocausal explanation" in *Stepchildren of Nature*, 105.
126. Ibid., 414, 419.
127. Ibid., 320.
128. Cited in Weeks, *Sexuality and Its Discontents*, 80.
129. Ellis, *Sexual Inversion*, 1.
130. Ibid., 22.
131. Ibid., 101.

132. Marie Stopes, *Married Love*, 2nd ed. (New York: Critic and Guide Company, 1918), 177.
133. Bloch, *Sexual Life in England*, 29.
134. Bloch, *Anthropological Studies*, 205.
135. Bloch, *Sexual Life of Our Time*, 489.
136. Bloch, *Anthropological Studies*, 215.
137. Ibid., 226.
138. Bloch, *The Sexual Life of Our Time*, 489.
139. Bloch, *Anthropological Studies*, 6, 244. Bloch wrote that "this general concept of the sexual anomalies as universal human, ubiquitous phenomena makes it necessary to recognize as physiologic much that previously had been regarded as pathologic" (6).
140. Bloch, *Sexual Life*, 489–90.
141. A. von Schrenck-Notzing, *Therapeutic Suggestion Psychopathia Sexualis*, trans. Charles Gilbert Chaddock (F. A. Davis, 1895), 147.
142. Schrenck-Notzing, *Therapeutic Suggestion Psychopathia Sexualis*, 155.
143. Ibid., 146.
144. Ellis, *Sexual Inversion*, 144.
145. Bloch, *Anthropological Studies*, 24.
146. Ibid., 6. Sigmund Freud, *Three Essays on the Theory of Sexuality*, trans. James Strachey (1905; repr. New York: Basic Books, 1962), 26.
147. Bullough, *Science in the Bedroom*, 69.
148. Hirschfeld, *Homosexuality of Men and Women*, 76.
149. Ibid., 455.

Chapter 4

1. On Steinach's life and career, see Harry Benjamin, "Eugen Steinach, 1861–1944: A Life of Research," *Scientific Monthly* 61 (1945): 427–42; Eugen Steinach, *Sex and Life: Forty Years of Biological and Medical Experiments* (New York: Viking, 1940); Chandak Sengoopta, *The Most Secret Quintessence of Life: Sex, Glands, and Hormones, 1850–1950* (Chicago: University of Chicago Press, 2006); and Heiko Stoff, *Ewige Jugend: Konzepte der Verjüngung vom späten 19. Jahrhundert bis ins Dritte Reich* (Cologne, Germany: Böhlau, 2004). All translations in this essay, unless otherwise mentioned, are my own.
2. Eugen Steinach, *Verjüngung durch experimentelle Neubelebung der alternden Pubertätsdrüse* (Berlin: Springer, 1920), 54–55. Since the operation of vasectomy had been performed for many years to relieve prostatic hypertrophy, some critics wondered why the rejuvenative potential of vasectomy had not been noticed before Steinach. See, for instance, Carl R. Moore, "The Regulation of Production and the Function of the Male Sex Hormone," *Journal of the American Medical Association* 97 (1931): 519. Surgeon Kenneth Walker in London remarked that the earlier vasectomists had not observed the rejuvenation of their patients simply because they had never taken sufficient care to prevent damage to the blood vessels and the nerves. See Kenneth M. Walker and J. Lumsden Cook, "Steinach's Rejuvenation Operation," *Lancet* 2 (February 1924): 224.
3. For reviews and excerpts of case reports, see Norman Haire, *Rejuvenation: The Work of Steinach, Voronoff, and Others* (London: Allen and Unwin, 1924); Peter

Schmidt, *The Theory and Practice of the Steinach Operation with a Report on One Hundred Cases* (London: Heinemann, 1924); and Peter Schmidt, *The Conquest of Old Age: Methods to Effect Rejuvenation and to Increase Functional Activity* (New York: Dutton, 1931).

4. "Gland Treatment Spreads in America," *New York Times*, April 8, 1923, sec. 9.
5. John de B. Limley, "Science Can Make You Grow Younger: Dr. Lorenz Illustrates from His Own Experience How a Simple Operation Renews Youthful Energy," *Liberty*, March 13, 1926, 22.
6. Harry Benjamin, "Preliminary Communication Regarding Steinach's Method of Rejuvenation," *New York Medical Journal* 114 (1921): 688.
7. Stephen Lock, " 'O That I Were Young Again': Yeats and the Steinach Operation," *British Medical Journal* 287 (1983): 1964–68; and Diana Wyndham, "Versemaking and Lovemaking—W. B. Yeats's 'Strange Second Puberty': Norman Haire and the Steinach Rejuvenation Operation," *Journal of the History of the Behavioral Sciences* 39 (2003): 25–50.
8. Steinach, *Sex and Life*, 170–71.
9. Edwin Clarke and L.S. Jacyna, *Nineteenth-Century Origins of Neuroscientific Concepts* (Berkeley: University of California Press, 1987), 312–13.
10. Paul F. Cranefield, "The Organic Physics of 1847 and the Biophysics of Today," *Journal of the History of Medicine and the Allied Sciences* 12 (1957): 407–23; and Laura Otis, "The Metaphoric Circuit: Organic and Technological Communication in the Nineteenth Century," *Journal of the History of Ideas* 63 (2002): 105–28.
11. On the telegraph, which was often compared with the nervous system, see Tom Standage, *The Victorian Internet* (London: Phoenix, 1998); and on the economic and cultural impact of electricity, Thomas P. Hughes, *Networks of Power: Electrification in Western Society, 1880–1930* (Baltimore: Johns Hopkins University Press, 1993).
12. George M. Beard, *American Nervousness: Its Causes and Consequences* (New York: Putnam's, 1881); and Janet Oppenheim, *"Shattered Nerves": Doctors, Patients and Depression in Victorian England* (New York: Oxford University Press, 1991), 79–180.
13. George M. Beard and A.D. Rockwell, *A Practical Treatise on the Medical and Surgical Uses of Electricity* (New York: William Wood, 1871); and Paula Bertucci and Giuliano Pancaldi, eds., *Electric Bodies: Episodes in the History of Medical Electricity* (Bologna, Italy: Università di Bologna, 2001).
14. Anson Rabinbach, *The Human Motor: Energy, Fatigue, and the Origins of Modernity* (Berkeley: University of California Press, 1992).
15. Rudolf Virchow, "Der puerperale Zustand: Das Weib und die Zelle," in *Gesammelte Abhandlungen zur wissenschaftlichen Medizin* (Frankfurt am Main, Germany: Meidinger, 1856), 735–79.
16. On Virchow's opposition to humoralism, see L.J. Rather, "Virchow's Review of Rokitansky's *Handbuch* in the *Preussische Medizinal-Zeitung*, Dec. 1846," *Clio Medica* 4 (1969): 127–40. My thanks to Harry Marks for this reference.
17. Russell C. Maulitz, "The Pathological Tradition," in *Companion Encyclopaedia of the History of Medicine*, vol. 1, ed. W. F. Bynum and Roy Porter (London: Routledge, 1993), 169–89.
18. Virchow, "Der puerperale Zustand," 739–40, 746.
19. Ibid., 747.

20. See Benedikt Stilling, *Physiologisch-pathologische und medicinisch-praktische Untersuchung über Spinal-Irritation* (Leipzig, Germany: Wiegand, 1840); and Moritz Romberg, *Lehrbuch der Nervenkrankheiten des Menschen*, vol. 1, pt. 2, 2nd ed. (Berlin: Duncker, 1851), 198–223, especially 210–11. See also Mark Micale, "Hysteria Male/Hysteria Female: Reflections on Comparative Gender Construction in Nineteenth-Century France and Britain," in *Science and Sensibility: Gender and Scientific Enquiry, 1780–1945*, ed. Marina Benjamin (Oxford: Blackwell, 1991), 200–39.
21. William Yarrell, "On the Influence of the Sexual Organ in Modifying External Character," *Journal of the Proceedings of the Linnean Society (Zoology)* 1 (1857): 81; Charles Darwin, *The Variation of Animals and Plants under Domestication*, vol. 2, 2nd ed. (New York: Appleton, 1897), 26–27; and August Weismann, *The Germ-Plasm: A Theory of Heredity*, trans. W. Newton Parker and Harriet Rönnfeldt (New York: Scribner's, 1893), 358–59.
22. See H. Ribbert, "Ueber die compensatorische Hypertrophie der Geschlechtsdrüsen," *Archiv der pathologische Anatomie und Physiologie* 120 (1890): 271; S. Samuel, "Das Gewebswachsthum bei Störungen der Innervation," *Archiv der pathologische Anatomie und Physiologie* 113 (1888): 313–14; and S. Samuel, "Trophoneurosen," in *Real-Encyclopädie der gesammten Heilkunde*, vol. 20, ed. Albert Eulenburg (Vienna: Urban & Schwarzenberg, 1890), 223–24.
23. A.A. Berthold, "The Transplantation of Testes," trans. D. P. Quiring, *Bulletin of the History of Medicine* 16 (1944): 399–401.
24. F. Goltz and A. Freusberg, "Ueber den Einfluss des Nervensystems auf die Vorgänge während der Schwangerschaft und des Gebärakts," *Archiv für gesamte Physiologie* 9 (1874): 552–65. Goltz—a student of Hermann von Helmholtz and teacher of Jacques Loeb—was better known in his time for his experimental investigation of reflex phenomena in frogs and for his critical stance on the theory of cerebral localization. For details, see Philip J. Pauly, "The Political Structure of the Brain: Cerebral Localization in Bismarckian Germany," *International Journal of Neuroscience* 21 (1983): 145–50.
25. H.H. Simmer, "Bilaterale Oophorektomie der Frau im späten 19. Jahrhundert: Zum methodologischen Wert der Kastration für die Entdeckung ovarieller Hormone," *Geburtshilfe und Frauenheilkunde* 43 (Sonderheft 1983): 57.
26. Michael J. Aminoff, *Brown-Séquard: A Visionary of Science* (New York: Raven Press, 1993), 163–70; and R.E. Billingham and W.B. Neaves, "Paratransplantation and Tissue Therapy," *Perspectives in Biology and Medicine* 22 (1979): 320–32.
27. On the spermatic economy, see Ben Barker-Benfield, "The Spermatic Economy: A Nineteenth Century View of Sexuality," *Feminist Studies* 1 (1972): 45–74.
28. Merriley Borell, "Brown-Séquard's Organotherapy and its Appearance in America at the End of the Nineteenth Century," *Bulletin of the History of Medicine* 50 (1976): 310; Borell, "Origins of the Hormone Concept: Internal Secretions and Physiological Research, 1889–1905" (PhD diss., Yale University, 1976).
29. Benjamin Harrow, *Glands in Health and Disease* (New York: Dutton, 1922), 185–86.
30. Merriley Borell, "Organotherapy and the Emergence of Reproductive Endocrinology," *Journal of the History of Biology* 18 (1985): 1–30. See also V. C. Medvei, *The History of Clinical Endocrinology* (Carnforth, UK: Parthenon, 1993).
31. F. Leydig, "Zur Anatomie der männlichen Geschlechtsorgane und Analdrüsen der Säugethiere," *Zeitschrift für wissenschaftliche Zoologie* 2 (1850): 1–57.

32. See, for example, Julius Tandler and Siegfried Grosz, "Untersuchungen an Skopzen," *Wiener klinische Wochenschrift* 21 (1908): 277–82.
33. Bouin and Ancel published extensively on the subject between 1903 and 1926. See the following examples: P. Ancel and P. Bouin, "Recherches sur les cellules interstitielles du testicule des mammifères," *Archives de zoologie expérimentale et générale* 4th ser., 1 (1903): 437–523; and Ancel and Bouin, "Les cellules séminales ont-elles une action sur les caractères sexuels? Discussion et nouvelles recherches," *Comptes rendus hebdomadaires des séances et mémoires de la Societé de Biologie (Paris)* 75 (1923): 175–78. See also Marc Klein, "Sur les interférences des sciences fondamentales et de la clinique dans l'essor de l'endocrinologie sexuelle," *Clio Medica* 8 (1973): 31–52.
34. E. Steinach, "Geschlechtstrieb und echt sekundäre Geschlechtsmerkmale als Folge der innersekretorischen Funktion der Keimdrüse," *Zentralblatt für Physiologie* 24 (1910): 551–66. On the persistence of sexuality in human eunuchs and castrates, see Alexander Lipschütz, *Internal Secretions of the Sex Glands: The Problem of the "Puberty Gland"* (Cambridge: Heffer, 1924), 17–18.
35. E. Steinach, "Willkürliche Umwandlung von Säugetier-Männchen in Tiere mit ausgeprägt weiblichen Geschlechtscharakteren und weiblicher Psyche," *Pflügers Archiv für die gesamte Physiologie* 144 (1912): 75. Many orthodox scientists opposed this claim, and the resulting disputes filled many pages in the medical journals of the time; it was only after the coming of electron microscopy that the question was settled definitively in favor of Steinach. For details, see Sengoopta, *The Most Secret Quintessence of Life*, 240–41.
36. See Artur Biedl, *Innere Sekretion: Ihre physiologischen Grundlagen und ihre Bedeutung für die Pathologie*, vol. 2, 2nd ed. (Berlin: Urban & Schwarzenberg, 1913), 199–343; and Francis H.A. Marshall, *The Physiology of Reproduction*, 2nd ed. (London: Longmans, 1922), 320–92. For a recent endorsement of Steinach's historical importance, see Anne Fausto-Sterling, *Sexing the Body: Gender Politics and the Construction of Sexuality* (New York: Basic Books, 1999).
37. For numerous examples, see Lipschütz, *The Internal Secretions of the Sex Glands.*
38. Steinach was nominated for the prize in 1921 by J.H. Zaaijer (Leiden); in 1922 by Y. Sakaki (Fukuoka); in 1927 by L. Haberlandt (Innsbruck); in 1930 by H.H. Meyer, S. Klein, and E. Pick (Vienna); in 1934 by A. Durig (Vienna); and in 1938 by J. Bock and colleagues. (Copenhagen). My thanks to the Nobel Committee for supplying me with this information.
39. See Steinach's letter to Harry Benjamin dated February 28, 1923, in Eugen Steinach–Harry Benjamin Correspondence, The New York Academy of Medicine Historical Collections.
40. Knud Sand, "Die Physiologie des Hodens," *Handbuch der inneren Sekretion*, vol. 2, pt. 2, ed. Max Hirsch (Leipzig, Germany: Kabbitzsch, 1933), 2073–74, 2109.
41. Steinach, *Sex and Life*, 18; and generally on the preoccupation with glands, Sengoopta, *Most Secret Quintessence of Life*, 69–75.
42. Chandak Sengoopta, "Glandular Politics: Experimental Biology, Clinical Medicine, and Homosexual Emancipation in Fin-de-Siècle Central Europe," *Isis* 89 (1998): 445–73; and Heiko Stoff, "Degenerierte Nervenkörper und regenerierte Hormonkörper: Eine kurze Geschichte der Verbesserung des Menschen zu Beginn des 20. Jahrhunderts," *Historische Anthropologie: Kultur, Gesellschaft, Alltag* 11 (2003): 225–39.

43. For details on Steinach, Voronoff, and the cultural impact of rejuvenation, see Sengoopta, *Most Secret Quintessence of Life*, 69–115. A leading expert on Cummings argues that "E. E. Cummings" is to be preferred to the uncapitalized version. See Norman Friedman, "Not e. e. cummings," *Spring* 1 (1992): 114–21 http://www.gvsu.edu/english/cummings/caps.htm.
44. Vivian Nutton, "Humoralism," in *Companion Encyclopedia of the History of Medicine*, vol. 1, ed. W. F. Bynum and Roy Porter (London: Routledge, 1993), 281–91.
45. See Toby Gelfand, *Professionalizing Modern Medicine: Paris Surgeons and Medical Science and Institutions in the Eighteenth Century* (Westport, CT: Greenwood Press, 1980); John Harley Warner, *Against the Spirit of System: The French Impulse in Nineteenth-Century American Medicine* (Princeton, NJ: Princeton University Press, 1998); and Elizabeth Haigh, *Xavier Bichat and the Medical Theory of the Eighteenth Century* (London: Wellcome Institute for the History of Medicine, 1984).
46. Michael Sappol, *A Traffic of Dead Bodies: Anatomy and Embodied Social Identity in Nineteenth-Century America* (Princeton, NJ: Princeton University Press, 2002), 53, 209–10, 270–71.
47. Brian Bracegirdle, "The Microscopical Tradition," in *Companion Encyclopedia of the History of Medicine*, vol. 1, ed. W. F. Bynum and Roy Porter, 102–119.
48. W. H. McMenemy, "Cellular Pathology," in *Medicine and Science in the 1860s*, ed. F.N.L. Poynter (London: Wellcome Institute for the History of Medicine, 1968), 13–43.
49. Paul Weindling, "Theories of the Cell State in Imperial Germany," in *Biology, Medicine and Society 1840–1940*, ed. Charles Webster (Cambridge: Cambridge University Press, 1981), 99–155.
50. Schäfer added, however, that the nervous system remained crucial in bioregulation—the glands acted in concert with or parallel to the nerves. See E. Sharpey-Schafer, "Endocrine Physiology," *Irish Journal of Medical Science* 6th ser., no. 69 (September 1931): 484. See also Merriley Borell, "Setting the Standards for a New Science: Edward Schäfer and Endocrinology," *Medical History* 22 (1978): 282–90.
51. See Sengoopta, *Most Secret Quintessence of Life* for an extended discussion of these factors.
52. Paul Weindling, *Health, Race, and German Politics Between National Unification and Nazism 1870–1945* (Cambridge: Cambridge University Press, 1989), 325.
53. Arne Hessenbruch, "A Brief History of X-Rays," *Endeavour* 26 (2002): 138.
54. See Bettyann Holtzmann Kevles, *Naked to the Bone: Medical Imaging in the Twentieth Century* (New Brunswick, NJ: Rutgers University Press, 1997), 39–43, 70–80; and Rebecca Herzig, "Removing Roots: North American 'Hiroshima Maidens' and the X-Ray," *Technology and Culture* 40 (1999): 723–45.
55. Lisa Cartwright, *Screening the Body: Tracing Medicine's Visual Culture* (Minneapolis: University of Minnesota Press, 1995), 110.

Chapter 5

1. Herbert Spencer, Essay, *The Leader*, March 20, 1852, http://www.clarehall.cam.ac.uk/userpages/djhc2/contbib/spencer.htm.

2. There is a good deal of slippage between the terms "English" and "British." Although Britain referred to the entire United Kingdom—England, Wales, Scotland, and Ireland—the term British as an adjective is often used loosely to refer to values and traits most closely identified with southern England. Although "English" and "British" could often be used interchangeably, for example, "Irish" or "Scottish" could not be so used: those terms designated specific differences from what was perceived as the dominant British culture.
3. M. Jeanne Peterson notes that thirty-three medical journals appeared between 1640 and 1749, whereas between 1800 and 1899, 458 medicine-related periodicals appeared in Britain (29). This proliferation was common across fields of expert knowledge; literary and news periodicals exploded even more dramatically. M. Jeanne Peterson, "Medicine," in *Victorian Periodicals and Victorian Society*, ed. J. Don Vann and Rosemary T. VanArsdel (Toronto: University of Toronto Press, 1994), 22–44.
4. Society for the Diffusion of Useful Knowledge (SDUK), *The Working Man's Companion. The Physician. 1. The Cholera* (London: Charles Knight, 1832), 160.
5. Ibid., 164–65.
6. William Farr, *Vital Statistics. A Memorial Volume of Selections from the Reports and Writings of William Farr*, intro. Mervyn Susser and Abraham Adelstein (Metuchen, NJ: Scarecrow Press, 1975), 166.
7. See Norbert Elias, *The Civilizing Process: The Development of Manners, Changes in the Code of Conduct and Feeling in Early Modern Times*, trans. Edmund Jephcott (New York: Urizen Books, 1978).
8. See, for example, Thomas Laqueur, *Making Sex: Body and Gender from the Greeks to Freud* (Cambridge, MA: Harvard University Press, 1990).
9. Mikhail Bakhtin, *Rabelais and His World* (Bloomington: Indiana University Press, 1984).
10. See Philippa Levine, *Prostitution, Race and Politics: Policing Venereal Disease in the British Empire* (London: Routledge, 2003).
11. Nadja Durbach, *Bodily Matters: The Anti-Vaccination Movement in England, 1853–1907* (Durham, NC: Duke University Press, 2005), 56.
12. Ibid., 6.
13. Ibid., 7.
14. Alfred Power, *Sanitary Rhymes . . . Personal Precautions against Cholera, and all Kinds of Fever, etc.* (London: T. Richards, 1871).
15. George Meredith, *The Ordeal of Richard Feverel* (New York: Penguin, 1999), 390.
16. Edwin Chadwick and Thomas Southwood Smith, "Report," *Irish University Press Parliamentary Papers 1849, Cholera I,* August 14, 1850, 608–9.
17. William Farr, "Influence of Elevation on Fatality of Cholera," *Journal of the Royal Statistical Society* 15 (1852): 174.
18. Ibid., 178.
19. Quoted in George Godwin, *Town Swamps and Social Bridges: The Sequel to a Glance at the Homes of the Thousands* (London: Routledge, Warnes and Routledge, 1859), 21.
20. "The Public Health Bill," *The Times,* (July 28, 1866), 7–8.
21. Herbert Spencer, *Social Statics*, 1851, http://oll.libertyfund.org/Texts/LFBooks/Spencer0236/SocialStatics/0331_Bk.html#LF_BK0331in01ch01.
22. For extensive discussions of bathing, sea bathing, and spas in this period, see the following: Charles F. Mullett, *Public Baths and Health in England, 16th–18th Century* (Baltimore: Johns Hopkins Press, 1946); Pamela Horn, *Pleasures and Pastimes*

in Victorian Britain (Stroud, UK: Sutton, 1999); John Lowerson and John Myerscough, *Time to Spare in Victorian England* (Hassocks, UK: Harvester Press, 1977); Malcolm R. Shifrin, *Victorian Turkish Baths*. www.victorianturkishbath.org; and Thomas Wright, *Clean and Decent: The Fascinating History of the Bathroom and the Water Closet* (New York: Viking Press, 1960).

23. In, *Time to Spare*, 31, Lowerson and Myerscough cite a fare of 9 shillings for a full-day round trip to Brighton for a family of four and a high point of 30,000 trippers at a time in the 1860s.
24. George Elson, *The Last of the Climbing Boys: An Autobiography*. London: John Long, 1900, quoted in Malcolm Shifrin, "George Elson: Chimney Sweep and Shampooer," Victorian Turkish Baths: Their Origin, Development, and Gradual Decline Web site, http://www.victorianturkishbath.org/_4PERSONS/AtoZPerson/Elson/ElsonEng.htm.
25. Malcolm Shifrin, "Rate Card for J H Faulkner's Hotels and Turkish Baths in London and Hastings," Victorian Turkish Baths: Their Origin, Development, and Gradual Decline Web site, http://www.victorianturkishbath.org/6directory/atozestab/england/hastwhite/aapix/faulknerchargecard_w.htm.
26. See Wright, *Clean and Decent,* for a detailed study.
27. Malcolm, Shifrin, "Cabinet and Lamp Baths: The DIY Turkish Bath," Victorian Turkish Baths: Their Origin, Development, and Gradual Decline Web site, http://www.victorianturkishbath.org/3topics/atozarts/cabinetsf.htm.
28. Charles Kingsley, *The Water-Babies* (Oxford: Oxford University Press, 1995).
29. Ouida, *Under Two Flags* (Oxford: Oxford University Press, 1995), 178.
30. Anthony Trollope, "The Turkish Bath," in *An Editor's Tales* (New York: Penguin, 1993).
31. Alan Bewell, *Romanticism and Colonial Disease* (Baltimore: Johns Hopkins University Press, 1999), 49, 270, and passim.
32. Bewell, *Romanticism and Colonial Disease*, 245.
33. Mark Harrison, *Climates and Constitutions* (New Delhi: Oxford University Press, 1999), 3.
34. Mark Harrison charts the admiration of Whitelaw Ainslie and others, who, in the 1820s, studied Indian indigenous medicine and found much to respect in it. Yet, by the mid-century, the British distrusted the local as quacks who retarded the progress of scientific medicine. See *Public Health in British India: Anglo-Indian Preventive Medicine, 1859–1914* (Cambridge: Cambridge University Press, 1994), 16, 41, 53. Harrison attributes some of this change to the Indian Rebellion of 1857 and some to the changes in European medicine from a humoral system more closely resembling Indian schemas to the scientific medical model. David Arnold, however locates the shift earlier, in 1835, and notes that it was not so much a significant change in attitude as the further manifestation of an existing tendency to privilege Western medicine while taking what was useful from native traditions. See *Colonizing the Body: State Medicine and Epidemic Disease in Nineteenth-Century India* (Berkeley: University of California Press, 1993), chap. 2.
35. George Johnson, *On Cholera and Choleraic Diarrhoea; Their Nature, Cause and Treatment. Two Lectures, Delivered at The Church Missionary College, Islington* (London, John Churchill and Sons, 1870), 26–27.
36. In Mrinalini Sinha, *The 'Manly Englishman' and the 'Effeminate Bengali' in the Late Nineteenth Century* (Manchester, UK: Manchester University Press; New York: St. Martin's Press, 1995), 15.
37. Joseph Furphy, *Such is Life*, http://www.gutenberg.org/etext/3470.

38. Thomas Hughes, *Tom Brown's Schooldays* (Oxford: Oxford University Press, 1999), 34–36.
39. Richard J. Holt, *Sport and the British: A Modern History* (Oxford: Clarendon Press, 1989), 71.
40. For a detailed study of athletics and the male body, especially on the Continent, see George L. Mosse, *The Image of Man: The Creation of Modern Masculinity* (Oxford: Oxford University Press, 1996).
41. For extensive discussions of the role of sport at this time, see the following: Peter Bailey, *Leisure and Class in Victorian England: Rational Recreation and the Contest for Control, 1830–1885* (London: Routledge and Kegan Paul, 1978); Gary Cross, *A Social History of Leisure since 1600* (State College, PA: Venture Publishing, 1990); Holt, *Sport and the British*; Horn, *Pleasures and Pastimes in Victorian England*; Lowerson and Myerscough, *Time to Spare in Victorian England*; J.A. Mangan, *Athleticism in the Victorian and Edwardian Public School* (London: Frank Cass, 2000) and *The Games Ethic and Imperialism: Aspects of the Diffusion of an Ideal* (New York: Viking, 1986).
42. Hughes, *Tom Brown's Schooldays*, 97.
43. Ibid., 98.
44. Thomas E. Jordan points out that thirteen-year-old London street boys measured by the Marine Society from 1753 to 1780 averaged 51.4 inches in height—ten inches below the height of thirteen year olds in London in the 1960s. In 1835, an Edinburgh physician recommended that a child be forty-eight inches tall to be considered equivalent to a thirteen year old and therefore eligible for factory work; in the 1990s, that was the average height of a British seven year old. *The Degeneracy Crisis and Victorian Youth* (Albany: State University of New York Press, 1993), 6.
45. Jordan, *The Degeneracy Crisis and Victorian Youth*, 23.
46. Hughes, *Tom Brown's Schooldays*, 104.
47. Bailey, *Leisure and Class in Victorian England*, 143–45.
48. Richard J. Holt, "Working Class Football and the City: The Problem of Continuity," *Proceedings of the Third Annual Conference of the British Society of Sports History* (n.p.: 1985), 68.
49. R.J. Holt, "Working Class Football and the City," 72.
50. This is not to say that there were no distinctions within Britain between English and other, especially Celtic, sports. Rugby was championed as a Welsh game, and various traditional sports survived in the north and in Scotland that were specifically championed by their players and fans as culturally distinctive. See Holt, *Sport and the British*, for the fullest discussion of these differences. However, there was a distinct tendency for Britishness to be subsumed within an English ideal, especially when Britishness was being communicated to an outside audience.
51. On polo, see Holt, *Sport and the British*, 215–217; for detailed discussions of the role of empire and British sport, and for British schools in India, see both Holt, *Sport and the British* and Mangan, *The Games Ethic and Imperialism.*
52. Mangan, *The Games Ethic and Imperialism*, 176.

Chapter 6

Research for this chapter was made possible in part by an Australian Research Council discovery grant (DP0558404) in 2005 and 2006 and a three-month research stay at the

Max Planck Institute for the History of Science (Department of Hans Jörg Rheinberger) in Berlin. I would also like to thank my colleagues Barbara Caine and Maria Nugent for comments on earlier versions of this chapter.

1. Carl Heinrich Stratz, *Die Schönheit des weiblichen Körpers*, 2nd ed. (Stuttgart, Germany: Ferdinand Enke, 1899), 13. All translations from German are my own.
2. Christopher E. Forth, *The Dreyfus Affair and the Crisis of French Manhood* (Baltimore: Johns Hopkins University Press, 2004), 178 and 184.
3. Lois W. Banner, *American Beauty* (New York: Knopf, 1983), 5, 154–59, and 226–40.
4. Michael Hau, *The Cult of Health and Beauty in Germany, 1890–1930* (Chicago: University of Chicago Press, 2003), 173ff; Banner, *American Beauty*, 278ff.
5. On Philhellenism in Germany, see: Esther S. Sünderhauf, *Griechensehnsucht und Kulturkritik. Die deutsche Rezeption von Winkelmanns Antikenideal* (Berlin: Akademie-Verlag, 2004); and Suzanne Marchand, *Down from Olympus: Archaeology and Philhellenism in Germany, 1750–1970* (Princeton, NJ: Princeton University Press, 1996). On France and Britain, see Athena S. Leoussi, *Nationalism and Classicism: The Classical Body as National Symbol in Nineteenth-Century England and France* (London: Macmillan, 1998).
6. Joan B. Landes, *Visualizing the Nation: Gender, Representation, and Revolution in Eighteenth-Century France* (Ithaca, NY: Cornell University Press, 2001), 50ff and 82; Antoine de Baecque, *The Body Politic: Corporeal Metaphor in Revolutionary France* (Stanford, CA: Stanford University Press, 1997), 309–23.
7. Adolf Zeising, *Neue Lehre von den Proportionen des menschlichen Körpers* (Leipzig, Germany: Rudolph Weigel, 1854), V.
8. Ibid., V.
9. Ibid., Xff. On the aesthetic aspects of Humboldt's *Cosmos*, see Andreas Daum, *Wissenschaftspopularisierung im 19. Jahrhundert. Bürgerliche Kultur, naturwissenschaftliche Bildung und die deutsche Öffentlichkeit, 1848–1914* (Munich: Oldenbourg, 1998), 269–79.
10. Robert Richards, *The Romantic Conception of Life. Science and Philosophy in the Age of Goethe* (Chicago: University of Chicago Press, 2002), 455, 470, and 477f.
11. Zeising, *Neue Lehre*, VII and 160.
12. Christoph Kockerbeck, *Die Schönheit des Lebendigen. Ästhetische Naturwahrnehmung im 19. Jahrhundert* (Vienna: Böhlau, 1997), 24ff. This was also quite different from the work of Karl Rosenkranz, who developed an "Aesthetic of the Ugly" from a dialectical relationship between the "beautiful" and "ugly" based on Hegel's philosophy. See Karl Rosenkranz, *Ästhetik des Hässlichen* (1853; repr. Darmstadt, Germany: Wissenschaftliche Buchgesellschaft, 1973).
13. Gottfried Schadow, *Polyclet oder von den Maassen des Menschen nach dem Geschlechte und Alter*, 5th ed. (Berlin: Ernst Wasmuth 1886). See also: Peter Gerlach, "Schadow's Poyclet (1834). *Die Bedeutung der Vermessung antiker Statuen für die Proportionslehre*," in *Beiträge zur Theorie der Künste im 19. Jahrhundert*, vol. 1, ed. Helmut Koopmann and J. Adolf Schmoll (Frankfurt, Germany: Vittorio Klostermann, 1971), 161–92. For a general survey of such systems by the same author, see *Proportionen, Körper, Leben. Quellen; Entwürfe, Kontroversen* (Cologne. Germany: Apex, 1990), especially 3–50.
14. Leoussi, *Nationalism and Classicism*, 42.

15. Carl Gustav Carus, *Symbolik der menschlichen Gestalt: Ein Handbuch der Menschenkenntnis* [Symbolism of Human Form: A Handbook of Knowledge of Human Nature], 2nd ed. (1858; repr. Hildesheim, Germany: Georg Olms, 1962), 55–60.
16. Ibid., 71.
17. Theodore M. Porter, *The Rise of Statistical Thinking 1820–1920* (Princeton, NJ: Princeton University Press, 1986), citations 101ff.
18. Carus, *Symbolik der menschlichen Gestalt*, 57–60.
19. Ibid., 28–36.
20. Ibid., 100ff.
21. Ibid., 102ff.
22. This four-partite racial taxonomy was based on older natural philosophical ideas of the quadruplicity of nature, according to which natural phenomena and humans had four different kinds of relations to the sun as the source of life. Day, night, morning, and evening dawn were analogous to different psychological tribal characteristics. Reinhard Mocek, "Der Naturphilosoph Carl Gustav Carus," in *Dresdner Hefte* 7 (1989), 15–29, especially 16ff.
23. Carus, *Symbolik der menschlichen Gestalt*, 104f. On Morton, see: Stephen Jay Gould, *The Mismeasure of Man* (New York: Norton, 1981).
24. Carus, *Symbolik der menschlichen Gestalt*, 379.
25. Ibid., 398ff.
26. Lucy Hartley, "A Science of Beauty? Femininity, Fitness and the Nineteenth-Century Physiognomic Tradition," *Women: A Cultural Review* 12, no. 1 (2001): 19–34, especially 27–31.
27. Eveleen Richards, "Darwin and the Descent of Woman," in *The Wider Domain of Evolutionary Thought*, ed. David Oldroyd and Ian Langham, 57–111 (Dordrecht, Netherlands: Reidel, 1983), 65ff.
28. Ernst Haeckel, *Kunstformen der Natur* [Art Forms of Nature] (Leipzig, Germany: Verl. des Bibliogr. Inst., 1899–1904).
29. Kockerbeck, *Schönheit des Lebendigen*, 82ff and 90–99.
30. Leoussi, *Nationalism and Classicism*, 35–55.
31. On Fritsch, see Annette Lewerentz, "Der Mediziner Gustav Fritsch als Fotograf. Dokumentation seiner anthropologisch-ethnographischen Untersuchungen in Fotografien der Berliner Gesellschaft für Anthropologie, Ethnologie und Urgeschichte," *Baessler-Archiv* Neue Folge 48 (2000): 271–309.
32. Gustav Fritsch, *Unsere Körperform im Lichte der modernen Kunst* (Berlin: Carl Habel, 1893), 4; Fritsch, *Ne sutor supra crepidam* (Berlin: Carl Habel, 1894), 21; Maria Makela, *The Munich Secession: Art and Artists in Turn-of-the-Century Munich* (Princeton, NJ: Princeton University Press, 1990), 104–13.
33. Fritsch, *Unsere Körperform*, citations on 4 and 17.
34. Gustav Fritsch, "Die graphischen Methoden zur Bestimmung der Verhältnisse des menschlichen Körpers," *Zeitschrift für Ethnologie* 27 (1895): 172–88, citation on 173. Michael Hau, "The Holistic Gaze in German Medicine, 1890–1930," *Bulletin of the History of Medicine* 74 (2000): 495–524, especially 498–506.
35. Fritsch, "Graphischen Methoden zur Bestimmung," citation on 174.
36. Gustav Fritsch, *Die Gestalt des Menschen: Mit Benutzung der Werke von E. Harless und C. Schmidt für Künstler und Anthropologen dargestellt* (Stuttgart, Germany: Paul Neff, 1899), citation on 147.
37. Paul Schultze-Naumburg, *Die Kultur des weiblichen Körpers als Grundlage der Frauenkleidung* (Jena, Germany: Diederichs, 1905), 21 and 61ff.

38. Anna Fischer-Dückelmann, *Das Geschlechtsleben des Weibes. Eine pysiologisch-soziale Studie mit ärztlichen Ratschlägen*, 13th. ed. (Berlin: Hugo Bermühler, 1908), 145f. Fischer-Dückelmann, *Die Frau als Hausärztin*, 2nd special ed. (Stuttgart, Germany: Süddeutsches Verlags-Institut, 1905), 144 and 151ff; Hau, *Cult of Health and Beauty*, 70–75.
39. Fischer-Dückelmann, *Frau als Hausärztin*, 198.
40. Hau, *Cult of Health and Beauty*, 44–54. There are a number of recent books on German and German-Jewish physical culture and nudism that explore the significance of embodied aesthetic ideals. See, for example: Todd Samuel Presner, *Muscular Judaism: The Jewish Body and the Politics of Regeneration* (London: Routledge, 2007); Chad Ross, *Naked Germany. Health, Race and the Nation* (Oxford: Berg, 2005): Maren Möhring, *Marmorleiber: Körperbildung in der deutschen Nacktkultur* (Cologne, Germany: Böhlau, 2004).
41. Forth, *Dreyfus Affair*, especially 33ff. and 58.
42. Ibid., 77.
43. Ibid., 190–95.
44. Ibid., 175–78.
45. Ibid., 202–29.
46. Claude Sigaud and Leon Vincent, *Les Origines de la Maladie: Essai sur l'Evolution de la Forme du Corps humain*, 2nd ed. (Paris: A Maloine, 1912), 46–64. On Sigaud, see: Jacques Bertholon, *L'Oeuvre de Claude Sigaud. Clinicien Lyonnais* (Paris: Vigot Frère, 1957).
47. Roberta J. Park, "Healthy, Moral, and Strong: Educational Views of Physical Exercise and Athletics in Nineteenth-Century America," in *Fitness in American Culture: Images of Health, Sport, and the Body, 1830–1940*, ed. Kathryn Grover (Amherst: University of Massachusetts Press, 1989), 123–68; Harvey Green, *Fit for America. Health, Fitness, Sport, and American Society* (New York: Pantheon Books, 1986), 181–215.
48. Roberta J. Park, "Muscles, Symmetry and Action: 'Do You Measure Up?' Defining Masculinity in Britain and America from the 1860s to the early 1900s," *The International Journal of the History of Sport* 22, no. 3 (2005): 365–95, especially 369–77; Clifford Putney, *Muscular Christianity: Manhood and Sports in Protestant America* (Cambridge, MA: Harvard University Press, 2001), 33–39.
49. Bernarr MacFadden, *Keeping Fit: Pep, Poise, Power for Men and Women* (New York: MacFadden Publications, 1929), preface.
50. Leoussi, *Nationalism and Classicism*, 113ff.
51. *Wege zu Kraft und Schönheit*, director Wilhelm Prager (1925) in *Bundesarchiv-Filmarchiv Berlin*.
52. Tamar Garb, *Bodies of Modernity. Figures and Flesh in Fin de Siècle France* (London: Thames and Hudson, 1998), 78. Forth, *Dreyfus Affair*, 203–34.
53. Svenja Goltermann, "Exercise and Perfection. Embodying the Nation in 19th Century Germany," *European Review of History* 11 (2004): 333–46.
54. Ferdinand Hueppe, *Hygiene der Körperübungen* (Leipzig, Germany: S. Hirzel, 1910), 5ff.
55. Ibid., 5–10.
56. Ibid., 30ff. and 60ff.
57. Fritsch, *Gestalt des Menschen*, 147ff. Some of Fritsch's colleagues, like Rudolf Virchow and Felix von Luschan, measured the performers of commercial exhibitions of colonial people (*Völkerschauen*) to assess human diversity: Sierra Ann Bruckner,

"The Tingle-Tangle of Modernity: Popular Anthropology and the Cultural Politics of Identity in Imperial Germany" (PhD diss., University of Iowa, 1999), 282–92 and 456–71.

58. Andrew Zimmerman, "Adventures in the Skin Trade: German Anthropology and Colonial Corporeality," in *Worldly Provincialism. German Anthropology in the Age of Empire*, ed. H. Glenn Penny and Matti Bunzl (Ann Arbor: University of Michigan Press, 2003), 156–78.
59. Hau, *Cult of Health and Beauty*, 91–100.
60. Sierra A. Bruckner, "Spectacles of (Human) Nature: Commercial Ethnography between Leisure, Learning, and *Schaulust*," in *Worldly Provincialism*, ed. Penny and Bunzl, 127–55.
61. Lewerentz, *Fritsch*, 284ff.
62. Hau, *Cult of Health*, 88–92. Images from: Carl Heinrich Stratz, *Die Rassenschönheit des Weibes*, 5th ed. (Stuttgart, Germany: Ferdinand Enke, 1904), 119, 304, and 326.
63. Erwin J. Haeberle, "Der verbotene Akt: Unzüchtige Fotos von 1850–1950," in *Das Aktfoto: Ansichten vom Körper im fotografischen Zeitalter*, ed. Michael Köhler and Gisela Barche (Munich: Bucher, 1985), 212–24.
64. Möhring, *Marmorleiber*, 205 and 256.
65. Hueppe, *Hygiene der Körperübungen*, 51ff.
66. Ibid., 22ff and 47ff.
67. Wolfgang Kohlrausch, "Zusammenhänge von Körperform und Leistung. Ergebnisse der anthropometrischen Messungen an den Athleten der Amsterdamer Olympiade," *Arbeitsphysiologie* 2 (1930), 187–204. On Kohlrausch, see Angelika Uhlmann, " 'Der Sport ist der praktische Arzt am Krankenlager des deutschen Volkes': Wolfgang Kohlrausch und die Geschichte der deutschen Sportmedizin" (Diss., Freiburg, 2004), http://freidok.ub.uni-freiburg.de/volltexte/1590/. On the cultural and economic background for Weimar sports science, see Michael Hau, "Sports in the Human Economy: '*Leibesübungen*,' Medicine, Psychology, and Performance Enhancement in the Weimar Republic," *Central European History* 41 (2008): 381–412.
68. Hau, *Cult of Health and Beauty*, 165ff.
69. Patricia Vertinsky, "Embodying Normalcy: Anthropometry and the Long Arm of William H. Sheldon's Somatotyping Project," *Journal of Sports History* 29, no. 1 (2002), 95–133, especially 114ff. Park, "Muscles, Symmetry and Action," 384–88.
70. Nikolas Rose, *Inventing Our Selves: Psychology, Power, and Personhood* (Cambridge: Cambridge University Press, 1996), 156ff.
71. Michael Anton Budd, *The Sculpture Machine. Physical Culture and Body Politics in the Age of Empire* (London: MacMillan, 1997), 37–57. The most famous of these health entrepreneurs was Eugen Sandow, who marketed his exercise system in the United States, Europe, and the British Empire beginning in the 1890s: David L. Chapman, *Sandow the Magnificent: Eugen Sandow and the Beginnings of Body Building* (Urbana: University of Illinois Press, 1994). For a compilation of the activities of health entrepreneurs, see Bernd Wedemeyer-Kolwe, *Der neue Mensch. Körperkultur im Kaiserreich und der Weimarer Republik* (Würzburg, Germany: Königshausen & Neumann, 2004), 301–31.
72. Hau, *Cult of Health and Beauty*, 176–98.
73. Heiko Stoff, *Ewige Jugend: Konzepte der Verjüngung vom späten 19. Jahrhundert bis ins Dritte Reich* (Cologne, Germany: Böhlau, 2004), 43–55, 130–44, and 374ff.

74. Cornelie Usborne,"The New Woman and Generation Conflict: Perceptions of Young Women's Sexual Mores in the Weimar Republic," in *Generations in Conflict: Youth Revolt and Generation Formation in Germany, 1770–1968*, ed. Mark Roseman (Cambridge: Cambridge University Press, 1995), 137–63; Hau, *Cult of Health and Beauty*, 70–75 and 193ff.
75. *Wege zu Kraft und Schönheit*, Bundesarchiv-Filmarchiv Berlin.
76. Modris Eksteins, *Rites of Spring: The Great War and the Birth of the Modern Age* (Boston: Houghton, 1989), 256ff.
77. Elizabeth Haiken, *Venus Envy. A History of Cosmetic Surgery* (Baltimore: Johns Hopkins University Press, 1997), 91–130; Stoff, *Ewige Jugend*, 290–301.
78. Hau, *Cult of Health and Beauty*, 182–87.
79. T. J. Jackson Lears, "American Advertising and the Reconstruction of the Body," in *Fitness in American Culture*, ed. Grover, 47–66, citation 62.

Chapter 7

I would like to thank Michael Sappol, who was a stimulating sounding board for earlier drafts of this chapter. I am also grateful to staff at the National Archives of Mauritius and India Office Library, London, and Jordanna Bailkin, James Bradley, Jane Caplan, Marina Carter, Ian Duffield, Tony Gorman, Hamish Maxwell-Stewart, and Peter Stanley, for so generously sharing their knowledge of tattooing. As ever, the late Tony Farrington was a mine of information.

1. Abby Schrader, "Branding the Other/Tattooing the Self: Bodily Inscription Among Convicts in Russia and the Soviet Union," in *Written on the Body: The Tattoo in European and American History*, ed. Jane Caplan (Princeton, NJ: Princeton University Press, 2000), 175.
2. David Arnold, *Colonizing the Body: State Medicine and Epidemic Disease in Nineteenth-Century India* (Berkeley: University of California Press, 1993); Tony Ballantyne and Antoinette Burton, eds., *Bodies in Contact: Rethinking Colonial Encounters in World History* (Durham, NC: Duke University Press, 2005); E. M. Collingham, *Imperial Bodies* (London: Polity, 2001); Steven Pierce and Anupama Rao, eds., *Discipline and the Other Body: Correction, Corporeality, Colonialism* (Durham, NC: Duke University Press, 2006); Ann Laura Stoler, *Race and the Education of Desire: Foucault's* History of Sexuality *and the Colonial Order of Things* (Durham, NC: Duke University Press, 1995).
3. For an overview of mainly anthropological literature on bodily inscription, see Enid Schildkrout, "Inscribing the Body," *Annual Review of Anthropology* 33 (2004): 319–44. See also Juniper Ellis, *Tattooing the World: Pacific Designs in Print and Skin* (New York: Columbia University Press, 2008).
4. Ann Laura Stoler and Frederick Cooper, "Between Metropole and Colony: Rethinking a Research Agenda," in *Tensions of Empire: Colonial Cultures in a Bourgeois World*, ed. Frederick Cooper and Ann Laura Stoler (Berkeley: University of California Press, 1997), 7.
5. Ann Laura Stoler, "Tense and Tender Ties: The Politics of Comparison in North American History and (Post) Colonial Studies," in *Haunted By Empire: Geographies of Intimacy in North American History*, ed. Ann Laura Stoler (Durham, NC: Duke University Press, 2006), 1–22.

6. The colonial discourses that accompanied the physical marks left by tropical diseases have been explored elsewhere. See Arnold, *Colonizing the Body*, chapter 3; Collingham, *Imperial Bodies*, 177–85. Note also the contention that late nineteenth-century fears about leprosy in Europe emerged in tandem with racial thinking that constructed lepers as colonial others. See Z. Gussow and G.S. Tracy, "Stigma and the Leprosy Phenomenon: The Social History of the Disease in the Nineteenth and Twentieth Centuries," *Bulletin of the History of Medicine* 44, no. 5 (1970): 425–49. For studies of colonial Australia and Africa, respectively, see Alison Bashford and Maria Nugent, "Leprosy and the Management of Race, Sexuality and Nation in Tropical Australia," in *Contagion*, ed. Alison Bashford and Claire Hooker (London: Routledge, 2001), 106–28; Megan Vaughan, *Curing Their Ills: Colonial Power and African Illness* (Cambridge: Polity Press, 1991), 77–99.
7. Jane Caplan, "Introduction," in *Written on the Body*, xv.
8. Juliet Fleming, "The Renaissance Tattoo," in *Written on the Body*, 61–82; Nicholas Thomas, "Introduction," in *Tattoo: Bodies, Art and Exchange in the Pacific and the West*, ed. Nicholas Thomas, Anna Cole, and Bronwen Douglas (London: Reaktion, 2005), 21.
9. Juliet Fleming, "The Renaissance Tattoo," in *Written on the Body*, 68.
10. Clare Anderson, *Legible Bodies: Race, Criminality and Colonialism in South Asia* (Oxford: Berg, 2004), 79.
11. James Bradley, "Body Commodification? Class and Tattoos in Victorian Britain," in *Written on the Body*, 136–55; James Bradley, "Embodied Explorations: Investigating Tattoos and the Transportation System," in *Representing Convicts: New Perspectives on Convict Forced Labour Migration*, ed. Ian Duffield and James Bradley (London: Leicester University Press, 1997), 183–203; James Bradley and Hamish Maxwell-Stewart, "'Behold the Man': Power, Observation and the Tattooed Convict," *Australian Studies* 12, no. 1 (1997): 71–97; Hamish Maxwell-Stewart and Ian Duffield, "Skin Deep Devotions: Religious Tattoos and Convict Transportation to Australia," in *Written on the Body*, 118–35. See also Joanna White, "Marks of Transgression: The Tattooing of Europeans in the Pacific Islands," in *Tattoo*, 77.
12. Bradley, "Body Commodification?" 145; Anna Cole, "Governing Tattoo: Reflections on a Colonial Trial," in *Tattoo*, 114. See also Christine M.E. Guth's account of Charles Longfellow's 1871–1873 sojourn in Japan: *Longfellow's Tattoos: Tourism, Collecting, and Japan* (Seattle: University of Washington Press, 2004).
13. Cole, "Governing Tattoo," 110–14.
14. Letter from G.E. Gascoigne to *The Lancet*, May 18, 1872, cited in A.W. Franks, "Note on the Tattooed Man from Burmah," *Journal of the Anthropological Institute of Great Britain and Ireland* 2 (1873): 231–32.
15. Alan Govenar, "The Changing Image of Tattooing in American Culture, 1846–1966," in *Written on the Body*, 213–14.
16. Schildkrout, "Inscribing the Body," 327.
17. George Bruce, *The Life of a Chelsea Pensioner by Himself Being the Memoirs of George Bruce Native of London and First European Resident in New Zealand upon whom on his adoption into the Native Race the Rank of a Chief was conferred in 1806*, Greenwich, ca. 1818. National Library of Australia, Canberra, MS3608. I thank Ian Duffield for this reference.
18. Linda Colley, *Captives: Britain, Empire and the World, 1600–1850* (London: Jonathan Cape, 2002).

19. White, "Marks of Transgression," 87–89 (quotes 88, 108). On O'Connell, see also Stephan Oettermann, "On Display: Tattooed Entertainers in America and Germany," in *Written on the Body*, 199. On tattooed beachcombers in nineteenth-century Samoa, see Sean Mallon, "Samoan *Tatau* as Global Practice," in *Tattoo*, 156.
20. Oettermann, "On Display," 200–201.
21. Franks, "Note on the Tattooed Man"; Dr Kaposi, "Der Tätowirte von Birma," *Wiener Medizinische Wochenschrift* 2 (1872): 39–45; *The Lancet*, June 1, 1872, 777. Later medical investigations centered on the communication of diseases like syphilis through tattoo needles and tattoo removal. See Bradley, "Body Commodification?," 143–45; Govenar, "The Changing Image," 213.
22. Oettermann, "On Display," 199–211. On circuses and carnivals in North America, see also Govenar, "The Changing Image," 222–26. On the de Burghs, see Gambier Bolton, "Pictures on the Human Skin," *Strand Magazine* 8 (1897): 431–32.
23. Bradley, "Body Commodification?" 146. For more details of tattooing practices in Japan, see Guth, *Longfellow's Tattoos*.
24. For instance, images of Thomas Cadell's tattoos (on both arms) are reproduced in Peter Stanley, *White Mutiny: British Military Culture in India, 1825–1875* (London: Hurst, 1998), 27. Lt. Thomas Cadell (1835–1919) became commander of the 2nd European Bengal Fusiliers in 1854 and received the Victoria Cross in 1857 for his part in suppressing the mutiny-rebellion. He was chief commissioner of the Andaman and Nicobar Islands (an Indian penal colony), 1879–1892.
25. Govenar, "The Changing Image," 215–17.
26. Bradley, "Body Commodification?" 147, 149, 154.
27. Schildkrout, "Inscribing the Body," 328.
28. Govenar, "The Changing Image," 213–14.
29. Bradley, "Body Commodification?" 145–55 (quote, 150).
30. Jordanna Bailkin, "Making Faces: Tattooed Women and Colonial Regimes," *History Workshop Journal* 59 (2005): 33–56 (quotes, 51, 49).
31. An indicative publication is W. D. Hambly, *The History of Tattooing and Its Significance With Some Account of Other Forms of Corporal Marking* (London: H.F. and G. Witherby, 1925).
32. Thomas, "Introduction," 12.
33. A. Gell, *Wrapping in Images: Tattooing in Polynesia* (Oxford: Clarendon Press, 1993), 38–39.
34. Anne D'Alleva, "Christian Skins: *Tatau* and the Evangelization of the Society Islands and Samoa," in *Tattoo*, 90–108 (quote. 91). See also Mallon, "Samoan *Tatau*," 149–51.
35. D'Alleva, "Christian Skins," 92–94.
36. Heidi Gengenbach, "Tattooed Secrets: Women's History in Magude District, Southern Mozambique," in *Bodies in Contact*, 253–73. The acquisition of such modern images by British aristocratic women in the late nineteenth century is also noted by Bailkin, "Making Faces," 49.
37. Edward Alpers, "Becoming Mozambique: Diaspora and Identity in Mozambique," in *History, Memory and Identity*, ed. Vijayalakshmi Teelock and Edward A. Alpers (Port Louis: Nelson Mandela Centre for African Culture/University of Mauritius, 2001), 123–31; Gengenbach, "Tattooed Secrets"; Michael A. Gomez, *Exchanging Our Country Marks: The Transformation of African Identities in the Colonial and Antebellum South* (Chapel Hill: University of North Carolina Press, 1998), 39–40, 97–98; Schildkrout, "Inscribing the Body," 332.

38. Megan Vaughan, "Scarification in Africa: Re-reading Colonial Evidence," *Cultural and Social History* 4, no. 3 (2007): 392.
39. Megan Vaughan, *Creating the Creole Island: Slavery in Eighteenth-Century Mauritius* (Durham, NC: Duke University Press, 2005), 117.
40. Bronwen Douglas, "Cureous Figures: European Voyagers and Tatau/Tattoo in Polynesia, 1595–1800." in *Tattoo*, 39–45; Linda Waimarie Nikora, Mohi Rua, and Ngahuia Te Awekotuku, "Wearing Moko," in *Tattoo*, 191–94.
41. Anderson, *Legible Bodies*, 65–67.
42. On Maori tattooing practices see Douglas, "Cureous Figures" and Nikora, Rua, and Awekotuku, "Wearing Moko." On collecting and the colonial body see Anderson, *Legible Bodies*, 183; Douglas M. Peers, "Colonial Knowledge and the Military in India, 1780–1860," *Journal of Imperial and Commonwealth History* 33, no. 2 (2005): 157–80.
43. "Scots to Decide on Return of Maori Artefacts," *The New Zealand Herald*, January 13, 2005. Robley was the author of *Moko; or Maori Tattooing* (London: Chapman and Hall, 1896).
44. Anderson, *Legible Bodies*, 68–73. See also Nicholas Dirks, *Castes of Mind: Colonialism and the Making of Modern India* (Princeton, NJ: Princeton University Press, 2001).
45. Anderson, *Legible Bodies*, 78–79.
46. Ibid., 79–80.
47. Hamish Maxwell-Stewart and Ian Duffield, "Skin Deep Devotions," 120.
48. Clare Anderson, "The Genealogy of the Modern Subject: Indian Convicts in Mauritius, 1814–1853," in *Representing Convicts*, 171.
49. Ibid., 146–47.
50. Simon Cole, *Suspect Identities: A History of Fingerprinting and Criminal Identification* (London: Harvard University Press, 2001), 29.
51. M[egalos] Caloyanni, "Étude des tatouages sur les criminals d'Egypte," *Bulletin de l'Institut D'Égypte* (ca. 1923): 115–28. I thank Tony Gorman for this reference.
52. Abby Schrader, "Branding the Other/Tattooing the Self: Bodily Inscription Among Convicts in Russia and the Soviet Union," in *Written on the Body*, 187–89.
53. Jane Caplan, " 'National Tattooing': Traditions of Tattooing in Nineteenth-century Europe," in *Written on the Body*, 156–73.
54. M. Kennedy, *Notes on Criminal Classes in the Bombay Presidency with Appendices regarding some Foreign Criminals who occasionally visit the Presidency, including Hints on the Detection of Counterfeit Coins* (Bombay: Government Central Press, 1908). Kennedy's work was used in other regional manuals, for example: F. C. Daly, *Manual of the Criminal Classes Operating in Bengal* (Calcutta: Bengal Secretariat Press, 1916); S. T. Hollins, *The Criminal Tribes of the United Provinces* (Allahabad, India: Government Press, 1914).
55. G. W. Gayer, *Some Criminal Tribes of India and Religious Mendicants* (Saugor, India: Police Officers' Training School, 1907), i.
56. William Crooke, *The Tribes and Castes of the North-West Provinces and Oudh*, 4 vols (Calcutta, Superintendent Government Printing, 1896), quote vol. II, 96.
57. Anderson, *Legible Bodies*, 73–78. The supposedly disfiguring qualities of facial tattoos have also been noted in the nineteenth-century Burmese context by Bailkin, "Making Faces," 37–38.
58. Unless otherwise indicated, this section is drawn from Anderson, *Legible Bodies*, chap. 2.

59. There are interesting parallels with penal tattooing in Japan, from the early eighteenth century to 1870, though there is no evidence of shared practice. See Daniel V. Botsman, *Punishment and Power in the Making of Modern Japan* (Princeton, NJ: Princeton University Press, 2005), 27, 28, 144.
60. Interesting comparative practices of resistance to penal tattooing can be found in Japan: Botsman, *Punishment and Power*, 28, 237 note 56.
61. Such concerns probably also explain the appeal of contiguous lithographs of slaves being branded—most famously by Nathaniel Currier—in abolitionist America.
62. Cole, "Governing Tattoo," 114–18; Bailkin, "Making Faces," 33–56.
63. Schrader, "Branding the Other," 174–92 (quote, 174).

Chapter 8

Parts of this chapter rework material from my book *Bodily Matters: The Anti-Vaccination Movement in England, 1853–1907*. I would like to thank Michael Sappol and Stephen Rice for their insightful comments on an earlier draft of this chapter and for their patience. Ben Cohen encouraged me to revisit this subject with new eyes and offered great advice along the way. Miles Cohen received his six-month immunizations during the process of revising this chapter. His reaction and my own to the process and the aftermath served as an important reminder of the importance of historicizing bodily experience and the continuities that exist between past and present anxieties around the infant's body. Luckily, his alarming carbuncle turned out to be merely dried snot.

1. *The News*, March 22, 1884, in Alfred Milnes Collection, vol. 5, 5, London School of Hygiene and Tropical Medicine. On Craigen see Sandra Stanley Holton, "Silk Dresses and Lavender Kid Gloves: The Wayward Career of Jessie Craigen, Working Suffragist," *Women's History Review* 5, no. 1 (1996): 129–49.
2. Nadja Durbach, *Bodily Matters: The Anti-Vaccination Movement in England, 1853–1907* (Durham, NC: Duke University Press, 2005); Ann Beck, "Issues in the Anti-Vaccination Movement in England," *Medical History* 4 (1960): 310–21; R.M. MacLeod, "Law, Medicine and Public Opinion: The Resistance to Compulsory Health Legislation 1870–1907," *Public Law* Summer (1967): 107–28, 189–211; Dorothy Porter and Roy Porter, "The Politics of Prevention: Anti-Vaccinationism and Public Health in Nineteenth-Century England," *Medical History* 32 (1988): 231–52; R.J. Lambert, "A Victorian National Health Service: State Vaccination 1855–71," *The Historical Journal* 5, no. 1 (1962): 1–18; Naomi Williams, "The Implementation of Compulsory Infant Smallpox Vaccination in England and Wales, 1840–1890," *Journal of Historical Geography* 20, no. 4 (1994): 396–412; Martin Fichman and Jennifer Keelan, "Resister's Logic: The Anti-vaccination Arguments of Alfred Russel Wallace and their Role in the Debates over Compulsory Vaccination in England, 1870–1907," *Studies in History and Philosophy of Science Part C: Studies in History and Philosophy of Biological and Biomedical Sciences* 38, no. 3 (2007): 585–607; Logie Barrow, "In the Beginning Was the Lymph: The Hollowing of Stational Vaccination in England and Wales, 1840–98," in *Medicine, Health and the Public Sphere in Britain, 1600–2000*, ed. Steve Sturdy (London: Routledge, 2002).
3. Walter Hadwen, *The Case against Vaccination* (London: National Anti-Vaccination League, 1896), 5.

4. Dorothy Porter and Roy Porter, "The Enforcement of Health: The British Debate," in *AIDS: The Burdens of History*, ed. Elizabeth Fee and Daniel M. Fox (Berkeley: University of California Press, 1988); Michel Foucault, *Discipline and Punish* (New York: Vintage Books, 1979); Michel Foucault, *The History of Sexuality: An Introduction* (New York: Vintage, 1990); Michel Foucault, *The Birth of the Clinic* (London: Tavistock, 1976).
5. Christopher Hamlin, *Public Health and Social Justice in the Age of Chadwick, Britain 1800–1854* (Cambridge: Cambridge University Press, 1998), 17.
6. Royston Lambert, *Sir John Simon 1816–1904 and English Social Administration* (London: Macgibbon and Kee, 1963), 618.
7. Deborah Brunton, "Pox Britannica: Smallpox Inoculation in Britain, 1721–1830" (Ph.D. diss., University of Pennsylvania, Philadelphia, 1990).
8. Hamlin, *Public Health and Social Justice*.
9. Epidemiological Society Small-Pox and Vaccination Committee. *Report on the State of Small-Pox and Vaccination in England and Wales and Other Countries, and on Compulsory Vaccination, With Tables and Appendices, Presented to the President and Council of the Epidemiological Society by the Small-Pox and Vaccination Committee, the 26th day of March 1853.* PP 1852–53 XVIII (HL 256), 5.
10. John Gibbs, *More Words on Vaccination* (London: Willis and Sotheran, 1856), 30.
11. *National Disease. Remarks upon the Prevailing Epidemic of Smallpox* (London: Longmans, Green, Reader and Dyer, 1871), 3–10.
12. *East London Observer*, October 28, 1876, 5; *Vaccination Inquirer*, March 1, 1887, 183.
13. R. H. Blakewell, "Is it Expedient to Make Vaccination Compulsory?" *Transactions of the New Zealand Institute* 24 (1891): 640.
14. John Shortt, *A Popular Lecture on Vaccination* (Madras, India: Gantz Brothers, 1865), 1–2.
15. *British Medical Journal*, December 19, 1863, 661.
16. Mary Spongberg, *Feminizing Venereal Disease: The Body of the Prostitute in Nineteenth-Century Medical Discourse* (New York: NYU Press, 1997), 143.
17. *Vaccination Inquirer*, December 1, 1886, 142; *Some Leading Arguments Against Compulsory Vaccination* (London: E.W. Allen, 1887), 17.
18. *Gloucester Epidemic of Smallpox, 1895–96. Report of the Committee Appointed By the Board of Guardians to Organize and Carry Out the General Vaccination of the City and District* (1896), 14.
19. William C. Dabney, "Maternal Impressions" in *Cyclopaedia of the Diseases of Children*, ed. John M. Keating (Edinburgh: Young J. Petland, 1889), 191–216; Jan Bondeson, *A Cabinet of Medical Curiosities* (London: I. B. Tauris, 1997), 144–69; Marie-Hélène Huet, *Monstrous Imagination* (Cambridge, MA: Harvard University Press, 1993).
20. Michael Worboys, *Spreading Germs* (Cambridge: Cambridge University Press, 2000); Terrie M. Romano, "The 1865 Cattle Plague and the Reception of the 'Germ Theory' in Mid-Victorian Britain," *Journal of the History of Medicine and Allied Sciences* 52, no. 1 (1997): 51–80.
21. *Evening Standard*, February 4, 1891, Alfred Milnes Collection, vol. 23, 52, London School of Hygiene and Tropical Medicine.
22. *Vaccination Inquirer*, January 1, 1898, 129.
23. Hilda Kean, *Animal Rights* (London: Reaktion Books, 1998); John Belchem, " 'Temperance in all Things': Vegetarianism, the Manx Press and the Alternative Agenda

of Reform in the 1840s," in *Living and Learning: Essays in Honour of J.F.C. Harrison*, ed. Malcolm Chase and Ian Dyck (Aldershot, UK: Scholar Press, 1996); Leah Leneman, "The Awakened Instinct: Vegetarianism and the Women's Suffrage Movement in Britain," *Women's History Review* 6, no. 2 (1997): 271–87.

24. *Vaccination Inquirer*, August 1, 1901, 94.
25. W.J. Furnival, *The Conscientious Objector: Who He Is! What He Has! What He Wants! And Why!* (privately published, 1902), 16.
26. *Vaccination Inquirer*, April 1, 1890, 2.
27. R. D. French, *Anti-Vivisection and Medical Science in Victorian Society* (Princeton, NJ: Princeton University Press, 1975).
28. Hilda Kean, "The 'Smooth Cool Men of Science': The Feminist and Socialist Response to Vivisection," *History Workshop Journal* 40 (1995): 16–38; Coral Lansbury, *The Old Brown Dog: Women, Workers, and Vivisection in Edwardian England* (Madison: University of Wisconsin Press, 1985); Mary Ann Elston, "Women and Anti-Vivisection in Victorian England 1870–1900," in *Vivisection in Historical Perspective*, ed. Nicholas A. Rupke (London: Croom Helm, 1987), 259–86; Anne L. Scott, "Physical Purity Feminism and State Medicine in Late Nineteenth-Century England," *Women's History Review* 8, no. 4 (1999): 625–53.
29. On the broader humanitarian movement, see Dan Weinbren, "Against *All* Cruelty: The Humanitarian League, 1891–1919," *History Workshop Journal* 38 (1994): 86–105.
30. *Vaccination Inquirer*, April 1, 1905, 10. On the relationship between Victorians and their animals, see James Turner, *Reckoning with the Beast: Animals, Pain, and Humanity in the Victorian Mind* (Baltimore: Johns Hopkins University Press, 1980).
31. *National Anti-Compulsory Vaccination Reporter*, October 7, 1880, 3, August 1, 1877, 2; *National Anti-Compulsory Vaccination League Occasional Circular*, February 1, 1876, 12; *East London Observer*, October 21, 1876, 6.
32. Quoted in *Vaccination Inquirer*, March 1, 1904, 236.
33. *The Anti-Vaccinator and Public Health Journal*, July 1, 1872, 89.
34. *National Anti-Compulsory Vaccination Reporter*, September 1, 1881, 202–3; September 1, 1882, 196; October 1, 1876, 8.
35. "The Vaccination Laws a Scandal to Public Honesty and Religion," in *Vaccination Tracts* (William Young, 1879), 6; Mary Hume-Rothery, *150 Reasons for Disobeying the Vaccination Law, by Persons Prosecuted Under It* (Cheltenham, UK: George F. Poole, 1871), 12.
36. I have been unable to locate either of these pamphlets, but the first is cited in George Sexton, *Vaccination Useless and Injurious* (London: George Howe, 1870), 5; the second is listed in London Society for the Abolition of Compulsory Vaccination, *Catalogue of Anti-Vaccination Literature* (London: E.W. Allen, 1894), 12. See also [E. Foster], *Murder and Mutilation! A Trial in a British Court of Justice* (John Moss, ca. 1873), 10; *Vaccination Inquirer*, April 1884, 4.
37. Logie Barrow, *Independent Spirits* (London: Routledge and Kegan Paul, 1986), 228.
38. Joy Dixon, *Divine Feminine: Theosophy and Feminism in England* (Baltimore: Johns Hopkins University Press, 2001), 121–51.
39. David Arnold, *Colonizing the Body: State Medicine and Epidemic Disease in Nineteenth Century India* (Berkeley: University of California Press, 1993), 116–58; Sanjoy Bhattacharya, Mark Harrison, and Michael Worboys, *Fractures States:*

Smallpox, Public Health and Vaccination Policy in British India 1800–1947 (Hyderabad, India: Orient Longman, 2005).

40. Arnold, *Colonizing the Body*, 143–44.
41. Giacomo Scarpelli, "'Nothing in Nature That Is Not Useful': The Anti-Vaccination Crusade and the Idea of *Harmonia Naturae* in Alfred Russel Wallace," *Nuncius* 7, no. 1 (1992): 109–30.
42. P. S. Brown, "Nineteenth-Century American Health Reformers and the Early Nature Cure Movement in Britain," *Medical History* 32 (1988): 174–75.
43. *National Disease: Remarks upon the Prevailing Epidemic of Smallpox* (London: Longmans, Green, Reader and Dyer, 1871), 48.
44. *The People*, May 25, 1884, Alfred Milnes Collection, vol. 5, 30, London School of Hygiene and Tropical Medicine.
45. *Herald of Health,* March 1878, 4.
46. Alison Bashford, *Purity and Pollution: Gender, Embodiment and Victorian Medicine* (Houndmills, UK: Macmillan, 1998), 136; David S. Barnes, *The Making of a Social Disease: Tuberculosis in Nineteenth Century France* (Berkeley: University of California Press, 1995), 41.
47. Walter S. Moon, *The Secret of Perfect Health: Disease Rendered Preventable and Removable by Washing its Germs Out of the Body* (London: Sanitary Engineering, ca. 1890).
48. J. J. Garth Wilkinson, *Smallpox and Vaccination* (London: F. Pitman, 1871), 42.
49. Quoted in Kean, *Animal Rights*, 133.
50. Porter and Porter, "The Politics of Prevention," 252; MacLeod, "Law, Medicine and Public Opinion," 211; Lloyd G. Stevenson, "Science Down the Drain," *Bulletin of the History of Medicine* 29, no. 1 (1955): 1–26.
51. Nadja Durbach, "Class, Gender, and the Conscientious Objector to Vaccination, 1898–1907," *Journal of British Studies* 41, no. 1 (2002): 58–83.

Chapter 9

1. William A. Alcott, "On the Study of Physiology as a Branch of General Education," *American Annals of Education and Instruction* 3 (September 1833): 403.
2. For Alcott's use of the language of "race" in this context, see ibid., 389.
3. Rosemarie Garland Thomson, *Extraordinary Bodies: Figuring Physical Disability in American Culture and Literature* (New York: Columbia University Press, 1997), 5.
4. See Georges Canguilhem, *The Normal and the Pathological* (New York: Zone Books, 1991).
5. This idea is a primary focus of Martha Banta in "Medical Therapies and Body Politics," *Prospects* 8 (1983): 59–128. See also Jacquelyn C. Miller, "The Body Politic and the Body Somatic: Benjamin Rush's Fear of Social Disorder and His Treatment for Yellow Fever," in *A Centre of Wonders: The Body in Early America*, ed. Janet Moore Lindman and Michele Lise Tarter (Ithaca, NY: Cornell University Press, 2001), 61–74; Dorinda Outram, *The Body and the French Revolution: Sex, Class and Political Culture* (New Haven, CT: Yale University Press, 1989), chap. 4; Joan Burbick, *Healing the Republic: The Language of Health and the Culture of Nationalism in Nineteenth-Century America* (New York: Cambridge University Press, 1994); and Joan Burbick, "'Intervals of Tranquility': The Language of Health in Antebellum America," *Prospects* 12 (1987): 175–99.

6. Catherine A. Holland, *The Body Politic: Foundings, Citizenship, and Difference in the American Political Imagination* (New York: Routledge, 2001), xxv.
7. This is what Catherine A. Holland argues. See ibid., xix.
8. Antoine de Baecque, *The Body Politic: Corporeal Metaphor in Revolutionary France, 1770–1800* (Stanford, CA: Stanford University Press, 1997), 1, 6, 8. See also Outram, *The Body and the French Revolution.*
9. Carolyn Sorisio, *Fleshing Out America: Race, Gender, and the Politics of the Body in American Literature, 1833–1879* (Athens: University of Georgia Press, 2002). The passage is on 2.
10. Much of this paragraph, including the final point about the changes to the United States Constitution by 1870, relies on Karen Sánchez-Eppler, *Touching Liberty: Abolition, Feminism, and the Politics of the Body* (Berkeley: University of California Press, 1993). See also Sorisio, *Fleshing Out America*, and Lauren Berlant, "National Brands/National Body: *Imitation of Life*," in *Comparative American Identities: Race, Sex, and Nationality in the Modern Text*, ed. Hortense J. Spillers (New York: Routledge, 1991), 110–40. Catherine Holland disputes the argument about disembodied personhood in early national America in *The Body Politic*. On the problem the body posed for claims to civic virtue, see Michael Warner, *The Letters of the Republic: Publication and the Public Sphere in Eighteenth-Century America* (Cambridge, MA: Harvard University Press, 1991).
11. On Buckland's death, see Rosamond Randall Beirne and John Henry Scarff, *William Buckland: Architect of Virginia and Maryland* (Baltimore: Maryland Historical Society, 1958), 105.
12. For an argument about the display of civic virtue in a portrait painted by Peale two years earlier, see Joseph Manca, "Cicero in America: Civic Duty and Private Happiness in Charles Willson Peale's Portrait of 'William Paca,'" *American Art* 17 (Spring 2003): 68–89. Karol Ann Peard Lawson has argued that Peale's support for the patriot cause led him to create a number of "allegorical portraits celebrating the moral rectitude and political actions of colonial sympathizers." I am suggesting that the portrait of Buckland might be added to this number. See Karol Ann Peard Lawson, "Charles Willson Peale's *John Dickinson*: An American Landscape as Political Allegory," *Proceedings of the American Philosophical Society* 136 (December 1992): 455–86. See also David C. Ward, *Charles Willson Peale: Art and Selfhood in the Early Republic* (Berkeley: University of California Press, 2004): 35–43. Ward suggests the symbolic meaning of homespun clothing in Peale's portraits of this period. Finally, see Sidney Hart, "A Graphic Case of Transatlantic Republicanism," in *New Perspectives on Charles Willson Peale: A 250th Anniversary Celebration*, ed. Lillian B. Miller and David C. Ward (Pittsburgh: University of Pittsburgh Press, 1991): 73–81. For a discussion of similar issues in early national American literature and portraiture more generally, see Christopher J. Lukasik, "The Face of the Public," *Early American Literature* 39 (2004): 413–64.
13. These two paintings are currently displayed together in the Yale University Art Gallery; the Neagle portrait hangs directly above the Peale portrait.
14. On Strickland's life and work, see Agnes Addison Gilchrist, *William Strickland, Architect and Engineer, 1788–1854* (Philadelphia: University of Pennsylvania Press, 1950).
15. In his analysis of a portrait Neagle painted in 1840, Robert Torchia similarly points to the "almost impossibly contorted manner" in which the sitter's hand comes

into view in the painting. See Robert W. Torchia, "No Strangers to the Ravages of Death: John Neagle's Portrait of Dr. John Abraham Elkinton," *Nineteenth Century* 18 (Fall 1998): 12.

16. My thanks to Ellen Ross for her help in making sense of this image.
17. On phrenology, see Roger Cooter, *The Cultural Meaning of Popular Science: Phrenology and the Organization of Consent in Nineteenth-Century Britain* (Cambridge: Cambridge University Press, 1984) and David de Giustino, *Conquest of Mind: Phrenology and Victorian Social Thought* (London: Croom Helm, 1975). On phrenology in the United States, see Madeleine B. Stern, *Heads and Headlines: The Phrenological Fowlers* (Norman: University of Oklahoma Press, 1971) and Robert E. Riegel, "The Introduction of Phrenology to the United States," *The American Historical Review* 39 (October 1933): 73–78.
18. Charles E. Rosenberg has argued that "the middle third of the nineteenth century was a formative period in the development of teaching about the body and its workings." See Charles E. Rosenberg, "Catechisms of Health: The Body in the Prebellum Classroom," *Bulletin of the History of Medicine* 69 (Summer 1995): 177. On popular anatomy and physiology in nineteenth-century America, see Michael Sappol, *A Traffic of Dead Bodies: Anatomy and Embodied Social Identity in Nineteenth-Century America* (Princeton, NJ: Princeton University Press, 2002). See also Stephen P. Rice, *Minding the Machine: Languages of Class in Early Industrial America* (Berkeley: University of California Press, 2004), chap. 4.
19. See John F. Kasson, *Rudeness and Civility: Manners in Nineteenth-Century Urban America* (New York: Hill and Wang, 1990), and Karen Halttunen, *Confidence Men and Painted Women: A Study of Middle-Class Culture in America, 1830–1870* (New Haven, CT: Yale University Press, 1982).
20. By far the richest study of these questions in America is Charles Colbert, *A Measure of Perfection: Phrenology and the Fine Arts in America* (Chapel Hill: University of North Carolina Press, 1997). For similar (although less wide-ranging) studies on France and Great Britain, respectively, see Judith Wechsler, *A Human Comedy: Physiognomy and Caricature in 19th-Century Paris* (London: Thames and Hudson, 1982) and Mary Cowling, *The Artist as Anthropologist: The Representation of Type and Character in Victorian Art* (Cambridge: Cambridge University Press, 1989). On American sculpture, see also John S. Crawford, "Physiognomy in Classical and American Portrait Busts," *The American Art Journal* 9 (May 1977): 49–60. On literary works, see, for instance, Graeme Tytler, *Physiognomy in the European Novel: Faces and Fortunes* (Princeton, NJ: Princeton University Press, 1982); Michael Hollington, "Dickens, 'Phiz' and Physiognomy," in *Imagination on a Long Rein: English Literature Illustrated*, ed. Joachim Möller (Marburg, Germany: Jonas Verlag, 1988): 125–35; and Taylor Stoehr, "Physiognomy and Phrenology in Hawthorne," *Huntington Library Quarterly* 37 (August 1974): 355–400.
21. Andrew Bank notes that phrenology quickly moved from the science of specifying individual traits to the science of generalizing group traits, especially according to race. "Although initially framed as a key to individual psychology, the phrenological system soon became widely applied as a theory of racial difference," he writes. "The leading proponents of the new discipline almost uniformly adapted their science of the brain to issues of racial differentiation." See Andrew Bank, "Of 'Native Skulls' and 'Noble Caucasians': Phrenology in Colonial South Africa," *Journal of Southern African Studies* 22 (September 1996): 389.

22. The very term "stereotype," which dates to the end of the eighteenth century, refers in its original meaning to the method by which type—including any relief illustrations—was cast in plaster or a similar material to be reproduced for printing. Once stereotyped, a printing plate could not be changed without considerable difficulty, but it could be used, stored, sold, and reused for years. By the middle of the decade the term was being used in a more general sense for any idea or expression that was widely repeated without change, but it was not until the early twentieth century that the term took on its current meaning. See Sander L. Gilman, *Difference and Pathology: Stereotypes of Sexuality, Race, and Madness* (Ithaca, NY: Cornell University Press, 1985): 15–16. On the accretion of meaning in the term, see "Stereotype," *Oxford English Dictionary Online* (Oxford: Oxford University Press, 2006), www.askoxford.com.
23. For a psychological approach to stereotyping, see Gilman, *Difference and Pathology*.
24. Joshua Brown, *Beyond the Lines: Pictorial Reporting, Everyday Life, and the Crisis of Gilded Age America* (Berkeley: University of California Press, 2002), 78. Brown goes on to discuss physiognomy and phrenology in wood-engraved illustration, noting that "these pseudo-sciences offered a way to comprehend and represent an increasingly complex and perplexing social reality." See ibid., 80.
25. See Rice, *Minding the Machine*.
26. Abraham Lincoln, "From Address to the Wisconsin State Agricultural Society, Milwaukee, Wisconsin," in *Selected Speeches and Writings by Abraham Lincoln*, ed. Don E. Fehrenbacher (New York: Vintage Books, Library of America ed., 1992), 235.
27. "Machinery, for Machine Making," *Graham's Magazine* 41 (November 1852), 475. Roger Cooter argues that phrenology as an entire system of thought served as a "representation and naturalization in human nature of hierarchical order," in part because of its "validation of mental over manual labor." See Cooter, *The Cultural Meaning of Popular Science*, 267–68. Allan Sekula similarly argues of both physiognomy and phrenology that "these disciplines would serve to legitimate on organic grounds the dominion of intellectual over manual labor." See Allan Sekula, "The Body and the Archive," *October* 39 (Winter 1986): 12.
28. The English engraver William J. Linton, who moved to the United States in 1866 to continue his work there, deplored the state that the craft had reached by mid-century, calling engravers who worked under the new industrialized system "two-legged, cheap machines for engraving—scarcely mechanics, mere machines, badly geared and ineffective." See W. J. Linton, "Art in Engraving on Wood," *The Atlantic Monthly* 43 (June 1879): 708. On the industrialization of wood engraving in the middle of the nineteenth century, see Brown, *Beyond the Lines*.
29. On the widening concern with energy and fatigue at the end of the nineteenth century and the beginning of the twentieth century, see Anson Rabinbach, *The Human Motor: Energy, Fatigue, and the Origins of Modernity* (Berkeley: University of California Press, 1990). See also Daniel Pick, *Faces of Degeneration: A European Disorder, c. 1848–c. 1918* (Cambridge: Cambridge University Press, 1989) and Kelly Hurley, *The Gothic Body: Sexuality, Materialism, and Degeneration at the* Fin de Siècle (Cambridge: Cambridge University Press, 1996).
30. John F. Kasson, *Houdini, Tarzan, and the Perfect Man: The White Male Body and the Challenge of Modernity in America* (New York: Hill and Wang, 2001), 19. See also Athena Devlin, *Between Profits and Primitivism: Shaping White Middle-Class Masculinity in the United States, 1880–1917* (New York: Routledge, 2005).

31. In the first four decades of the century, for instance, education reformers on both sides of the Atlantic established dozens of so-called manual labor schools that required students to combine physical labor with study for the purpose of preserving their health in the face of demanding and potentially debilitating mental labor. In Great Britain at mid-century a muscular Christianity movement among white, Protestant, middle-class, and reform-minded men that celebrated so-called manly courage and fortitude—especially in the service of imperialist aims—came to inform a variety of cultural forms, including novels, sermons, and paintings. On manual labor schools, see Rice, *Minding the Machine*, chap. 3. On the muscular Christianity (or "Christian manliness") movement in Great Britain, see Norman Vance, *The Sinews of the Spirit: The Ideal of Christian Manliness in Victorian Literature and Religious Thought* (Cambridge: Cambridge University Press, 1985) and Donald E. Hall, ed., *Muscular Christianity: Embodying the Victorian Age* (Cambridge: Cambridge University Press, 1994). On the movement's later manifestation in the United States, see Clifford Putney, *Muscular Christianity: Manhood and Sports in Protestant America, 1880–1920* (Cambridge, MA: Harvard University Press, 2001).
32. Jack London, *The Iron Heel* (1907; repr. Chicago: Lawrence Hills Books, 1981), 8, 7. On masculinity in the paintings of Thomas Eakins, see Martin A. Berger, *Man Made: Thomas Eakins and the Construction of Gilded Age Manhood* (Berkeley: University of California Press, 2000). On ideas about bodily vitality in American literature during this period, including discussions of Jack London, see Mark Seltzer, *Bodies and Machines* (New York: Routledge, 1992). For a fascinating discussion of some of Tilden's work, and of manliness in American sculpture more generally at the end of the nineteenth and the beginning of the twentieth centuries, see Melissa Dabakis, *Visualizing Labor in American Sculpture: Monuments, Manliness, and the Work Ethic, 1880–1935* (New York: Cambridge University Press, 1999).
33. James Weldon Johnson, *The Autobiography of an Ex-Coloured Man* (1912; repr. New York: Vintage, 1989), 186. Leon Litwack writes of American lynching photographs that "not only did photographers capture the execution itself, but also the carnival-like atmosphere and the expectant mood of the crowd." See Leon Litwack, "Hellhounds," in *Without Sanctuary: Lynching Photography in America*, James Allen (Santa Fe, NM: Twin Palms Publishers, 2004), 11. Amy Louise Wood argues somewhat differently that lynching "created its own photographic conventions," conventions that juxtaposed "stalwart and controlled" white mob members with "captive and defiled" black victims. See Amy Louise Wood, "Lynching Photography and the Visual Reproduction of White Supremacy," *American Nineteenth Century History* 6 (September 2005): 375. See also Jacqueline Goldsby, *A Spectacular Secret: Lynching in American Life and Literature* (Chicago: University of Chicago Press, 2006), chap. 5; Dora Apel, *Imagery of Lynching: Black Men, White Women, and the Mob* (Piscataway, NJ: Rutgers University Press, 2004); and Alison Piepmeier, *Out in Public: Configurations of Women's Bodies in Nineteenth-Century America* (Chapel Hill: University of North Carolina Press, 2004), chap. 4.
34. On the power of lynching photographs in their circulation, see Goldsby, *A Spectacular Secret*, 247–53.
35. Eadweard Muybridge, *Animal Locomotion: Prospectus and Catalogue of Plates* (Philadelphia, 1887), 12, published in facsimile in Eadweard Muybridge, *Muybridge's Complete Human and Animal Locomotion*, vol. 3 (Mineola, NY: Dover Publications, 1979), 1585–97.

36. The prospectus offers insight into Muybridge's gender order as well, identifying most of the male models by their occupations ("teachers in their respective professions," "public acrobats," "mechanics [who] are experts in their particular trades"), while identifying the female models by their marital status, age, and body size ("slender," "medium height," "heavily built"). See Muybridge, *Animal Locomotion*, 12. On some of the gendered meanings of these photographs, see Marta Braun, "Muybridge's Scientific Fictions," *Studies in Visual Communication* 10 (Summer 1984): 15.
37. Jayne Morgan, "Eadweard Muybridge and W. S. Playfair: An Aesthetics of Neursathenia," *History of Photography* 23 (Autumn 1999): 229. Morgan shows that Muybridge probably knew Mitchell's work on neurasthenia, and while working in Philadelphia was personally acquainted with some of the leading "nerve doctors" of the day.
38. On the zoopraxiscope, see Robert Bartlett Haas, *Muybridge: Man in Motion* (Berkeley: University of California Press, 1976), 116–20.
39. See, for instance, Janine A. Mileaf, "Poses for the Camera: Eadweard Muybridge's Studies of the Human Figure," *American Art* 16 (Autumn 2002), 31–53, and Sarah Gordon, "Prestige, Professionalism, and the Paradox of Eadweard Muybridge's *Animal Locomotion* Nudes," *The Pennsylvania Magazine of History and Biography* 130 (January 2006): 79–104.
40. A reviewer for the *Century Magazine*, for instance, referred to Muybridge's photographic technique as "an important addition to the instruments of scientific research," comparing Muybridge's camera with the microscope in that it "put before the eye what it could not see unaided." See Talcott Williams, "Animal Locomotion in the Muybridge Photographs," *The Century Magazine* 34 (July 1887): 356, 358.
41. Mileaf, 35. Sarah Gordon also argues that the publication plan for *Animal Locomotion* served the purpose of guaranteeing the respectability of Muybridge's project. See Gordon, "Prestige, Professionalism."
42. Marta Braun notes of *Animal Locomotion* that "although carried out under the rubric of science, Muybridge's method and conclusions are inconsistent with what we understand to be scientific experiment in the analysis of locomotion." See Braun, "Muybridge's Scientific Fictions," 2.
43. Marta Braun, "Muybridge's *Animal Locomotion*: The Director's Cut," *History of Photography* 24 (Spring 2000): 54. On some of the proto-cinematic aspects of Muybridge's work, see Rebecca Solnit, *River of Shadows: Eadweard Muybridge and the Technological Wild West* (New York: Viking, 2003), 220–38. See also Marta Braun, "The Expanded Present: Photographing Movement," in *Beauty of Another Order: Photography in Science*, ed. Ann Thomas (New Haven, CT: Yale University Press, 1997), 150–85.

Chapter 10

1. Bruce Mazlish, *The Fourth Discontinuity: The Co-Evolution of Humans and Machines* (New Haven, CT: Yale University Press, 1993), 4.
2. A good example is Tom Seigfried, *The Bit and the Pendulum: From Quantum Physics to M-theory—The New Physics of Information* (New York: John Wiley, 2000).
3. Quoted in A. Doyon and L. Liagre, *Jacques Vaucanson, mécanicien de genie* (Paris: Presses universitaires de France, 1967), 148l.

4. See Jessica Riskin, "The Defecating Duck, or, the Ambiguous Origins of Artificial Life," *Critical Inquiry* 24, no. 4 (Summer 2003): 599–633.
5. On the automata and the application of biomechanics to the "animal economy," see A. Doyon and L. Liagre, "Méthodolgie comparée du biomécanisme et de la mécanique comparée," *Dialectica* 10, no. 4 (December 1956): 319.
6. Walter Benjamin, "On the Mimetic Faculty," *Reflections: Essays, Aphorisms, Autobiographical Writings*, ed., Peter Demetz, trans., Edmund Jephcott (New York: Harcourt, Brace, Jovanovich, 1979), 336.
7. James Lastra, *Sound Technology and the American cinema: Perception, Representation, Modernity* (New York: Columbia University Press, 2000), 16–24.
8. Julien Offray de La Mettrie, *L'Homme-machine* (1748; repr. Paris: Denoël-Gonthier, 1981).
9. See my *The Human Motor: Energy, Fatigue, and the Origins of Modernity* (New York: Basic Books, 1990), chap. 2.
10. Hermann von Helmholtz, "Über die Wechselwirkung der Naturkräfte und die darauf Bezüglichen neuesten Ermittelungen der Physik," in *Populäre Wissenschaftliche Vorträge. Drittes Heft* (Braunschweig, Germany: Druck und Verlag Friedrich Vieweg und Sohn, 1876), 102.
11. Mark Seltzer, *Bodies and Machines* (New York: Routledge, 1992), 29.
12. Michel Serres, *Hermès III: La Traduction* (Paris: Minuit, 1974), 258.
13. Ibid., 258.
14. Hermann von Helmholtz, "Über die Wechselwirkung der Naturkräfte, 104. Thomas Kuhn discovered that "Helmholtz used the terms, *Arbeitskraft*, *bewegende Kraft*, *mechanische Arbeit* and *Arbeit* interchangeably for his fundamental measurable force." See Thomas S. Kuhn, "Energy Conservation as an Example of Simultaneous Discovery," in *The Essential Tension: Selected Studies in Scientific Tradition and Change* (Chicago: University of Chicago Press, 1977), 66–104.
15. Von Helmholtz, "Über die Wechselwirkung der Naturkrafte, 125.
16. See the important discussion in Peter Beilharz, *Labor's Utopias: Bolshevism, Fabianism, Social Democracy* (London: Routledge, 1992), 108.
17. Von Helmhotz, 1876, 103.
18. Ibid.
19. Karl Marx, *Das Kapital: A Critique of Political Economy. Vol. 1, Pelican Marx Library*, trans. Ben Fowkes (Harmondsworth, UK: Penguin, New Left Review, 1976), 647
20. Albrecht Wellmer, "The Latent Positivism of Marx's Philosophy of History," in *Critical Theory of Society* (New York: Seabury Press, 1974), 113.
21. Marx, *Das Kapital*, vol. III, 820.
22. Marx, *Das Kapital*, vol. I, 134.
23. Ibid., 137.
24. Charles S. Maier, *In Search of Stability: Explorations in Historical Political Economy* (Cambridge: Cambridge University Press, 1987), 26.
25. Angelo Mosso, *Fatigue*, trans. Margaret Drummond and William Blackley Drummond (New York: G.P. Putnam's Sons, 1904). On Mosso, see Rabinbach, *The Human Motor*, chap. 5.
26. Jules Amar, *The Human Motor, or, the Scientific Foundations of Labor and Industry*, (London: Routledge, 1920), 393.
27. Jules Amar, *The Physiology of Industrial Organisation and the Re-Employment of the Disabled*, trans. Bernard Miall and Albert Frank Stanley Kent (New York: Macmillan, 1919), 129.

28. Amar, *Human Motor*, 421.
29. Amar, *Physiology of Industrial Organisation*, 101.
30. Ibid., 127.
31. On Marey, see Rabinbach, *Human Motor*, chap. 4; Marta Braun, *Picturing Time: The Work of Étienne-Jules Marey (1830–1904)* (Chicago: University of Chicago Press, 1994).
32. Amar, *Physiology of Industrial Organisation*, 160, 161, 204.
33. Amar, *Human Motor*, 469.
34. Amar, *Physiology of Industrial Organisation*, 127.
35. Amar, *Human Motor*, 393.
36. Jules Amar, "L'organisation scientifique du travail humain," *La Technique Moderne* 7, no. 4. (August 15, 1913): 118; Jean-Marie Lahy, *Le Système Taylor: Analyse et commentaires* (Paris: Gauthier-Villars 1921), 171.
37. Mia Fineman, "Ecce Homo Prostheticus," *New German Critique* 76 (Winter 1999): 85–115.
38. Walther Moede, "Kraftfahrer-Eignungsprüfungen beim deutschen Heer 1915–1918," *IndustrielllePsychotechnik* 3, no. 1 (1926): 79.
39. Fritz Giese, *Psychologie der Arbeitshand*, in *Handbuch der Biolgischen Arbeitsmethoden*, ed. Emil Abderhalden (Berlin: Urban & Schwarzenberg, 1928), 804.
40. S. Jaeger and I. Stäuble, "Die Psychotechnik und ihre gesellschaftlichen Entwicklungsbedingungen," In *Anwendungeniim Berufsleben: Arbeits-Wirtschafts-Und Verkehrspsychologie, Kindlers "Psychologie Des 20. Jahrhunderts,"* Bd. 13, ed. François Stoll, (Zurich: Kindler, 1981), 53–94.
41. See Andreas Killen, *Berlin Electropolis Shock, Nerves and German Modernity* (Berkeley: University of California Press, 2006), 209, 210.
42. Ulfried Geuter, *The Professionalization of German Psychology in Nazi Germany* (Cambridge: Cambridge University Press, 1992), 86.
43. On Schlesinger, see Peter Hinrichs, Lothar Peter, and G. Briefs, *Industrieller Friede? Arbeitswissenschaft, Rationalisierung Und Arbeiterbewegung in Der Weimarer Republik, Kleine Bibliothek* (Cologne, Germany: Pahl-Rugenstein, 1976), 84.
44. Edgar Atzler, *Körper Und Arbeit: Handbuch der. Arbeitsphysiologie* (Leipzig, Germany: G. Thieme, 1927), 419–420.
45. See Anson Rabinbach, "Betriebspsychologie zwischen Psychotechnik und Politik während der Weimarer Republik: Der Fall Otto Lipmann," in *Betriebsärzte und produktionsbezogene Gesundheitspolitik*, ed. Dietrich Milles (Bremerhaven, Germany: Verlag für neue Wissenschaft, 1992), 41–64.
46. Killen, *Berlin Electropolis*, 183–86.
47. I am indebted to Andreas Killen for this insight into Giese. See Fritz Giese, *Handbuch Der Arbeitswissenschaft.* 2nd expanded and edited edition, Aufl. "Eignungsprüfungen an Erwachsenen" (Halle, Germany: Marhold, 1925), 209–10.
48. "Philosophen an der Front!" was the rallying cry in Fritz Giese, *Nietzsche—Die Erfüllung* (Tübingen, Germany: Mohr, 1934).
49. See Rabinbach, *Human Motor*, 284.
50. Ibid., for example, "Betriebspsychologie zwischen Psychotechnik und Politik während der Weimarer Republik: Der Fall Otto Lipmann," in *Betriebsärzte und produktionsbezogene Gesundheitspolitik*, ed. Dietrich Milles (Bremerhaven, Germany: Verlag für neue Wissenschaft, 1992), 41–64.
51. Peter Beilharz, *Labor's Utopias: Bolshevism, Fabianism, Social Democracy* (London: Routledge, 1992), 30, 31.

52. Sheila Fitzpatrick, "War and Society in Soviet Context: Soviet Labor Before, During, and After World War II," *International Labor and Working-Class History* 35 (Spring 1989): 37–52.
53. History of the Communist party of the Soviet Union (Bolsheviks) Short course. Edited by a commission of the Central Committee of the C.P.S.U. (B.) Authorized by the Central Committee of the C.P.S.U. (B.) [Moscow: Foreign Language Publishing House, 1954], 311.
54. Stephen Kotkin, *Magnetic Mountain* (Berkeley: University of California Press, 1995).
55. See Charles F. Heckscher, *The New Unionism: Employee Involvement in the Changing Corporation* (New York: Basic Books, 1987); Peter F. Drucker, "Are Unions becoming Irrelevant?" *Wall Street Journal*, September 22, 1982.
56. Lee Sproul and Sara Kiesler, *Connections: New Ways of Working in the Networked Organization* (Cambridge: MIT Press, 1992), 175; See also, Michael Schrage, *Shared Minds: The New Technologies of Collaboration* (New York: Random House, 1990); Peter M. Senge, *The Fifth Discipline: The Art & Practice of The Learning Organization* (New York: Currency Books, 1990).
57. Juliet Schor, *The Overworked American: The Unexpected Decline of Leisure* (New York: Basic Books, 1991).
58. Mazlish, *Fourth Discontinuity*, 228.
59. Shoshana Zuboff, *In the Age of the Smart Machine: The Future of Work and Power* (New York: Basic Books, 1988), 393.
60. Bob Davis and Dana Milbank, "'Employee Ennui': If the US Work Ethic Is Fading, 'Laziness' May Not Be the Reason," *Wall Street Journal*, February 7, 1992, 1.
61. Kevin Kelly, *Out of Control: The New Biology of Machines* (London: Fourth Estate, 1994), 25.
62. Ibid., 349.
63. Claus Emmeche, *The Garden in the Machine: The Emerging Science of Artificial Life*, trans. Steven Sampson (Princeton, NJ: Princeton University Press, 1994), 146.

BIBLIOGRAPHY

Ackerman, D. *An Alchemy of Mind*. New York: Scribner, 2004.

Acton, W. *The Contagious Diseases Act*. London: John Churchill & Sons, 1870.

Adams, Rachel, and David Savran, eds. *The Masculinity Studies Reader*. London: Blackwell, 2002.

Alcott, William A. "On the Study of Physiology as a Branch of General Education." *American Annals of Education and Instruction* 3 (September 1833): 385–403.

Alfred Milnes Collection of Newspaper Clippings, London School of Hygiene and Tropical Medicine.

Allen, James, ed. *Without Sanctuary: Lynching Photography in America*. Santa Fe, NM: Twin Palms Publishers, 2004.

Alpers, Edward. "Becoming Mozambique: Diaspora and Identity in Mozambique." In *History, Memory and Identity*. Edited by Vijayalakshmi Teelock and Edward A. Alpers, 117–55. Port Louis, Mauritius: Nelson Mandela Centre for African Culture/ University of Mauritius, 2001.

Altick, Richard. *The Shows of London*. Cambridge: Belknap Press, 1978.

Amar, Jules. *The Human Motor, or, the Scientific Foundations of Labor and Industry*. London: Routledge, 1920.

Amar, Jules. "L'organisation scientifique du travail humain." *La Technique Moderne* 7, no. 4. (August 15, 1913): 118.

Amar, Jules. *The Physiology of Industrial Organisation and the Re-Employment of the Disabled*. Translated by Bernard Miall and Albert Frank Stanley Kent. New York: Macmillan, 1919.

Aminoff, M.J. *Brown-Séquard: A Visionary of Science*. New York: Raven Press, 1993.

Ancel, P., and Bouin, P. "Les cellules séminales ont-elles une action sur les caractères sexuels? Discussion et nouvelles recherches." *Comptes rendus hebdomadaires des séances et mémoires de la Societé de Biologie* (Paris) 75 (1923): 175–78.

Ancel, P., and Bouin, P. "Recherches sur les cellules interstitielles du testicule des mammifères." *Archives de zoologie expérimentale et générale* 4th series, 1 (1903): 437–523.

Anderson, Amanda. *The Way We Argue Now.* Princeton, NJ: Princeton University Press, 2006.

Anderson, Benedict. *Imagined Communities: Reflections on the Origin and Spread of Nationalism.* Rev. ed. London: Verso, 1991.

Anderson, Clare. *Legible Bodies: Race, Criminality and Colonialism in South Asia.* Oxford: Berg, 2004.

Anderson, Clare. "The Genealogy of the Modern Subject: Indian Convicts in Mauritius, 1814–1853." In *Representing Convicts.* Edited by Ian Duffield and James Bradley, 164–82. London: Leicester University Press, 1997.

Angelides, S. *A History of Bisexuality.* Chicago: University Chicago Press, 2001.

Anon. "Burlesque Trust Shaky." *Washington Post*, February 18, 1905, 9.

Anon. "The Discussion at Glasgow on the Pathology of Phthisis Pulmonalis." *British Medical Journal* i (1881): 818.

Anon. "Dr Koch's Investigations upon the Treatment of Tuberculosis." *Lancet* ii (1890), 933.

Anon. "Gland Treatment Spreads in America." *New York Times*, April 8, 1923, 2.

Anon. "Hommes serpents et femmes couleuvres." *Lecture pour tous*, February 1908.

Anon. "Intramural Internment." *Pharmaceutical Journal and Transactions* 9 (1850): 446–48.

Anon. "Kernan Endows Hospital; Theatrical Man Gives $100,000 for Benefit of Crippled Children." *Washington Post*, November 29, 1910, 3.

Anon. *National Disease. Remarks Upon the Prevailing Epidemic of Smallpox.* London: Longmans, Green, Reader and Dyer, 1871.

Anon. "A New Theatre Circuit." *New York Times*, May 28, 1906, 9.

Anon. *The Lancet*, June 1, 1872, 777.

Anon. "Machinery, for Machine Making." *Graham's Magazine* 41 (November 1852): 469–76.

Anon. *The Phoenix of Sodom, or the Vere Street Coterie.* London: Sold by J. Cook, 1813.

Anon. "The Public Health Bill." *The Times*, July 28, 1866, 7–8.

Anon. "Purchases Two Theatres; James L. Kernan Buys Houses in Washington and Baltimore, May Be in Stair-Hanlin Syndicate." *New York Times*, February 12, 1903, 9.

Anon. "Review IV, Papers on Sanitary Science." *British and Foreign Medico-Chirurgical Review* 40 (1867): 65.

Anon. "Scots to Decide on Return of Maori Artefacts," *The New Zealand Herald*, January 13, 1905.

Anon. "Sells His Theaters; James L. Kernan Relinquishes Burlesque Holdings. . . ," *Washington Post*, May 12, 1906, 12.

Anon. *Some Leading Arguments Against Compulsory Vaccination.* London: E. W. Allen, 1887.

Anon. "Tuberculosis as a Contagious Disease." *British Medical Journal* i (1880): 704.

Anon. *Vaccination.* Reprinted from the *Herald of Health*, March 1878.

Anon. "The Vaccination Laws a Scandal to Public Honesty and Religion." In *Vaccination Tracts.* Edited by W. Young. London: William Young, 1879.

Apel, Dora. *Imagery of Lynching: Black Men, White Women, and the Mob.* Piscataway, NJ: Rutgers University Press, 2004.

Arbuthnot, F. F., trans. *The Kama Sutra of Vatsyayana.* Cosmopoli: the Kama Shastra Society of London and Benares, 1883.

Ariès, Phillippe. *Centuries of Childhood: A Social History of Family Life*. Translated by Robert Baldick. London: Jonathan Cape, 1962.

Arnold, David. *Colonizing the Body: State Medicine and Epidemic Disease in Nineteenth Century India*. Berkeley: University of California Press, 1993.

Arnold, David. "Crisis and Contradiction in India's Public Health." In *The History of Public Health and the Modern State*. Edited by Dorothy Porter. London: Wellcome Institute Series in the History of Medicine, 1994.

Arnold, David, ed. *Imperial Medicine and Indigenous Societies*. Manchester, UK: Manchester University Press, 1988.

Atzler, Edgar. *Körper Und Arbeit: Handbuch der. Arbeitsphysiologie*. Leipzig, Germany: G. Thieme, 1927.

Augustine, P. Charles. *A Commentary on the New Code of Canon Law*. London: B. Herder, 1921.

Badinter, Elisabeth. *Mother Love: Myth and Reality: Motherhood in Modern History*. New York: Macmillan, 1981.

Bailey, Peter. *Leisure and Class in Victorian England: Rational Recreation and the Contest for Control, 1830–1885*. London: Routledge and Kegan Paul, 1978.

Bailkin, Jordanna. "Making Faces: Tattooed Women and Colonial Regimes." *History Workshop Journal* 59 (2005): 33–56.

Bakhtin, M. *Rabelais and His World*. Bloomington: Indiana University Press, 1984.

Baldwin, P. *Contagion and the State in Europe, 1830–1930*. Cambridge: Cambridge University Press, 1999.

Ballantyne, Tony. *Orientalism and Race Aryanism in the British Empire*. Houndmills, UK: Palgrave, 2002.

Ballantyne, Tony, and Antoinette Burton, eds. *Bodies in Contact: Rethinking Colonial Encounters in World History*. Durham, NC: Duke University Press, 2005.

Bank, Andrew. "Of 'Native Skulls' and 'Noble Caucasians': Phrenology in Colonial South Africa." *Journal of Southern African Studies* 22 (September 1996): 387–403.

Banner, L. *American Beauty*. New York: Knopf, 1983.

Banta, Martha. "Medical Therapies and Body Politics." *Prospects* 8 (1983): 59–128.

Barker-Benfield, B. "The Spermatic Economy: A Nineteenth Century View of Sexuality." *Feminist Studies* 1 (1972): 45–74.

Barnes, D.S. *The Making of a Social Disease: Tuberculosis in Nineteenth Century France*. Berkeley: University of California Press, 1995.

Barrow, Logie. *Independent Spirits*. London: Routledge and Kegan Paul, 1986.

Barrow, Logie. "In the Beginning Was the Lymph: The Hollowing of Stational Vaccination in England and Wales, 1840–98." In *Medicine, Health and the Public Sphere in Britain, 1600–2000*. Edited by S. Sturdy. London: Routledge, 2002.

Bashford, A. *Purity and Pollution: Gender, Embodiment and Victorian Medicine*. Houndmills, UK: Macmillan, 1998.

Bashford, Alison, and Maria Nugent. "Leprosy and the Management of Race, Sexuality and Nation in Tropical Australia." In *Contagion*. Edited by Alison Bashford and Claire Hooker, 106–28. London: Routledge, 2001.

Beard, G.M. and A.D. Rockwell. *A Practical Treatise on the Medical and Surgical Uses of Electricity*. New York: William Wood, 1871.

Beard, G.M., *American Nervousness: Its Causes and Consequences*. New York: Putnam's, 1881.

Beck, A. "Issues in the Anti-Vaccination Movement in England." *Medical History* 4 (1960): 310–21.

Beck, T. *Elements of Medical Jurisprudence*. 2 vols. Albany, NY: Webster and Skinners, 1823.

Beer, Gillian. *Darwin's Plots: Evolutionary Narrative in Darwin, George Eliot and 19th-century Fiction*. Cambridge: Cambridge University Press, 2000.

Beilharz, Peter. *Labor's Utopias: Bolshevism, Fabianism, Social Democracy*. London: Routledge, 1992.

Beirne, Rosamond Randall, and John Henry Scarff. *William Buckland: Architect of Virginia and Maryland*. Baltimore: Maryland Historical Society, 1958.

Belchem, J. "'Temperance in all Things': Vegetarianism, the Manx Press and the Alternative Agenda of Reform in the 1840s." In *Living and Learning: Essays in Honour of J.F.C. Harrison*. Edited by M. Chase and I. Dyck. Aldershot, UK: Scholar Press, 1996.

Benjamin, Harry. "Eugen Steinach, 1861–1944: A Life of Research." *Scientific Monthly* 61 (1945): 427–42.

Benjamin, Harry. "Preliminary Communication Regarding Steinach's Method of Rejuvenation." *New York Medical Journal* 114 (1921): 687–92.

Benjamin, Walter. "On the Mimetic Faculty." In *Reflections: Essays, Aphorisms, Autobiographical Writings*. Edited by Peter Demetz, Translated by Edmund Jephcott, 333–36. New York: Harcourt, Brace, Jovanovich, 1979.

Bennet, J.H. "The Treatment of Pulmonary Consumption." *British Medical Journal* ii (1867): 137.

Bennett, Hughes. "Phthisis pulmonalis." In *A System of Medicine*. Edited by J. Reynolds, vol. 3, 571. Macmillan, 1871.

Benson, Susan Porter. *Counter-cultures: Saleswomen, Managers and Customers in American Department Stores, 1890–1940*. Chicago: University of Illinois Press, 1986.

Berger, Martin A. *Man Made: Thomas Eakins and the Construction of Gilded Age Manhood*. Berkeley: University of California Press, 2000.

Berlant, Lauren. "National Brands/National Body." In *Comparative American Identities: Race, Sex, and Nationality in the Modern Text*. Edited by Hortense J. Spillers, 110–40. New York: Routledge, 1991.

Berthold, A.A. "The Transplantation of Testes." Translated by D.P. Quiring. *Bulletin of the History of Medicine* 16 (1944): 399–401.

Bertholon, J. *L'Oeuvre de Claude Sigaud. Clinicien Lyonnais*. Paris: Vigot Frère, 1957.

Bertucci, Paula, and Giuliano Pancaldi, eds. *Electric Bodies: Episodes in the History of Medical Electricity*. Bologna. Italy: Università di Bologna, 2001.

Bewell, A. *Romanticism and Colonial Disease*. Baltimore: Johns Hopkins University Press, 1999.

Bhattacharya, Sanjoy, MarkHarrison, and Michael Worboys. *Fractures States: Smallpox, Public Health and Vaccination Policy in British India 1800–1947*. Hyderabad, India: Orient Longman, 2005.

Bichat, X. "Physiological Researches on Life and Death." In *Significant Contributions to the History of Psychology, 1750–1920*. Vol. II. Edited by Daniel N. Robinson. Washington, DC: University Publications of America, 1978.

Biedl, A. *Innere Sekretion: Ihre physiologischen Grundlagen und ihre Bedeutung für die Pathologie*. 2nd ed. Berlin: Urban & Schwarzenberg, 1913.

Billingham, R.E. and W.B. Neaves. "Paratransplantation and Tissue Therapy." *Perspectives in Biology and Medicine* 22 (1979): 320–32.

Birken, L. *Consuming Desire*. Ithaca, NY: Cornell University Press, 1988.

Blakewell, R.H. "Is it Expedient to Make Vaccination Compulsory?" *Transactions of the New Zealand Institute* 24 (1891): 636–41.

Blanchard, Laman. "A Visit to the General Cemetery at Kensal Green." *Ainsworth's Magazine* 2 (August 1842): 177.

Bleys, R. *The Geography of Perversion, 1750–1918.* London: Cassell, 1996.

Bloch, Iwan. *Anthropological Studies in the Strange Sexual Practices of All Races in All Ages.* New York: Anthropological Press, n.d.

Bloch, Iwan. *Betrage zur Aetiologie der Psychopathia Sexualis, Erster Teil.* Dresden, Germany: H.R. Dorn, 1902.

Bloch, Iwan. *Sexual Life in England.* Translated by William Forstern. London: Corgi Books, 1965.

Bloch, Iwan. *The Sexual Life of Our Time.* Translated by M. Paul Eden. New York: Allied Book Company, 1928.

Bogdan, Robert. *Freak Show: Presenting Human Oddities for Amusement and Profit.* Chicago: University of Chicago Press, 1990.

Bolton, Gambier. "Pictures on the Human Skin." *Strand Magazine* 8 (1897): 427–34.

Bondeson, J. *A Cabinet of Medical Curiosities.* London: I.B. Tauris, 1997.

Borell, M. "Brown-Séquard's Organotherapy and its Appearance in America at the End of the Nineteenth Century." *Bulletin of the History of Medicine* 50 (1976): 309–20.

Borell, M. "Organotherapy and the Emergence of Reproductive Endocrinology." *Journal of the History of Biology* 18 (1985): 1–30.

Borell, M. "Origins of the Hormone Concept: Internal Secretions and Physiological Research, 1889–1905." PhD diss., Yale University, 1976.

Borell, M. "Setting the Standards for a New Science: Edward Schäfer and Endocrinology." *Medical History* 22 (1978): 282–90.

Bornside, G.H. "Waldemar Haffkine's Cholera Vaccines and the Ferran-Haffkine Priority Dispute." *Journal of the History of Medicine and Allied Sciences* 37 (1982): 399–422.

Botsman, Daniel V. *Punishment and Power in the Making of Modern Japan.* Princeton, NJ: Princeton University Press, 2005.

Bourdieu, P. *Distinction: A Social Critique of the Judgement of Taste.* Translated by Richard Nice. Cambridge, MA: Harvard University Press, 1984.

Bowler, Peter J. *Evolution: The History of an Idea.* Berkeley: University of California Press, 2003.

Bowler, Peter J. *The Non-Darwinian Revolution: Reinterpreting an Historical Myth.* Baltimore: Johns Hopkins University Press, 1988.

Bracegirdle, B. "The Microscopical Tradition." In *Companion Encyclopedia of the History of Medicine.* Vol. 1. Edited by W.F. Bynum and Roy Porter, 102–119. London: Routledge, 1993.

Bradley, James. "Body Commodification? Class and Tattoos in Victorian Britain." In *Written on the Body: The Tattoo in European and American History.* Edited by Jane Caplan, 136–55. Princeton, NJ: Princeton University Press, 2000.

Bradley, James. "Embodied Explorations: Investigating Tattoos and the Transportation System." In *Representing Convicts: New Perspectives on Convict Forced Labour Migration.* Edited by Ian Duffield and James Bradley, 183–203. London: Leicester University Press, 1997.

Bradley, James, and Hamish Maxwell-Stewart. " 'Behold the Man': Power, Observation and the Tattooed Convict." *Australian Studies* 12, no. 1 (1997): 71–97.

Braun, Marta. "The Expanded Present: Photographing Movement." In *Beauty of Another Order: Photography in Science*. Edited by Ann Thomas, 150–85. New Haven, CT: Yale University Press, 1997.

Braun, Marta. "Muybridge's *Animal Locomotion*: The Director's Cut." *History of Photography* 24 (Spring 2000): 52–54.

Braun, Marta. "Muybridge's Scientific Fictions." *Studies in Visual Communication* 10 (Summer 1984): 2–21.

Braun, Marta. *Picturing Time: The Work of Etienne-Jules Marey (1830–1904)*. Chicago: University of Chicago Press, 1994.

Bristow, Joseph. *Effeminate England*. New York: Columbia University Press, 1995.

Bristow, Joseph. *Sexuality*. London: Routledge, 1997.

Brontë, Charlotte. *Jane Eyre*. 1847. Reprint, London: Penguin, 2006.

Brown, Gillian. *Domestic Individualism: Imagining Self in 19th-century America*. Berkeley: University of California Press, 1992.

Brown, Joshua. *Beyond the Lines: Pictorial Reporting, Everyday Life, and the Crisis of Gilded Age America*. Berkeley: University of California Press, 2002.

Brown, Julie K. *Health and Medicine on Display: International Expositions in the United States, 1876–1904*. Cambridge: MIT Press, 2009.

Brown, P. S. "Nineteenth-Century American Health Reformers and the Early Nature Cure Movement in Britain." *Medical History* 32 (1988): 174–94.

Browne, Janet. *Darwin's* Origin of Species: *A Biography*. London: Allen & Unwin, 2006.

Bruce, George. *The Life of a Chelsea Pensioner by Himself Being the Memoirs of George Bruce Native of London and First European Resident in New Zealand upon whom on his adoption into the Native Race the Rank of a Chief was conferred in 1806*. Greenwich, ca. 1818. National Library of Australia, Canberra, MS3608

Bruckner, S. "Spectacles of (Human) Nature: Commercial Ethnography between Leisure, Learning, and *Schaulust*." In *Worldly Provincialism. German Anthropology in the Age of Empire*. Edited by Glenn Penny and Matti Bunzl, 127–55. Ann Arbor: University of Michigan Press, 2003.

Bruckner, S. "The Tingle-Tangle of Modernity: Popular Anthropology and the Cultural Politics of Identity in Imperial Germany." PhD diss., University of Iowa, 1999.

Brunton, Deborah. "Pox Britannica: Smallpox Inoculation in Britain, 1721–1830." PhD diss., University of Pennsylvania, 1990.

Budd, Michael Anton. *The Sculpture Machine: Physical Culture and Body Politics in the Age of Empire*. London: MacMillan, 1997.

Bullough, Vern. *Science in the Bedroom*. New York: Basic Books, 1994.

Burbick, Joan. *Healing the Republic: The Language of Health and the Culture of Nationalism in Nineteenth-Century America*. New York: Cambridge University Press, 1994.

Burbick, Joan. "'Intervals of Tranquility': The Language of Health in Antebellum America." *Prospects* 12 (1987): 175–99.

Bushman, Richard. *The Refinement of America: Persons, Houses, Cities*. New York: Vintage, 1993.

Butler, Judith. *Bodies That Matter: On the Discursive Limits of Sex*. New York: Routledge, 1997.

Butler, Judith. *Feminist Contentions: A Philosophical Exchange*. New York: Routledge, 1995.

Bynum, Caroline Walker. *Fragmentation and Redemption: Essays on Gender and the Human Body in Medieval Religion*. New York: Zone Books, 1991.

Bynum, W. F. "An Experiment That Failed: Malaria Control at Mian Mir." *Parassitologia* 36 (1994): 107–120.

Caloyanni, M[egalos]. "Étude des tatouages sur les criminals d'Egypte." *Bulletin de l'Institut D'Égypte* circa 1923, 115–28.

Calwell, W. "The Hygienic Treatment of Consumption Independently of Sanatoria." *British Medical Journal* ii (1898): 947.

Canguilhem, Georges. *The Normal and the Pathological.* New York: Zone Books, 1991.

Caplan, Jane. "Introduction." In *Written on the Body: The Tattoo in European and American History.* Edited by Jane Caplan, xv. Princeton, NJ: Princeton University Press, 2000.

Caplan, Jane. "'National Tattooing': Traditions of Tattooing in Nineteenth-century Europe." In *Written on the Body: The Tattoo in European and American History.* Edited by Jane Caplan, 156–73. Princeton, NJ: Princeton University Press, 2000.

Caplan, Jane, ed. *Written on the Body: The Tattoo in European and American History.* Princeton, NJ: Princeton University Press, 2000.

Cartwright, L. *Screening the Body: Tracing Medicine's Visual Culture.* Minneapolis: University of Minnesota Press, 1995

Carus, C.G. *Symbolik der menschlichen Gestalt. Ein Handbuch der Menschenkenntnis.* 2nd ed. 1858. Repr. Hildesheim, Germany: Georg Olms, 1962.

Cassidy, Tina. *Birth: The Surprising History of How We Are Born.* New York: Atlantic Monthly Press, 2006.

Castle, T. *The Apparitional Lesbian.* New York: Columbia University Press, 1993.

Caton, Donald. *What a Blessing She Had Chloroform: The Medical and Social Response to the Pain of Childbirth from 1800 to the Present.* New Haven, CT: Yale University Press, 1999.

Chadwick, E. *Report on the Labouring Population of Great Britain.* Edited with an introduction by M.W. Flinn. 1842. Repr. Edinburgh: Edinburgh University Press, 1965.

Chadwick, E., and Thomas Southwood S. Smith. "Report." *Irish University Press Parliamentary Papers 1849. Cholera I,* August 14, 1850.

Chakravarty, G. *The Indian Mutiny and the British Imagination.* Cambridge: Cambridge University Press, 2005.

Chamberlain, Alexander Francis. *The Child: A Study in the Evolution of Man.* London: W. Scott, 1900.

Chapman, David L. *Sandow the Magnficent. Eugen Sandow and the Beginnings of Body Building.* Urbana: University of Illinois Press, 1994.

Clark, Anna. *Scandal: The Sexual Politics of the British Constitution.* Princeton, NJ: Princeton University Press, 2005.

Clarke, E., and L.S. Jacyna. *Nineteenth-Century Origins of Neuroscientific Concepts.* Berkeley: University of California Press, 1987.

Coats, J.F. "The Spontaneous Healing of Tuberculosis." *British Medical Journal* 1891, ii, pp. 933–8.

Cody, Lisa Forman. *Birthing the Nation: Sex, Science, and the Conception of Eighteenth-Century Britons.* Oxford: Oxford University Press, 2005.

Cody, Lisa Forman. "The Politics of Illegitimacy." *Journal of Women's History* 11 (2000): 131–56.

Cohen, Patricia Cline. *The Murder of Helen Jewett.* New York: Knopf, 1998.

Colbert, Charles. *A Measure of Perfection: Phrenology and the Fine Arts in America.* Chapel Hill: University of North Carolina Press, 1997.

Cole, Anna. "Governing Tattoo: Reflections on a Colonial Trial." In *Tattoo*. Edited by Nicholas Thomas, Anna Cole, and Bronwen Douglas, 109–19. London: Reaktion, 2005.

Cole, Joshua. *The Power of Large Numbers: Population, Politics, and Gender in Nineteenth-Century France*. Ithaca, NY: Cornell University Press, 2000.

Cole, Simon. *Suspect Identities: A History of Fingerprinting and Criminal Identification*. London: Harvard University Press, 2001.

Coleman, Deidre. *Romantic Colonization and British Anti-Slavery.* Cambridge: Cambridge University Press, 2005.

Coleman, W. *Death Is a Social Disease: Public Health and Political Economy in Early Industrial France*. Madison: University of Wisconsin Press, 1982.

Coleman, W. "Koch's Comma Bacillus: The First Year." *Bulletin for the History of Medicine* 61 (1987): 315–42.

Colley, Linda. *Captives: Britain, Empire and the World, 1600–1850*. London: Jonathan Cape, 2002.

Collingham, E. M. *Imperial Bodies*. London: Polity, 2001.

Collins, W. W. *Man and Wife*. Oxford: Oxford University Press, 1995.

Condrau, F., and M. Worboys. "Epidemics and Infections in Nineteenth-Century Britain." *Social History of Medicine* 20 (2007): 80–91.

Cooter, Roger J. *The Cultural Meaning of Popular Science: Phrenology and the Organization of Consent in Nineteenth-Century Britain*. Cambridge: Cambridge University Press, 1984.

Cope, Richard. *Ploughshare of War: The Origins of the Anglo-Zulu War of 1879*. Pietermaritzburg: University of Natal Press, 1999.

Corbin, Alain. *The Foul and the Fragrant: Odor and the French Social Imagination*. Cambridge, MA: Harvard University Press, 1986.

Cowling, Mary. *The Artist as Anthropologist: The Representation of Type and Character in Victorian Art*. Cambridge: Cambridge University Press, 1989.

Crais, Clifton, and Pamela Scully. *Sara Baartman and the Hottentot Venus: A Ghost Story and a Biography*. Princeton, NJ: Princeton University, 2008.

Cranefield, P. F. "The Organic Physics of 1847 and the Biophysics of Today." *Journal of the History of Medicine and the Allied Sciences* 12 (1957): 407–23.

Crary, Jonathan, and Sanford Kwinter, eds. *Incorporations* [*Zone* 6]. New York: Zone Books, 1992.

Crawford, John S. "Physiognomy in Classical and American Portrait Busts." *The American Art Journal* 9 (May 1977): 49–60.

Crompton, L. *Byron and Greek Love*. Berkeley: University California Press, 1985.

Crooke, William. *The Tribes and Castes of the North-West Provinces and Oudh*. 4 vols. Calcutta, India: Superintendent Government Printing, 1896.

Cross, Gary. *A Social History of Leisure Since 1600*. State College, PA: Venture Publishing, 1990.

Cullen, M. J. *The Statistical Movement in Early Victorian Britain: The Foundations of Empirical Social Research*. Hassocks, UK: Harvester Press, 1975.

Cullimore, H. *Consumption as a Contagious Disease* (A translation of Professor Cohnheim's pamphlet "Die Tuberklose vom Standpunkte der Infections-Lehre'). London: Baillière, Tindall and Cox, 1880.

Cunningham, A. "Transforming Plague: The Laboratory and the Identity of the Infectious Disease." In *The Laboratory Revolution in Medicine*. Edited by A. Cunningham and P. Williams, 209–44. Cambridge: Cambridge University Press, 1992.

Cunningham, Andrew, and Perry Williams, eds. *The Laboratory Revolution in Medicine*. Cambridge: Cambridge University Press, 1992.

Cunningham, Hugh. *Children and Childhood in Western Society Since 1500*. 2nd ed. London: Longman, 2005.

Dabakis, Melissa. *Visualizing Labor in American Sculpture: Monuments, Manliness, and the Work Ethic, 1880–1935*. New York: Cambridge University Press, 1999.

Dabney, W. C. "Maternal Impressions." In *Cyclopaedia of the Diseases of Children*. Edited by J. M. Keating. Edinburgh: Young J. Petland, 1889.

D'Alleva, Anne. "Christian Skins: *Tatau* and the Evangelization of the Society Islands and Samoa." In *Tattoo*. Edited by Nicholas Thomas, Anna Cole, and Bronwen Douglas, 90–108. London: Reaktion, 2005.

Daly, F. C. *Manual of the Criminal Classes Operating in Bengal*. Calcutta, India: Bengal Secretariat Press, 1916.

Darwin, Charles. *On the Origin of the Species by Means of Natural Selection, or the Preservation of Favoured Races in the Struggle for Life*. London, 1859.

Darwin, Charles. *The Descent of Man and Selection in Relation to Sex*. London, 1871.

Darwin, Charles. *The Variation of Animals and Plants Under Domestication*. 2nd ed. New York: Appleton, 1897.

Darwin, Charles to Charles Lyell, January 10, 1860; letter 2647. Darwin Correspondence Project. http://www.darwinproject.ac.uk/darwinletters/calendar/entry-2647.html.

Daum, A. *Wissenschaftspopularisierung im 19. Jahrhundert. Bürgerliche Kultur, naturwissenschaftliche Bildung und die deutsche Öffentlichkeit, 1848–1914*. Munich: Oldenbourg, 1998.

Daunton, M. J. *House and Home in the Victorian City: Working Class Housing 1850–1914*. London: Edward Arnold, 1983.

Davidson, A. *The Emergence of Sexuality*. Cambridge, MA: Harvard University Press, 2001.

Davis, Bob, and Dana Milbank. "'Employee Ennui': If the US Work Ethic Is Fading, 'Laziness' May Not Be the Reason." *Wall Street Journal*, February 7, 1992, 1.

Dean, C. *The Frail Social Body*. Berkeley: University California Press, 2000.

de Baecque, Antoine. *The Body Politic: Corporeal Metaphor in Revolutionary France, 1770–1800*. Stanford, CA: Stanford University Press, 1997.

de Giustino, David. *Conquest of Mind: Phrenology and Victorian Social Thought*. London: Croom Helm, 1975.

de Groot, J. "'Sex' and 'Race': The Construction of Language and Image in the Nineteenth Century." In *Cultures of Empire*. Edited by Catherine Hall. London: Routledge, 2000.

DeLacy, Margaret. "Puerperal Fever in Eighteenth-Century Britain." *Bulletin of the History of Medicine* 63 (1989): 521–56.

Dellamora, R. *Masculine Desire; The Sexual Politics of Victorian Aestheticism*. Chapel Hill: University North Carolina Press, 1990.

Dennett, Andrea. *Weird and Wonderful: The Dime Museum in America*. New York: NYU Press, 1997.

Desmond, Adrian. *Darwin: The Life of a Tormented Evolutionist*. New York: Norton, 1994.

Desmond, Adrian. *The Politics of Evolution: Morphology, Medicine and Reform in Radical London*. Chicago: University of Chicago Press, 1989.

Devlin, Athena. *Between Profits and Primitivism: Shaping White Middle-Class Masculinity in the United States, 1880–1917*. New York: Routledge, 2005.

Dictionary of National Biography. Oxford: Clarendon Press, 2004.

Digby, A. *Madness, Morality and Medicine: A Study of the York Retreat, 1796–1914*. Cambridge: Cambridge University Press, 1985.

Digby, A. "Moral treatment at the Retreat, 1796–1846." In *The Anatomy of Madness: Essays in the History of Psychiatry*, vol. 2. Edited by R. Porter, W. Bynum, and Michael Shepherd, 52–72. London: Tavistock, 1985.

Dimock, George. "Child of the Mills: Re-Reading Lewis Hine's Child-Labour Photographs." *Oxford Art Journal* 16, no. 2 (1993): 37–54.

Dirks, Nicholas. *Castes of Mind: Colonialism and the Making of Modern India*. Princeton, NJ: Princeton University Press, 2001.

Dixon, J. *Divine Feminine: Theosophy and Feminism in England*. Baltimore: Johns Hopkins University Press, 2001.

Donnison, Jean. *Midwives and Medical Men: A History of the Struggle for the Control of Childbirth*. 2nd ed. London: Historical Publications, 1988.

Donoghue, E. *Passions Between Women: British Lesbian Culture 1668–1801*. London: Scarlet Press, 1993.

Donzelot, Jacques. *The Policing of Families*. New York: Pantheon, 1979.

Douglas, Ann. *The Feminization of American Culture*. New York: Knopf, 1977.

Douglas, Bronwen. "'Cureous Figures': European Voyagers and Tatau/Tattoo in Polynesia, 1595–1800." In *Tattoo*. Edited by Nicholas Thomas, Anna Cole, and Bronwen Douglas, 39–45. London: Reaktion, 2005.

Doyon A., and L. Liagre. *Jacques Vaucanson, mécanicien de genie*. Paris: Presses universitaires de France, 1967.

Doyon, A., and L. Liagre. "Méthodolgie comparée du biomécanisme et de la mécanique comparée." *Dialectica* 10, no. 4 (December 1956): 319.

Driver, Felix. *Power and Pauperism: The Workhouse System, 1834–1884*. Cambridge: Cambridge University Press, 1993.

Drucker, Peter F. "Are Unions Becoming Irrelevant?" *Wall Street Journal*, September 22, 1982.

Drysdale, C. R. *Alpine Heights and Change of Climate in the Prevention and Treatment of Pulmonary Consumption*. London: Odell & Ives, 1869.

Dudden, Faye. *Serving Women: Household Service in 19th-Century America*. Middletown, CT: Wesleyan University Press, 1983.

Durbach, Nadja. *Bodily Matters: The Anti-Vaccination Movement in England, 1853–1907*. Durham, NC: Duke University Press, 2005.

Durbach, Nadja. "Class, Gender, and the Conscientious Objector to Vaccination, 1898–1907." *Journal of British Studies* 41, no. 1 (2002): 58–83.

Dyer, G. "Thieves, Boxers, Sodomites, Poets: Being Flash to Byron's *Don Juan*." *PMLA* 116, no. 3 (2001): 562–78.

Eassie, William. *Healthy Houses: A Handbook to the History, Defects, and Remedies of Drainage, Ventilation, Warming, and Kindred Subjects*. London: Simpkins, Marshall, 1872.

Eassie, William. *Sanitary Arrangements for Dwellings: Intended for the Use of Officers of Health, Architects, Builders and Householders*. London: Smith, Elder and Co., 1874.

Ehrenreich, Barbara, and Deirdre English. *For Her Own Good: Two Centuries of the Experts' Advice to Women*. Rev. ed. New York: Random House, 2005.

Eksteins, M. *Rites of Spring: The Great War and the Birth of the Modern Age*. Boston: Houghton, 1989.

Elias, Norbert. *The Civilizing Process: The Development of Manners, Changes in the Code of Conduct and Feeling in Early Modern Times*. Translated by Edmund Jephcott. New York: Urizen Books, 1978.

Elias, Norbert. *The Civilizing Process*. Vol. 1, *The History of Manners*. 1939. Reprint, Oxford: Blackwell, 1969.

Eliot, George. *Adam Bede*. Harmondsworth, UK: Penguin, 1980.

Ellis, Havelock. *Studies in the Psychology of Sex*. Philadelphia: F.A. Davis Company, 1900.

Ellis, Havelock, and John Addington Symonds. *Sexual Inversion*. 1897. Repr. New York: Arno Press, 1975.

Ellis, Juniper. *Tattooing the World: Pacific Designs in Print and Skin*. New York: Columbia University Press, 2008.

Ellis, Mrs. [Sarah Stickney]. *The Mothers of England. Their Influence and Responsibility*. New York: Edward Walker, 1850.

Elson, George. *The Last of the Climbing Boys: An Autobiography*. London: John Long, 1900.

Elston, M.A. "Women and Anti-Vivisection in Victorian England 1870–1900." In *Vivisection in Historical Perspective*. Edited by N.A. Rupke. London: Croom Helm, 1987.

Emmeche, Claus. *The Garden in the Machine: The Emerging Science of Artificial Life*. Translated by Steven Sampson. Princeton, NJ: Princeton University Press, 1994.

Epidemiological Society Small-Pox and Vaccination Committee. "Report on the State of Small-Pox and Vaccination in England and Wales and Other Countries, and on Compulsory Vaccination, With Tables and Appendices." Presented to the President and Council the 26th day of March 1853. PP 1852–53 XVIII (HL 256).

Evans, Richard. *Death in Hamburg: Society and Politics in the Cholera Years 1830–1910*. New York: Oxford University Press, 1987.

Farr, W. "Influence of Elevation on Fatality of Cholera." *Journal of the Royal Statistical Society* 15 (1852):155–83.

Farr, W. *Report on the Mortality of Cholera in England 1848–49*. London, 1852.

Farr, W. *Vital Statistics. A Memorial Volume of Selections from the Reports and Writings of William Farr*. Introduction by Mervyn Susser and Abraham Adelstein. Metuchen, NJ: Scarecrow Press, 1975.

Fass, Paula, and Mary Mason, eds. *Childhood in America*. New York: NYU Press, 2000.

Fausto-Sterling, A. *Sexing the Body: Gender Politics and the Construction of Sexuality*. New York: Basic Books, 1999.

Feher, Michel, with Ramona Naddaff and Nadia Tazi, eds. *Fragments for a History of the Human Body* [*Zone* 3–5]. New York: Zone Books, 1989.

Feuer, L.S. "The Friendship of Edwin Ray Lankester and Karl Marx: The Last Episode in Marx's Intellectual Evolution." *Journal of the History of Ideas* 40 (1979): 633–48.

Fichman, M., and J. Keelan. "Resister's Logic: The Anti-vaccination Arguments of Alfred Russel Wallace and their Role in the Debates Over Compulsory Vaccination in England, 1870–1907." *Studies in History and Philosophy of Science Part C: Studies in History and Philosophy of Biological and Biomedical Sciences* 38, no. 3 (2007): 585–607.

Fineman, Mia. "Ecce Homo Prostheticus." *New German Critique* 76 (Winter 1999): 85–115.

Fischer-Dückelmann, A. *Das Geschlechtsleben des Weibes. Eine pysiologisch-soziale Studie mit ärztliche Ratschlägen.* 13th ed. Berlin: Hugo Bermühler, 1908.

Fischer-Dückelmann, A. *Die Frau als Hausärztin.* 2nd special ed. Stuttgart, Germany: Süddeutsches Verlags-Institut, 1905.

Fitzpatrick, Sheila. "War and Society in Soviet Context: Soviet Labor before, during, and after World War II." *International Labor and Working-Class History* 35 (Spring 1989): 37–52.

Flanders, Judith. *Inside the Victorian Home: A Portrait of Domestic Life in Victorian England.* New York: W. W. Norton, 2003.

Fleming, Juliet. "The Renaissance Tattoo." In *Written on the Body: The Tattoo in European and American History.* Edited by Jane Caplan, 61–82. Princeton, NJ: Princeton University Press, 2000.

Forth, Christopher E. *The Dreyfus Affair and the Crisis of French Manhood.* Baltimore: Johns Hopkins University Press, 2004.

Forth, Christopher E., and Ivan Crozier, eds. *Body Parts: Critical Explorations in Corporeality.* Lanham, MD: Lexington Books, 2005.

[Foster, E.] *Murder and Mutilation! A Trial in a British Court of Justice.* John Moss, ca. 1873.

Foucault, Michel. *Abnormal.* New York: Picador, 2004.

Foucault, Michel. *Discipline and Punish: The Birth of the Prison.* Translated by Alan Sheridan. New York: Pantheon, 1978.

Foucault, Michel. "Nietzsche, Genealogy, History." In *Language, Counter-Memory, Practice: Selected Essays and Interviews.* Translated by Donald F. Bouchard and Sherry Simon. Ithaca, NY: Cornell University Press, 1977.

Foucault, Michel. *The Archaeology of Knowledge.* Translated by A. M. Sheridan Smith. New York: Pantheon, 1972.

Foucault, Michel. *The Birth of the Clinic: An Archaeology of Medical Perception.* Translated by A. M. Sheridan Smith. New York: Pantheon 1973.

Foucault, Michel. *The History of Sexuality.* Vol. 1, *An Introduction.* Translated by Robert Hurley. New York: Pantheon, 1978.

Foucault, Michel. *The Order of Things: An Archaeology of the Human Sciences.* New York: Pantheon, 1970.

Fowler, J. K. *Arrested Pulmonary Tuberculosis.* London: Churchill, 1892.

Franks, A. W. "Note on the Tattooed Man from Burmah." *Journal of the Anthropological Institute of Great Britain and Ireland* 2 (1873): 231–32.

French, R. D. *Anti-Vivisection and Medical Science in Victorian Society.* Princeton, NJ: Princeton University Press, 1975.

Freud, Sigmund. *Standard Edition of the Works of Sigmund Freud,* 24 vols. Translated by James Strachey. London: Hogarth Press and the Institute of Psychoanalysis, 1953–1974.

Freud, Sigmund. *Three Essays on the Theory of Sexuality.* Translated by James Strachey. New York: Basic Books, 1962.

Freud, Sigmund, and Joseph Breuer. *Studies in Hysteria.* Translated by Nicola Luckhurst. Harmondsworth, UK: Penguin, 2004.

Fritsch, Gustav. *Die Gestalt des Menschen: Mit Benutzung der Werke von E. Harless und C. Schmidt für Künstler und Anthropologen dargestellt.* Stuttgart, Germany: Paul Neff, 1899.

Fritsch, Gustav. "Die graphischen Methoden zur Bestimmung der Verhältnisse des menschlichen Körpers." *Zeitschrift für Ethnologie* 27 (1895): 172–88.
Fritsch, Gustav. *Ne sutor supra crepidam*. Berlin: Carl Habel, 1894.
Fritsch, Gustav. *Unsere Körperform im Lichte der modernen Kunst*. Berlin: Carl Habel, 1893.
Fuchs, Rachel G. *Poor and Pregnant in Paris: Strategies for Survival in the Nineteenth Century*. New Brunswick, NJ: Rutgers University Press, 1992.
Fulford, Roger. *Dearest Child: Letters Between Queen Victoria and the Princess Royal, 1858–1861*. New York: Holt, Rinehart, and Winston, 1964.
Furnival, W. J. *The Conscientious Objector: Who He Is! What He Has! What He Wants! And Why!* privately published, 1902.
Furphy, J. [Tom Collins]. *Such Is Life*, http://www.gutenberg.org/etext/3470.
Fuss, D. *Inside/Out Lesbian Theories, Gay Theories*. Edited by Diana Fuss. New York: Routledge, 1991.
Garb, T. *Bodies of Modernity: Figures and Flesh in Fin de Siècle France*. London: Thames and Hudson, 1998.
Garton, S. *Histories of Sexuality*. London: Equinox, 2004.
Gay, Peter. *The Bourgeois Experience*, vol. 1, *The Education of the Senses*. New York: Oxford University Press, 1984.
Gayer, G. W. *Some Criminal Tribes of India and Religious Mendicants*. Saugor, India: Police Officers' Training School, 1907.
Gelfand, T. *Professionalizing Modern Medicine: Paris Surgeons and Medical Science and Institutions in the Eighteenth Century*. Westport, CT: Greenwood Press, 1980.
Gell, A. *Wrapping in Images: Tattooing in Polynesia*. Oxford: Clarendon Press, 1993.
Gengenbach, Heidi. "Tattooed Secrets: Women's History in Magude District, Southern Mozambique." In *Bodies in Contact: Rethinking Colonial Encounters in World History*. Edited by Tony Ballantyne and Antoinette Burton, 253–73. Durham, NC: Duke University Press, 2005.
Gerlach, P. *Proportionen, Körper, Leben. Quellen; Entwürfe, Kontroversen*. Cologne, Germany: Apex, 1990.
Gerlach, P. "Schadow's Poyclet (1834). Die Bedeutung der Vermessung antiker Statuen für die Proportionslehre." In *Beiträge zur Theorie der Künste im 19. Jahrhundert*, vol. 1. Edited by H. Koopmann and J. A. Schmoll, 161–92. Frankfurt, Germany: Vittorio Klostermann, 1971.
Geuter, Ulfried. *The Professionalization of German Psychology in Nazi Germany*. Cambridge: Cambridge University Press, 1992.
Gibbs, J. *More Words on Vaccination*. London: Willis and Sotheran, 1856.
Giese, Fritz. *Handbuch Der Arbeitswissenschaft*. Halle, Germany: Marhold, 1925.
Giese, Fritz. *Nietzsche—Die Erfüllung*. Tübingen, Germany: Mohr, 1934.
Giese, Fritz. "Psychologie der Arbeitshand." In *Handbuch der Biolgischen Arbeitsmethoden*. Edited by Emil Abderhalden, 804. Berlin: Urban & Schwarzenberg, 1928.
Gijswijt-Hofstra, M., and R. Porter. *Cultures of Neurasthenia from Beard to the First World War*. Amsterdam: Rodopi, 2001.
Gilchrist, Agnes Addison. *William Strickland, Architect and Engineer, 1788–1854*. Philadelphia: University of Pennsylvania Press, 1950.
Gilfoyle, Timothy J. *City of Eros: New York City, Prostitution, and the Commercialization of Sex, 1790–1920*. New York: Norton, 1992.
Gilman, Charlotte Perkins. *The Yellow Wallpaper*. London: Virago, 1981.

Gilman, Sander. "Black Bodies, White Bodies: Toward an Iconography of Female Sexuality in Late 19th-century Art, Medicine, And Literature." In *Race, Writing and Difference*. Edited by Henry Gates, 223–61. Chicago: University of Chicago Press, 1985.

Gilman, Sander. *Difference and Pathology: Stereotypes of Sexuality, Race, and Madness*. Ithaca, NY: Cornell University Press, 1985.

Gilman, Sander. *Disease and Representation: Images of Illness from Madness to AIDS*. Ithaca, NY: Cornell University Press, 1988.

Gilman, Sander. *Sexuality: An Illustrated History, Representing the Sexual in Medicine and Culture from the Middle Ages to the Age of AIDS*. New York: John Wiley & Sons, 1989.

Glendening, John. *The Evolutionary Imagination in Late-Victorian Novels: An Entangled Bank*. Burlington, VT: Ashgate, 2008.

Gloucester Epidemic of Smallpox, 1895–96. Report of the Committee Appointed By the Board of Guardians to Organize and Carry Out the General Vaccination of the City and District, 1896.

Godwin, G. *London Shadows. A Glance at the 'Homes' of the Thousands*. London: G. Routledge & Co., 1854.

Godwin, G. *Town Swamps and Social Bridges. The Sequel to a Glance at the Homes of the Thousands*. London: Routledge, Warnes and Routledge, 1859.

Goldsby, Jacqueline. *A Spectacular Secret: Lynching in American Life and Literature*. Chicago: University of Chicago Press, 2006.

Goldstein, Jan. *Console and Classify: The French Psychiatric Profession in the 19th Century*. Chicago: University of Chicago Press, 1987.

Goltermann, S. "Exercise and Perfection. Embodying the Nation in 19th Century Germany." *European Review of History* 11 (2004): 333–46.

Goltz, F., and A. Freusberg. "Ueber den Einfluss des Nervensystems auf die Vorgänge während der Schwangerschaft und des Gebärakts." *Archiv für gesamte Physiologie* 9 (1874): 552–65.

Gomez, Michael A. *Exchanging Our Country Marks: The Transformation of African identities in the Colonial and Antebellum South*. Chapel Hill: University of North Carolina Press, 1998.

Gordon, Sarah. "Prestige, Professionalism, and the Paradox of Eadweard Muybridge's *Animal Locomotion* Nudes." *The Pennsylvania Magazine of History and Biography* 130 (January 2006): 79–104.

Gould, Stephen Jay. "The Hottentot Venus." In *The Flamingo's Smile*. New York: Norton, 1985.

Gould, Stephen Jay. *The Mismeasure of Man*. New York: Norton, 1981.

Govenar, Alan. "The Changing Image of Tattooing in American Culture, 1846–1966." In *Written on the Body*. Edited by Jane Caplan, 212–33. Princeton, NJ: Princeton University Press, 2000.

Gradmann, C. *Krankheit im Labor: Robert Koch und die medizinische Bakteriologie*. Göttingen, Germany: Wallstein, 2005.

Gradmann, C. "Robert Koch and the Pressures of Scientific Research: Tuberculosis and Tuberculin." *Medical History* 45 (2001): 1–32.

Grant, Hague W. *The Eugenic Marriage: A Personal; Guide to the New Science of Better Living and Healthy Babies*. New York: Review of Review Companies, 1916.

Gray, Thomas. *Elegy Written in a Country Church Yard*. 4th ed. London: R. Dodsley, 1751.

Green, Harvey. *Fit for America: Health, Fitness, Sport, and American Society*. New York: Pantheon Books, 1986.

Green, T. H. *An Introduction to Pathology and Morbid Anatomy*. London: Henry Renshaw, 1871.

Grove, William Robert. *On the Correlation of Physical Forces*. 5th ed. London: Longmans, Green, 1867.

Gussow, Z., and G. S. Tracy. "Stigma and the Leprosy Phenomenon: The Social History of the Disease in the Nineteenth and Twentieth Centuries." *Bulletin of the History of Medicine* 44, no. 5 (1970): 425–49.

Guth, Christine M. E. *Longfellow's Tattoos: Tourism, Collecting, and Japan*. Seattle: University of Washington Press, 2004.

Guy, Jeff. *The Destruction of the Zulu Kingdom: The Civil War in Zululand 1879–1884*. London: Longman, 1979.

Guyot-Daubès. *Curiosités physiologiques: Les homme-phénomès: Force—agilité—adresse*, edited by G. Masson. Paris: Libraire de l'Académie de Médecine, 1885.

Haas, Robert Bartlett. *Muybridge: Man in Motion*. Berkeley: University of California Press, 1976.

Haddington, Earl of. *Select Poems on Several Occasions by the Right Honorable The Earl of H*****N*. London, 1824.

Hadwen, Walter. *The Case Against Vaccination*. London: National Anti-Vaccination League, 1896.

Haeberle, E. "Der verbotene Akt: Unzüchtige Fotos von 1850–1950." In *Das Aktfoto: Ansichten vom Körper im fotografischen Zeitalter*. Edited by Michael Köhler and Gisela Barche, 212–24. Munich: Bucher, 1985.

Haeckel, E. *Der Monismus als Band zwischen Religion und Wissenschaft*. Bonn, Germany: Strauss, 1892.

Haeckel, E. *Kunstformen der Natur* [Art Forms of Nature]. Leipzig, Germany: Verl. des Bibliogr. Inst., 1899–1904.

Haeckel, E. *The History of Creation: Or the Development of the Earth and its Inhabitants by the Action of Natural Causes*. New York: Appleton, 1876.

Haeckel, E. *The Wonders of Life*. New York: Harper, 1905.

Haigh, E. *Xavier Bichat and the Medical Theory of the Eighteenth Century*. London: Wellcome Institute for the History of Medicine, 1984.

Haiken, Elizabeth. *Venus Envy: A History of Cosmetic Surgery*. Baltimore: Johns Hopkins University Press, 1997.

Haire, N. *Rejuvenation: The Work of Steinach, Voronoff, and Others*. London: Allen and Unwin, 1924.

Hales, Peter Bacon. *Silver Cities: The Photography of American Urbanization, 1839–1915*. Philadelphia: Temple University Press, 1983.

Haley, B. *The Healthy Body and Victorian Culture*. Cambridge, MA: Harvard University Press, 1978.

Hall, C. *Civilizing Subjects: Colony and Metropole in the English Imagination 1830–1867*. Chicago: University Chicago Press, 2002.

Hall, Donald E., ed. *Muscular Christianity: Embodying the Victorian Age*. Cambridge: Cambridge University Press, 1994.

Hall, Lesley. "British fin de Siecle Sexology." In *New Sexual Agendas*. Edited by Lynne Segal. New York: NYU Press, 1997.

Halperin, D. *How to Do the History of Homosexuality*. Chicago: University of Chicago Press, 2002.

Halttunen, Karen. *Confidence Men and Painted Women: A Study of Middle-Class Culture in America, 1830–1870*. New Haven, CT: Yale University Press, 1982.

Hambly, W. D. *The History of Tattooing and Its Significance with Some Account of Other Forms of Corporal Marking*. London: H.F. and G. Witherby, 1925.

Hamlin, Christopher. *Public Health and Social Justice in the Age of Chadwick: Britain, 1800–1854*. Cambridge: Cambridge University Press, 1998.

Hamlin, Christopher. *A Science of Impurity: Water Analysis in Nineteenth Century England*. Bristol, UK: Adam Hilger, 1990.

Hanson, E. *Decadence and Catholicism*. Cambridge, MA: Harvard University Press, 1997.

Hardy, Anne. *Health and Medicine in Britain since 1860*. Basingstoke, UK: Macmillan, 2000.

Hardy, Anne. *The Epidemic Streets: Infectious Disease and the Rise of Preventive Medicine, 1856–1900*. Oxford, UK: Clarendon Press, 1993.

Harrison, M. *Climates and Constitutions*. New Delhi, India: Oxford University Press, 1999.

Harrison, M. *Public Health in British India: Anglo-Indian Preventive Medicine, 1859–1914*. Cambridge: Cambridge University Press, 1994.

Harrow, B. *Glands in Health and Disease*. New York: Dutton, 1922.

Hart, Sidney. "A Graphic Case of Transatlantic Republicanism." In *New Perspectives on Charles Willson Peale: A 250th Anniversary Celebration*. Edited by Lillian B. Miller and David C. Ward, 73–81. Pittsburgh: University of Pittsburgh Press, 1991.

Hartley, L. "A Science of Beauty? Femininity, Fitness and the Nineteenth-Century Physiognomic Tradition." *Women: A Cultural Review* 12, no. 1 (2001): 19–34.

Hau, M. "The Holistic Gaze in German Medicine, 1890–1930." *Bulletin of the History of Medicine* 74 (2000): 495–524.

Hau, M. "Sports in the Human Economy: '*Leibesübungen*.' Medicine Psychology and Performance Enhancement during the Weimar Republic." *Central European History* 41 (2008): 381–412.

Hau, M. *The Cult of Health and Beauty in Germany, 1890–1930*. Chicago: University of Chicago Press, 2003.

Heckscher, Charles F. *The New Unionism: Employee Involvement in the Changing Corporation*. New York: Basic Books, 1987.

Hegel, G. *Phenomenology of the Spirit*. Translated by A.V. Miller. Oxford: Oxford University Press, 1977.

Hellerstein, Erna Olafson Hellerstein, Leslie Parker Hume, and Karen M. Offen, eds. *Victorian Women: A Documentary Account of Women's Lives in Nineteenth-Century England, France, and the United States*. Stanford, CA: Stanford University Press, 1981.

Hemans, Felicia. "Casabianca." *Monthly Magazine or British Register* (1826).

Herzig, R. "Removing Roots: North American 'Hiroshima Maidens' and the X-Ray." *Technology and Culture* 40 (1999): 723–45.

Hessenbruch, A. "A Brief History of X-Rays." *Endeavour* 26 (2002): 137–41.

Hill, O. *Our Common Land, and Other Short Essays*. London: Sydenham, 1877.

Hilts, V.L. "Epidemiology and the Statistical Movement." *Times, Places, and Persons: Aspects of the History of Epidemiology, Bulletin of the History of Medicine Supplements*, NS 4 (1980): 43–60.

Hinrichs, Peter, Lothar Peter, and G. Briefs. *Industrieller Friede? Arbeitswissenschaft, Rationalisierung Und Arbeiterbewegung in Der Weimarer Republik, Kleine Bibliothek*. Cologne: Pahl-Rugenstein, 1976.

Hirschfeld, M. *Homosexuality of Men and Women*. Translated by Michael Lombardi-Nash. New York: Prometheus Books, 2000.

Hirschfeld, M. *Les Homosexuels de Berlin*. Paris: Librarie Medicale, 1908.

Holland, Catherine A. *The Body Politic: Foundings, Citizenship, and Difference in the American Political Imagination*. New York: Routledge, 2001.

Hollington, Michael. "Dickens, 'Phiz' and Physiognomy." In *Imagination on a Long Rein: English Literature Illustrated*. Edited by Joachim Möller, 125–35. Marburg, Germany: Jonas Verlag, 1988.

Hollins, S.T. *The Criminal Tribes of the United Provinces*. Allahabad, India: Government Press, 1914.

Holmes, Rachel. *The Hottentot Venus*. Bloomsbury, UK: Random House, 2006.

Holt, Richard J. *Sport and the British: A Modern History*. Oxford, UK: Clarendon Press, 1989.

Holt, Richard J. "Working Class Football and the City: The Problem of Continuity." *Proceedings of the Third Annual Conference of the British Society of Sports History* (1985): 64–85.

Holton, S.S. "Silk Dresses and Lavender Kid Gloves: The Wayward Career of Jessie Craigen, Working Suffragist." *Women's History Review* 5, no. 1 (1996): 129–49.

Horn, Pamela. *Pleasures and Pastimes in Victorian Britain*. Stroud, UK: Sutton, 1999.

Howard, John. *The State of the Prisons in England and Wales*. Warrington, UK: William Eyres, 1777.

Howard-Jones, N. *The Scientific Background to the International Sanitary Conferences, 1851–1938*. Geneva, Switzerland: World Health Organization, 1975.

Hueppe, F. *Hygiene der Körperübungen*. Leipzig, Germany: S. Hirzel, 1910.

Huet, M.H. *Monstrous Imagination*. Cambridge, MA: Harvard University Press, 1993.

Hughes, Thomas P. *Networks of Power: Electrification in Western Society, 1880–1930*. Baltimore: Johns Hopkins University Press, 1993.

Hughes, Thomas. *Tom Brown's Schooldays*. Oxford: Oxford University Press, 1999.

Hume-Rothery, M. *150 Reasons for Disobeying the Vaccination Law, by Persons Prosecuted Under It*. Cheltenham, UK: George F. Poole, 1871.

Hunt, L. "Pornography and the French Revolution." In *The Invention of Pornography*. Edited by Lynn Hunt. New York: Zone Books, 1996.

Hurley, Kelly. *The Gothic Body: Sexuality, Materialism, and Degeneration at the* fin de siècle. Cambridge: Cambridge University Press, 1996.

Hyam, R. *Empire and Sexuality: The British Experience*. Manchester, UK: Manchester University Press, 1990.

Isaacs, J.D. "D.D. Cunningham and the Aetiology of Cholera in British India, 1869–1897." *Medical History* 42 (1998): 279–305.

Jacob, François. *The Logic of Life: A History of Heredity*. New York: Random House, 1973.

Jaeger, S., and I. Stäuble. "Die Psychotechnik und ihre gesellschaftlichen Entwicklungsbedingungen." In *Anwendungeniim Berufsleben: Arbeits-Wirtschafts-Und Verkehrspsychologie, Kindlers "Psychologie Des 20. Jahrhunderts."* Bd. 13, edited by François Stoll, 53–94. Zürich: Kindler, 1981.

Jewson, N. "Disappearance of the *Sickman* from Medical Cosmologies 1770–1870." *Sociology* 10 (1976): 225–44.

Johnson, G. *On Cholera and Choleraic Diarrhoea; Their Nature, Cause and Treatment. Two Lectures, Delivered at the Church Missionary College, Islington*. London: John Churchill and Sons, 1870.

Johnson, James Weldon. *The Autobiography of an Ex-Coloured Man*. New York: Vintage, 1989.

Jordan, T.E. *The Degeneracy Crisis and Victorian Youth*. Albany: State University of New York Press, 1993.

Jordanova, Ludmilla. *Sexual Visions: Images of Gender in Science and Medicine Between the 18th and 20th Centuries*. Madison: University of Wisconsin Press, 1989.

Kaposi, Dr. "Der Tätowirte von Birma." *Wiener Medizinische Wochenschrift* 2 (1872): 39–45.

Kasson, John F. *Houdini, Tarzan, and the Perfect Man: The White Male Body and the Challenge of Modernity in America*. New York: Hill and Wang, 2001.

Kasson, John F. *Rudeness and Civility: Manners in Nineteenth-Century Urban America*. New York: Hill and Wang, 1990.

Kaye, Richard. *The Flirt's Tragedy*. Charlottesville: University of Virginia Press, 2002.

Kean, H. *Animal Rights*. London: Reaktion Books, 1998.

Kean, H. "The 'Smooth Cool Men of Science': The Feminist and Socialist Response to Vivisection." *History Workshop Journal* 40 (1995): 16–38.

Kelly, Kevin. *Out of Control: The New Biology of Machines*. London: Fourth Estate, 1994.

Kendrick, W. *The Secret Museum: Pornography in Modern Culture*. New York: Viking, 1987.

Kennedy, M. *Notes on Criminal Classes in the Bombay Presidency with Appendices regarding some Foreign Criminals who occasionally visit the Presidency, including Hints on the Detection of Counterfeit Coins*. Bombay, India: Government Central Press, 1908.

Kern, Stephen. *Anatomy and Destiny: A Cultural History of the Human Body*. Indianapolis: University of Indiana Press, 1975.

Kevles, B.H. *Naked to the Bone: Medical Imaging in the Twentieth Century*. New Brunswick, NJ: Rutgers University Press, 1997.

Kevles, Daniel. *In the Name of Eugenics: Genetics and the Uses of Human Heredity*. New York: Knopf, 1985.

Killen, Andreas. *Berlin Electropolis Shock, Nerves and German Modernity*. Berkeley: University of California Press, 2006.

Kingsley, Charles. *The Water-Babies*. Oxford: Oxford University Press, 1995.

Klein, M. "Sur les interférences des sciences fondamentales et de la clinique dans l'essor de l'endocrinologie sexuelle." *Clio Medica* 8 (1973): 31–52.

Kockerbeck, C. *Die Schönheit des Lebendigen. Ästhetische Naturwahrnehmung im 19. Jahrhundert*. Vienna: Böhlau, 1997.

Kohlrausch, W. "Zusammenhänge von Körperform und Leistung. Ergebnisse der anthropometrischen Messungen an den Athleten der Amsterdamer Olympiade." *Arbeitsphysiologie* 2 (1930): 187–204.

Kotkin, Stephen. *Magnetic Mountain: Stalinism as a Civilization*. Berkeley: University of California Press, 1995.

Koven, S. *Slumming: Sexual and Social Politics in Victorian London*. Princeton, NJ: Princeton University Press, 2004.

Krafft-Ebing, Richard von. *Psychopathia Sexualis with Especial Reference to Contrary Sexual Instinct, a Medico-legal Study*, trans. Charles Gilbert Chaddock. Philadelphia: F. A. Davis, 1893.

Kuhn, Thomas S. "Energy Conservation as an Example of Simultaneous Discovery." In *The Essential Tension: Selected Studies in Scientific Tradition and Change*. Edited by Thomas S. Kuhn, 66–104. Chicago: University of Chicago Press, 1977.

Lahy, Jean-Marie. *Le Système Taylor: Analyse et commentaries*. Paris: Gauthier-Villars, 1921.

Lambert, R. J. "A Victorian National Health Service: State Vaccination 1855–71." *The Historical Journal* 5, no. 1 (1962): 1–18.

Lambert, R. J. *Sir John Simon 1816–1904 and English Social Administration*. London: Macgibbon and Kee, 1963.

La Mettrie, Julien Offray de. *L'Homme-machine*. 1748; repr. Paris: Denoël-Gonthier, 1981.

Landes, J. *Visualizing the Nation: Gender, Representation, and Revolution in Eighteenth-Century France*. Ithaca, NY: Cornell University Press, 2001.

Lansbury, C. *The Old Brown Dog: Women, Workers, and Vivisection in Edwardian England*. Madison: University of Wisconsin Press, 1985.

Laqueur, Thomas. *Making Sex: Body and Gender from the Greeks to Freud*. Cambridge, MA: Harvard University Press, 1990.

Laqueur, Thomas, and Catherine Gallagher, eds. *The Making of the Modern Body: Sexuality and Society in the 19th Century*. Berkeley: University of California Press, 1987.

Lasch, Christopher. *The Culture of Narcissism: American Life in an Age of Diminishing Expectations*. New York: Norton, 1979.

Lastra, James. *Sound Technology and the American Cinema: Perception, Representation, Modernity*. New York: Columbia University Press, 2000.

Latour, Bruno. *Science in Action: How to Follow Scientists and Engineers Through Society*. Cambridge, MA: Harvard University Press, 1987.

Latour, Bruno. *The Pasteurization of France*. Cambridge, MA: Harvard University Press, 1987.

Lawlor, C. *Consumption and Literature: The Making of the Romantic Disease*. Basingstoke, UK: Palgrave, 2007.

Lawrence, C. *Medicine in the Making of Modern Britain, 1700–1920*. London: Routledge, 1994.

Lawson, Karol Ann Peard. "Charles Willson Peale's *John Dickinson*: An American Landscape as Political Allegory." *Proceedings of the American Philosophical Society* 136 (December 1992): 455–86.

Lears, J. "American Advertising and the Reconstruction of the Body." In *Fitness in American Culture: Images of Health, Sport, and the Body, 1830–1940*. Edited by Kathryn Grover, 47–66. Amherst: University of Massachusetts Press, 1989.

Leavitt, Judith Walzer. *Brought to Bed: Childbearing in America, 1750–1950*. New York: Oxford University Press, 1986.

Leneman, L. "The Awakened Instinct: Vegetarianism and the Women's Suffrage Movement in Britain." *Women's History Review* 6, no. 2 (1997): 271–87.

Leoussi, A. *Nationalism and Classicism: The Classical Body as National Symbol in Nineteenth-Century England and France*. London: Macmillan, 1998.

Levine, Philippa. *Prostitution, Race and Politics: Policing Venereal Disease in the British Empire*. London: Routledge, 2003.

Levine, Philippa. "Venereal Disease, Prostitution, and the Politics of Empire: The Case of British India." *Journal of the History of Sexuality* 4 (1993–1994): 579–602.

Lewerentz, A. "Der Mediziner Gustav Fritsch als Fotograf. Dokumentation seiner anthropologisch-ethnographischen Untersuchungen in Fotografien der Berliner Gesellschaft für Anthropologie, Ethnologie und Urgeschichte." *Baessler-Archiv, Neue Folge* 48 (2000): 271–309.

[Lewes, George H.] "Training in Relation to Health." *Cornhill Magazine* 9 (1864): 219–31.

Leydig, F. "Zur Anatomie der männlichen Geschlechtsorgane und Analdrüsen der Säugethiere." *Zeitschrift für wissenschaftliche Zoologie* 2 (1850): 1–57.

Limley, J de B. "Science Can Make You Grow Younger: Dr. Lorenz Illustrates from His Own Experience How a Simple Operation Renews Youthful Energy." *Liberty* 13 (March 1926): 19–22.

Lincoln, Abraham. "From Address to the Wisconsin State Agricultural Society, Milwaukee, Wisconsin." In *Selected Speeches and Writings by Abraham Lincoln*. Edited by Don E. Fehrenbacher, 233–37. New York: Vintage Books, Library of America Edition, 1992.

Linton, W.J. "Art in Engraving on Wood." *The Atlantic Monthly* 43 (June 1879): 705–15.

Lipschütz, A. *Internal Secretions of the Sex Glands: The Problem of the "Puberty Gland."* Cambridge: Heffer, 1924.

Litwack, Leon. "Hellhounds." In *Without Sanctuary: Lynching Photography in America*. Edited by James Allen, 8–37. Santa Fe, NM: Twin Palms Publishers, 2004.

Livi-Bacci, Massim. *The Population of Europe: A History of Europe*. Oxford: Blackwell, 2000.

Lock, S. "'O That I Were Young Again': Yeats and the Steinach Operation." *British Medical Journal* 287 (1983): 1964–68.

London, Jack. *The Iron Heel*. 1907. Repr. Chicago: Lawrence Hills Books, 1981.

London Society for the Abolition of Compulsory Vaccination. *Catalogue of Anti-Vaccination Literature*. London: E.W. Allen, 1894.

Lowe, L. *Critical Terrains: French and British Orientalism*. Ithaca, NY: Cornell University Press, 1991.

Lowerson, John, and John Myerscough. *Time to Spare in Victorian England*. Hassocks, UK: Harvester Press, 1977.

Luckin, W. "The Final Catastrophe: Cholera in London, 1866." *Medical History* 21 (1977): 32–42.

Lukasik, Christopher J. "The Face of the Public." *Early American Literature* 39 (2004): 413–64.

Lutz, Tom. *American Nervousness 1903: An Anecdotal History*. Ithaca, NY: Cornell University Press, 1993.

Lyttelton, W.H. *Scripture Revelation of the Life of Man After Death*. London: Daldy, Isbisther & Co., 1875.

MacFadden, B. *Keeping Fit: Pep, Poise, Power for Men and Women*. New York: MacFadden Publications, 1929.

MacKenzie, C. *Psychiatry for the Rich: A History of Ticehurst Private Asylum, 1792–1917*. London: Routledge, 1992.

MacLeod, R.M. "Law, Medicine and Public Opinion: The Resistance to Compulsory Health Legislation 1870–1907." *Public Law* Summer (1967): 107–28, 189–211.

Maier, Charles S. *In Search of Stability: Explorations in Historical Political Economy*. Cambridge: Cambridge University Press, 1987.

Makela, M. *The Munich Secession: Art and Artists in Turn-of-the-Century Munich*. Princeton, NJ: Princeton University Press, 1990.

Mallon, Sean. "Samoan *Tatau* as Global Practice." In *Tattoo: Bodies, Art, and Exchange in the Pacific and Europe*. Edited by Nicholas Thomas, Anna Cole, and Bronwen Douglas, 145–70. London: Reaktion, 2005.

Malthus, Thomas. *An Essay on the Principle of Population*. Edited by Philip Appelman. 1798. Repr. New York: Norton, 1976.

Manca, Joseph. "Cicero in America: Civic Duty and Private Happiness in Charles Willson Peale's Portrait of William Paca." *American Art* 17 (Spring 2003): 68–89.

Mangan, J.A. *Athleticism in the Victorian and Edwardian Public School.* London: Frank Cass, 2000.

Mangan, J.A. *The Games Ethic and Imperialism: Aspects of the Diffusion of an Ideal.* New York: Viking, 1986.

Marchand, S. *Down from Olympus: Archaeology and Philhellenism in Germany, 1750–1970.* Princeton, NJ: Princeton University Press, 1996.

Marshall, F.H.A. *The Physiology of Reproduction.* 2nd ed. London: Longmans, 1922.

Marx, Karl. *Das Kapital: A Critique of Political Economy,* vol. 1. Translated by Ben Fowkes. London: Penguin Books, 1976.

Mason, Michael. *The Making of Victorian Sexuality.* New York: Oxford University Press, 1994.

Maulitz, R.C. "The Pathological Tradition." In *Companion Encyclopaedia of the History of Medicine,* vol. 1. Edited by W.F. Bynum and Roy Porter, 169–89. London: Routledge, 1993.

Maxwell-Stewart, Hamish, and Ian Duffield. "Skin Deep Devotions: Religious Tattoos and Convict Transportation to Australia." In *Written on the Body: The Tattoo in European and American History.* Edited by Jane Caplan, 118–35. Princeton, NJ: Princeton University Press, 2000.

Maza, S. *The Myth of the French Bourgeoisie.* Cambridge, MA: Harvard University Press, 2003.

Mazlish, Bruce. *The Fourth Discontinuity: The Co-Evolution of Humans and Machines.* New Haven, CT: Yale University Press, 1993.

McClintock, Anne. *Imperial Leather: Race, Gender, and Sexuality in the Colonial Contest.* New York: Routledge, 1995.

McMenemy, W.H. "Cellular Pathology." In *Medicine and Science in the 1860s.* Edited by F.N.L. Poynter, 13–43. London: Wellcome Institute for the History of Medicine, 1968.

McNeeley, Ian F. *"Medicine on a Grand Scale": Rudolf Virchow, Liberalism, and the Public Health.* London: Wellcome, 2002.

Medvei, V.C. *The History of Clinical Endocrinology.* Carnforth, UK: Parthenon, 1993.

Meredith, George. *The Ordeal of Richard Feverel.* 1859. Repr. New York: Penguin, 1999.

Micale, M. "Hysteria Male/Hysteria Female: Reflections on Comparative Gender Construction in Nineteenth-Century France and Britain." In *Science and Sensibility: Gender and Scientific Enquiry, 1780–1945.* Edited by Marina Benjamin, 200–39. Oxford: Blackwell, 1991.

Michie, H. *The Flesh Made Word: Female Figures and Women's Bodies.* New York: Oxford University Press, 1987.

Mileaf, Janine A. "Poses for the Camera: Eadweard Muybridge's Studies of the Human Figure." *American Art* 16 (Autumn 2002): 31–53.

Miles, Robert. *Gothic Writing, 1750–1820: A Genealogy.* London: Routledge, 1993.

Miller, Jacquelyn C. "The Body Politic and the Body Somatic: Benjamin Rush's Fear of Social Disorder and His Treatment for Yellow Fever." In *A Centre of Wonders: The Body in Early America.* Edited by Janet Moore Lindman and Michele Lise Tarter, 61–74. Ithaca, NY: Cornell University Press, 2001.

Mintz, Steven. *Huck's Raft: A History of American Childhood.* Cambridge, MA: Harvard University Press, 2004.

Mocek, R. "Der Naturphilosoph Carl Gustav Carus." *Dresdner Hefte* 7 (1989): 15–29

Moede, Walther. "Kraftfahrer-Eignungsprüfungen beim deutschen Heer 1915–1918." *IndustrielllePsychotechnik* 3, no. 1 (1926): 79.

Moeller, Susan D. "The Cultural Construction of Urban Poverty: Images of Poverty in New York City, 1890–1917." *Journal of American Culture* 18 (Winter 1995): 1–16.

Möhring, Maren. *Marmorleiber: Körperbildung in der deutschen Nacktkultur.* Cologne: Böhlau, 2004.

Moon, W. S. *The Secret of Perfect Health: Disease Rendered Preventable and Removable by Washing its Germs Out of the Body.* London: Sanitary Engineering, ca. 1890.

Mooney, G. "Infectious Diseases and Epidemiologic Transition in Victorian Britain? Definitely." *Social History of Medicine* 20 (2007): 595–606.

Moor, E. *Hindu Infanticide: An Account of the Measures Adopted for Suppressing the Practice of the Systematic Murder by Their Parents of Female Infants.* London: J. Johnson, 1811.

Moore, C. R. "The Regulation of Production and the Function of the Male Sex Hormone." *Journal of the American Medical Association* 97 (1931): 518–22.

Morel, B. A. *Traité des dégénérescences physiques, intellectuelles et morales de l'espèce humaine et des causes qui produisent ces variétés maladives.* London: H. Baillière, 1857.

Morgan, Jayne. "Eadweard Muybridge and W. S. Playfair: An Aesthetics of Neurasthenia." *History of Photography* 23 (Autumn 1999): 225–31.

Morton, Peter. *The Vital Science: Biology and the Literary Imagination 1860–1900.* London: Allen & Unwin, 1984.

Moscucci, Ornella. *The Science of Woman: Gynaecology and Gender in England, 1800–1929.* Cambridge: Cambridge University Press, 1990

Mosse, George L. *The Image of Man: The Creation of Modern Masculinity.* Oxford: Oxford University Press, 1996.

Mosse, George L. *Nationalism and Sexuality Respectability and Abnormal Sexuality in Modern Europe.* New York: Howard Fertig, 1985.

Mosso, Angelo. *Fatigue.* Translated by Margaret Drummond and William Blackley Drummond. New York: G. P. Putnam's Sons, 1904.

Mullet, C. F. *Public Baths and Health in England, 16th–18th Century.* Baltimore: Johns Hopkins University Press, 1946.

Muri, Allison. *The Enlightenment Cyborg: A History of Communications and Control in the Human Machine, 1660–1830.* Toronto: University of Toronto Press, 2006.

Muybridge, Eadweard. *Animal Locomotion: Prospectus and Catalogue of Plates* (Philadelphia, 1887), published in facsimile in Eadweard Muybridge, *Muybridge's Complete Human and Animal Locomtion*, vol. 3. Mineola, NY: Dover Publications, 1979, 1585–97.

Nasaw, David. *Children of the City: At Work and at Play.* New York: Oxford University Press, 1986.

Nussbaum, F. *Torrid Zones: Maternity, Sexuality, and Empire in Eighteenth-Century English Narratives.* Baltimore: Johns Hopkins University Press, 1995.

Nutton, V. "Humoralism." In *Companion Encyclopedia of the History of Medicine*, vol. 1. Edited by W.F. Bynum and Roy Porter, 281–91. London: Routledge, 1993.
Nye, Robert A. *Crime, Madness, & Politics in Modern France: The Medical Concept of National Decline*. Princeton, NJ: Princeton University Press, 1984.
Nye, Robert A. "National Sexological Traditions." *Science in Context* 4, no. 2 (1991): 387–406.
O'Donnell, K., and M. O'Rourke, eds. *Love, Sex, Intimacy, and Friendship Between Men, 1550–1800*. Houndmills, UK: Palgrave, 2003.
Oettermann, Stephan. "On Display: Tattooed Entertainers in America and Germany." In *Written on the Body: The Tattoo in European and American History*. Edited by Jane Caplan, 193–211. Princeton, NJ: Princeton University Press, 2000.
Office of National Statistics. *The Health of Adult Britain, 1841–1994*, vol. 1. London: Stationery Office, 1997.
Oosterhuis, H. *Stepchildren of Nature: Krafft-Ebing, Psychiatry, and the Making of Sexual Identity*. Chicago: University Chicago Press, 2000.
Oppenheim, J. "*'Shattered Nerves': Doctors, Patients, and Depression in Victorian England*. New York: Oxford University Press, 1991.
Otis, L. "The Metaphoric Circuit: Organic and Technological Communication in the Nineteenth Century." *Journal of the History of Ideas* 63 (2002): 105–28.
Otis, Laura. *Müller's Lab*. New York: Oxford University Press, 2007.
Ott, K. *Fevered Lives: Tuberculosis in American Culture Since 1870*. Cambridge, MA: Harvard University Press, 1996.
Ouida [Louise de la Ramee]. *Under Two Flags*. Oxford: Oxford University Press, 1995.
Outram, Dorinda. *The Body and the French Revolution: Sex, Class, and Political Culture*. New Haven, CT: Yale University Press, 1989.
Oxford English Dictionary. Oxford: Clarendon Press, 1989.
Oxford English Dictionary Online. "Stereotype." Oxford: Oxford University Press, 2006.
Park, Mungo. *Travels in the Interior Districts of Africa*. London: W. Bulmer and Co., 1799.
Park, Roberta J. "Healthy, Moral, and Strong. Educational Views of Physical Exercise and Athletics in Nineteenth-Century America." In *Fitness in American Culture: Images of Health, Sport, and the Body, 1830–1940*. Edited by Kathryn Grover, 123–68. Amherst: University of Massachusetts Press, 1989
Park, Roberta J. "Muscles, Symmetry and Action: 'Do You Measure Up?' Defining Masculinity in Britain and America from the 1860s to the early 1900s." *International Journal of the History of Sport* 22, no. 3 (2005): 365–95.
Parkes, Edmund. *Manual of Practical Hygiene: Prepared Especially for the Use of Medical Service on the Army*. London: John Churchill and Sons, 1864.
Pauly, P.J. "The Political Structure of the Brain: Cerebral Localization in Bismarckian Germany." *International Journal of Neuroscience* 21 (1983): 145–50.
Peers, Douglas M. "Colonial Knowledge and the Military in India, 1780–1860." *Journal of Imperial and Commonwealth History*, 33, no. 2 (2005): 157–80.
Pelling, Margaret. *Cholera, Fever and English Medicine, 1825–1860*. Oxford: Oxford University Press, 1978.
Pemberton, N., and M. Worboys. *Mad Dogs and Englishmen: Rabies in Britain, 1830–2000*. Basingstoke, UK: Palgrave, 2007.
Pemble, J. *The Mediterranean Passion: Victorians and Edwardians in the South*. Oxford: Clarendon Press, 1987.

Perrot, Michelle. "Roles and Characters." In *From the Fires of Revolution to the Great War*. Edited by Michelle Perrot, 167–259, vol. 4. In *A History of Private Life*. Edited by Phillippe Ariès and Georges Duby. Cambridge, MA: Harvard University Press, 1990.

Peterson, M.J. "Medicine." *Victorian Periodicals and Victorian Society*. Edited by J. Don Vann and Rosemary T. VanArsdel, 22–44. Toronto: University of Toronto Press, 1994.

Pick, Daniel. *Faces of Degeneration: A European Disorder, c. 1848–c. 1918*. Cambridge: Cambridge University Press, 1989.

Piepmeier, Alison. *Out in Public: Configurations of Women's Bodies in Nineteenth-Century America*. Chapel Hill: University of North Carolina Press, 2004.

Pierce, Steven, and Anupama Rao, eds. *Discipline and the Other Body: Correction, Corporeality, Colonialism*. Durham, NC: Duke University Press, 2006.

Pike, David L. *Subterranean Cities: The World Beneath Paris and London, 1800–1945*. Ithaca, NY: Cornell University Press, 2005.

Playfair, Sir Lyon. *Subjects of Social Welfare*. 2nd ed. London: Cassell & Co., 1889.

Poovey, Mary. *Making a Social Body: British Cultural Formation, 1830–1864*. Chicago: University of Chicago Press, 1995.

Porter, Dorothy. *Health, Civilization, and the State: A History of Public Health from Ancient to Modern Times*. London: Routledge, 1999.

Porter, Dorothy, and Roy Porter. "The Politics of Prevention: Anti-Vaccinationism and Public Health in Nineteenth-Century England." *Medical History* 32 (1988): 231–52.

Porter, Dorothy, and Roy Porter. "The Enforcement of Health: The British Debate." In *AIDS: The Burdens of History*, edited by E. Fee and D. M. Fox. Berkeley: University of California Press, 1988.

Porter, Roy. *Bodies Politic: Disease, Death and Doctors in Britain, 1650–1900*. London: Reaktion, 2001.

Porter, Roy. *Madness: A Brief History*. Oxford: Oxford University Press, 2002.

Porter, Roy. "Madness and Its Institutions." In *Medicine in Society*. Edited by A. Wear, 277–301. Cambridge: Cambridge University Press, 1992.

Porter, Roy. "Perversion in the Past." *Communicationes de Historia Artis Medicinae* 44 (1999): 17–25.

Porter, Roy. *The Greatest Benefit to Mankind*. London: Harper Collins, 1997

Porter, Roy. and Lesley Hall. *The Facts of Life: The Creation of Sexual Knowledge in Britain, 1650–1950*. New Haven, CT: Yale University Press, 1995.

Porter, Roy, and W.F. Bynum, eds. *Medicine and the Five Senses*. Cambridge: Cambridge University Press, 1993.

Porter, T. *The Rise of Statistical Thinking 1820–1920*. Princeton, NJ: Princeton University Press, 1986.

Power, A. *Sanitary Rhymes . . . Personal Precautions against Cholera, and all Kinds of Fever, etc*. London: T. Richards, 1871.

Presner, Todd Samuel. *Muscular Judaism: The Jewish Body and the Politics of Regeneration*. London: Routledge, 2007.

Putney, C. *Muscular Christianity: Manhood and Sports in Protestant America*. Cambridge, MA: Harvard University Press, 2001.

Qureshi, Sadiah. "Displaying Sara Baartman, the 'Hottentot Venus'." *History of Science* 42 (2004): 233–57.

Rabinbach, Anson. "Betriebspsychologie zwischen Psychotechnik und Politik während der Weimarer Republik: Der Fall Otto Lipmann." In *Betriebsärzte und produk-*

tionsbezogene Gesundheitspolitik. Edited by Dietrich Milles, 41–64. Bremerhaven, Germany: Verlag für neue Wissenschaft, 1992.

Rabinbach, Anson. *The Human Motor: Energy, Fatigue, and the Origins of Modernity*. New York: Basic Books, 1990.

Ransome, A., "The Consumption Scare." *Medical Chronicle* 2 (1895): 241–49.

Rather, L. J. "Virchow's Review of Rokitansky's *Handbuch* in the *Preussische Medizinal-Zeitung*, Dec. 1846." *Clio Medica* 4 (1969):127–40.

Rawlinson, R. "Old Lessons in Sanitary Science Revived, and New Lessons Considered." *Transactions of the Sanitary Institute* 2 (1880–81): 127–28.

Registrar General of Births, Deaths and Marriages in England Forty-second Annual Report, British Parliamentary Papers, 1881 [C.2907], 27, p. 79.

Reis, João José. *Death Is a Festival: Funeral Rites and Rebellion in Nineteenth-Century Brazil*. Chapel Hill: University of North Carolina Press, 2003.

Riall, Lucy. *Garibaldi: Invention of a Hero*. New Haven, CT: Yale University Press, 2007.

Ribbert, H. "Ueber die compensatorische Hypertrophie der Geschlechtsdrüsen." *Archiv der pathologische Anatomie und Physiologie* 120 (1890): 247–72.

Rice, Stephen P. *Minding the Machine: Languages of Class in Early Industrial America*. Berkeley: University of California Press, 2004.

Richards, Eveleen. "Darwin and the Descent of Woman." In *The Wider Domain of Evolutionary Thought*. Edited by D. Oldroyd and I. Langham, 57–111. Dordrecht: Reidel, 1983.

Richards, Graham. *Mental Machinery: The Origins and Consequences of Psychological Ideas*. Baltimore: Johns Hopkins University Press, 1993.

Richards, Robert J. *The Romantic Conception of Life: Science and Philosophy in the Age of Goethe*. Chicago: University of Chicago Press, 2002.

Richards, Robert J. *The Tragic Sense of Life: Ernst Haeckel and the Struggle Over Evolutionary Thought*. Chicago: University of Chicago Press, 2008.

Richardson, Michael K., and Gerhard Keuck. "Haeckel's ABC of Evolution and Development." *Biological Reviews* 77 (2002): 495–528.

Richardson, Ruth. *Death, Dissection and the Destitute*. London: Routledge, 1987.

Ridley, Matt. *Genome: The Autobiography of a Species in 23 Chapters*. New York: Harper, 1999.

Riegel, Robert E. "The Introduction of Phrenology to the United States." *The American Historical Review* 39 (October 1933): 73–78.

Riley, J. C. *Rising Life Expectancy: A Global History*. Cambridge: Cambridge University Press, 2001.

Riley, J. C. *Sick, Not Dead: The Health of British Workingmen During the Mortality Decline*. Baltimore: Johns Hopkins University Press, 1997.

Riskin, Jessica. "The Defecating Duck, or, the Ambiguous Origins of Artificial Life." *Critical Inquiry* 24, no. 4 (Summer 2003): 599–633.

Roberts, H. *The Dwellings of the Labouring Classes, their arrangement and construction, illustrated by a reference to the Model Houses of the Society for Improving the Condition of the Labouring Classes, and other buildings recently erected: being An Essay, Read January 21, 1850, at the Royal Society of British Architects*. London: Society for Improving the Condition of the Labouring Classes, 1850.

Robertson, R. "Family History in Relation to Contagion in Phthisis Pulmonalis," *British Medical Journal* ii (1882): 624.

Robley, Horatio. *Moko; or Maori Tattooing*. London: Chapman and Hall, 1896.

Rockwood, George. "Photography as Legal Evidence." *The Manufacturer and Builder* 21, no. 4 (1889): 81.

Romano, Terrie M. "The 1865 Cattle Plague and the Reception of the 'Germ Theory' in Mid-Victorian Britain." *Journal of the History of Medicine and Allied Sciences* 52, no. 1 (1997): 51–80.

Romberg, M. *Lehrbuch der Nervenkrankheiten des Menschen*. 2nd ed. Berlin: Duncker, 1851.

Rose, Nikolas. *Inventing Our Selves: Psychology, Power, and Personhood*. Cambridge: Cambridge University Press, 1996.

Rosenberg, C. E. "Body and Mind in Nineteenth-century Medicine: Some Clinical Origins of the Neurosis Concept." In *Explaining Epidemics and Other Studies in the History of Medicine*. Edited by C. E. Rosenberg, 77. Cambridge: Cambridge University Press, 1992.

Rosenberg, C. E. "Catechisms of Health: The Body in the Prebellum Classroom." *Bulletin of the History of Medicine* 69 (Summer 1995): 175–97.

Rosenberg, C. E. *The Cholera Years: The United States in 1832, 1849, and 1866*. Chicago: University of Chicago Press, 1962.

Rosenberg, C. E. "Contested Boundaries: Psychiatry, Disease, and Diagnosis." *Perspectives in Biology and Medicine* 49 (2006): 407–24.

Rosenberg, C. E. *Explaining Epidemics and Other Studies in the History of Medicine*. Cambridge: Cambridge University Press, 1992.

Rosenberg, C. E. "Pathologies of Progress: The Idea of Civilization as Risk." *Bulletin of the History of Medicine* 72 (1998): 714–30.

Rosenberg, C. E. "The Place of George M. Beard in Nineteenth-Century Psychiatry." *Bulletin of the History of Medicine* 36 (1962): 245–59.

Rosenkranz, K. *Ästhetik des Hässlichen*. 1853. Repr. Darmstadt, Germany: Wissenschaftliche Buchgesellschaft, 1973.

Ross, Chad. *Naked Germany: Health, Race and the Nation*. Oxford: Berg, 2005.

Ross, Robert. *Status and Respectability in the Cape Colony, 1750–1870: A Tragedy of Manners*. Cambridge: Cambridge University Press, 1999.

Russell, Bertrand. *Marriage and Morals*. London: Allen Unwin, 1929.

Rycroft, Daniel. *Representing Rebellion: Visual Aspects of Counter-insurgency in Colonial India*. New Delhi, India: Oxford University Press, 2006.

Samuel, S. "Das Gewebswachsthum bei Störungen der Innervation." *Archiv der pathologische Anatomie und Physiologie* 113 (1888): 272–314.

Samuel, S. "Trophoneurosen." In *Real-Encyclopädie der gesammten Heilkunde*. 20 vols. Edited by Albert Eulenburg. Vienna: Urban & Schwarzenberg, 1885–1890.

Samuels, Shirley, ed. *The Culture of Sentiment: Race, Gender, and Sentimentality in 19th-century America*. New York: Oxford University Press, 1992.

Sánchez-Eppler, Karen. *Touching Liberty: Abolition, Feminism, and the Politics of the Body*. Berkeley: University of California Press, 1993.

Sand, K. "Die Physiologie des Hodens." In *Handbuch der inneren Sekretion*. 2 vols. Edited by M. Hirsch, 2017–2268. Leipzig, Germany: Kabbitzsch, 1928–1932.

Sanger, Margaret. *The Woman Rebel*. 1914.

Sappol, Michael. *Dream Anatomy*. Washington, DC: Government Printing Office, 2006.

Sappol, Michael. *Traffic of Dead Bodies: Anatomy and Embodied Social Identity in Nineteenth-Century America*. Princeton, NJ: Princeton University Press, 2002.

Saur, P. B. *Maternity: A Book for Every Wife and Mother*. Chicago: L. P. Miller, 1887.

Scarpelli, G. "'Nothing in Nature That Is Not Useful': The Anti-Vaccination Crusade and the Idea of *Harmonia Naturae* in Alfred Russel Wallace." *Nuncius* 7, no. 1 (1992): 109–30.
Schadow, G. *Polyclet oder von den Maassen des Menschen nach dem Geschlechte und Alter*. 5th. ed. Berlin: Ernst Wasmuth, 1886.
Schiebinger, L. "Colonial Science." *Isis* 96 (March 2005): 52–88.
Schiebinger, L. *Nature's Body*. Cambridge, MA: Harvard University Press, 1990.
Schiebinger, L. "Women and Gender in Science." In *The Cambridge History of Science*, vol. 4. Edited by Roy Porter. Cambridge: Cambridge University Press, 2003.
Schildkrout, Enid. "Inscribing the Body." *Annual Review of Anthropology*, 33 (2004): 319–44.
Schmidt, P. *The Conquest of Old Age: Methods to Effect Rejuvenation and to Increase Functional Activity*. New York: Dutton, 1931.
Schmidt, P. *The Theory and Practice of the Steinach Operation with a Report on One Hundred Cases*. London: Heinemann, 1924.
Schor, Juliet. *The Overworked American: The Unexpected Decline of Leisure*. New York: Basic Books, 1991.
Schrader, Abby. "Branding the Other/Tattooing the Self: Bodily Inscription among Convicts in Russia and the Soviet Union." In *Written on the Body: The Tattoo in European and American History*. Edited by Jane Caplan, 174–92. Princeton, NJ: Princeton University Press, 2000.
Schrage, Michael. *Shared Minds: The New Technologies of Collaboration*. New York: Random House, 1990.
Schultze-Naumburg, Paul. *Die Kultur des weiblichen Körpers als Grundlage der Frauenkleidung*. Jena, Germany: Diederichs, 1905.
Schweber, Libby. *Disciplining Statistics: Demography and Vital Statistics in France and England, 1830–1885*. Durham, NC: Duke University Press, 2006.
Scott, A.L. "Physical Purity Feminism and State Medicine in Late Nineteenth-Century England." *Women's History Review* 8, no. 4 (1999): 625–53.
Scull, A. "The Discovery of the Asylum Revisited: Lunacy Reform in the New American Republic." In *Madhouses, Mad-Doctors, and Madmen: The Social History of Psychiatry in the Victorian Era*. Edited by A. Scull, 144–165. London: Athlone Press, 1981.
Scull, A. *The Most Solitary of Afflictions: Madness and Society in Britain, 1700–1900*. New Haven, CT: Yale University Press, 1993.
Sedgwick, E. *Between Men: English Literature and Male Homosocial Desire*. New York: Columbia University Press, 1985.
Sedgwick, E. *Epistemology of the Closet*. Berkeley: University of California Press, 1990.
Seigfried, Tom. *The Bit and the Pendulum: From Quantum Physics to M-Theory—The New Physics of Information*. New York: John Wiley & Sons, 2000.
Sekula, Allan. "The Body and the Archive." *October* 39 (Winter 1986): 3–64.
Seltzer, Mark. *Bodies and Machines*. New York: Routledge, 1992.
Senge, Peter M. *The Fifth Discipline: The Art & Practice of the Learning Organization*. New York: Currency Books, 1990.
Sengoopta, C. "Glandular Politics: Experimental Biology, Clinical Medicine, and Homosexual Emancipation in Fin-de-Siècle Central Europe." *Isis* 89 (1998): 445–73.
Sengoopta, C. *The Most Secret Quintessence of Life: Sex, Glands, and Hormones, 1850–1950*. Chicago: University of Chicago Press, 2006.
Serres, Michel. *Hermès III: La Traduction*. Paris: Minuit, 1974.

Seton, William. "The Century's Progress in Science." *Catholic World* (May 1889): 146–67.

Sexton, George. *Vaccination Useless and Injurious*. London: George Howe, 1870.

Sha, R. "Romanticism and the Sciences of Perversion." *Wordsworth Circle* 36 (2006): 43–48.

Sharpey-Schafer, E. "Endocrine Physiology." *Irish Journal of Medical Science,* 6th ser., no. 69 (September 1931): 483–505.

Shelley, Mary. *Frankenstein*. Edited by J. Paul Hunter. 1818. Repr. New York: W.W. Norton, 1995.

Shields, David S. "The Science of Lying." In *Messy Beginnings: Post-coloniality and Early American Studies*. Edited by M.J. Schueller and Edward Watts, 223–36. New Brunswick, NJ: Rutgers University Press, 2003.

Shifrin, Malcolm R. Victorian Turkish Baths: Their Origin, Development, and Gradual Decline Web site, www.victorianturkishbath.org.

Shingleton Smith, R. "Phthisis." *Medical Annual* (1892): 366–84.

Shortt, J. *A Popular Lecture on Vaccination*. Madras, India: Gantz Brothers, 1865.

Sigaud, Claude, and Leon Vincent. *Les Origines de la Maladie. Essai sur l'Evolution de la Forme du Corps humain*. 2nd ed. Paris: A Maloine, 1912.

Sigel, Lisa. *Governing Pleasures: Pornography and Social Change in England, 1815–1914*. New Brunswick, NJ: Rutgers University Press, 2002.

Sigsworth, M. "Cholera in the East and West Ridings of Yorkshire, 1848–1893." PhD diss., Sheffield Hallam University, 1991.

Simmer, H.H. "Bilaterale Oophorektomie der Frau im späten 19. Jahrhundert: Zum methodologischen Wert der Kastration für die Entdeckung ovarieller Hormone." *Geburtshilfe und Frauenheilkunde* 43 (1983; Sonderheft): 54–59.

Simon, J. "Proceedings Against Cholera under the Disease Prevention Acts and Otherwise." *Ninth Annual Report of the Medical Officer of the Privy Council for 1866*, British Parliamentary Papers, 1867, [3949], xxxvii, 1.

Sinha, M. *Colonial Masculinity*, Manchester, UK: Manchester University Press, 1995.

Sinha, M. *The "Manly Englishman" and the "Effeminate Bengali" in the Late Nineteenth Century*. Manchester, UK: Manchester University Press, 1995.

Sloane, David Charles. *The Last Great Necessity: Cemeteries in American History*, Baltimore: Johns Hopkins University Press, 1995.

Smith, F.B. *The Retreat of Tuberculosis*. London: Croom Helm, 1988.

Smith, Jonathan. *Charles Darwin and Victorian Visual Culture*. Cambridge: Cambridge University Press, 2006.

Smith-Rosenberg, Carroll. *Disorderly Conduct: Visions of Gender in Victorian America*. New York: Oxford University Press, 1986.

Snow, John. *On the Mode of Communication of Cholera*. 2nd ed. London: John Churchill, 1855.

Society for the Diffusion of Useful Knowledge (SDUK). *The Working Man's Companion: The Physician. 1. The Cholera*. London: Charles Knight, 1832.

Söder, H-P. "Disease and Health as Contexts of Modernity: Max Nordau as a Critic of Fin-de-Siècle Modernism." *German Studies Review* 14 (1991): 473–87.

Solnit, Rebecca. *River of Shadows: Eadweard Muybridge and the Technological Wild West*. New York: Viking, 2003.

Somerville, S. *Queering the Color Line: Race and the Invention of Homosexuality in American Culture*. Durham, NC: Duke University Press, 2000.

Sorisio, Carolyn. *Fleshing Out America: Race, Gender, and the Politics of the Body in American Literature, 1833–1879*. Athens: University of Georgia Press, 2002.

Spargo, John. *The Bitter Cry of the Children*. New York: Macmillan, 1906.

Spencer, H. Essay, *The Leader*, March 20, 1852, http://www.clarehall.cam.ac.uk/user-pages/djhc2/contbib/spencer.htm.

Spencer, H. *Social Statics*, 1851. http://oll.libertyfund.org/Texts/LFBooks/Spencer0236/SocialStatics/0331_Bk.html#LF_BK0331in01ch01.

Spongberg, M. *Feminizing Venereal Disease: The Body of the Prostitute in Nineteenth-Century Medical Discourse*. New York: NYU Press, 1997.

Sproul, Lee, and Sara Kiesler. *Connections: New Ways of Working in the Networked Organization*. Cambridge: MIT Press, 1992.

Squire, J. E. *The Hygienic Prevention of Consumption*. London: C. Griffin, 1893.

Stalin, Joseph. *Problems of Leninism*. Moscow: Cooperative Publishing Society of Foreign Workers in the USSR, 1933.

Standage, T. *The Victorian Internet*. London: Phoenix, 1998.

Stange, Maren. "Jacob Riis and Urban Visual Culture: The Lantern Slide Exhibition as Entertainment and Ideology." *Journal of Urban History* 15 (1989): 274–303.

Stange, Maren. *Symbols of Ideal Life: Social Documentary Photography in America, 1890–1950*. Cambridge: Cambridge University Press, 1989.

Stanley, Peter. *White Mutiny: British Military Culture in India, 1825–1875*. London, Hurst, 1998.

Stansell, Christine. *City of Women: Sex and Class in New York, 1789–1860*. Chicago: University of Illinois Press, 1987.

Steedman, Carolyn. *Strange Dislocations: Childhood and the Idea of Human Interiority, 1780–1930*. Cambridge, MA: Harvard University Press, 1995.

Stein, George J. "Biological Science and the Roots of Nazism." *American Scientist* 76 (1988): 50–58.

Steinach, Eugen. "Geschlechtstrieb und echt sekundäre Geschlechtsmerkmale als Folge der innersekretorischen Funktion der Keimdrüse." *Zentralblatt für Physiologie* 24 (1910): 551–66.

Steinach, Eugen, and Harry Benjamin Collection, Letter dated February 28, 1923. The New York Academy of Medicine Historical Collections.

Steinach, Eugen. "Willkürliche Umwandlung von Säugetier-Männchen in Tiere mit ausgeprägt weiblichen Geschlechtscharakteren und weiblicher Psyche." *Pflügers Archiv für die gesamte Physiologie* 144 (1912):71–108.

Steinach, Eugen. *Sex and Life: Forty Years of Biological and Medical Experiments*. New York: Viking, 1940.

Steinach, Eugen. *Verjüngung durch experimentelle Neubelebung der alternden Pubertätsdrüse*. Berlin: Springer, 1920.

Stepan, Nancy Leys. *The Hour of Eugenics: Race, Gender, and Nation in Latin America*. Ithaca, NY: Cornell University Press, 1991.

Stern, Madeleine B. *Heads and Headlines: The Phrenological Fowlers*. Norman: University of Oklahoma Press, 1971.

Stevenson, L G. "Science Down the Drain." *Bulletin of the History of Medicine* 29, no. 1 (1955): 1–26.

Stilling, B. *Physiologisch-pathologische und medicinisch-praktische Untersuchung über Spinal-Irritation*. Leipzig, Germany: Wiegand, 1840.

Stoehr, Taylor. "Physiognomy and Phrenology in Hawthorne." *Huntington Library Quarterly* 37 (August 1974): 355–400.

Stoff, H. "Degenerierte Nervenkörper und regenerierte Hormonkörper: Eine kurze Geschichte der Verbesserung des Menschen zu Beginn des 20. Jahrhunderts." *Historische Anthropologie: Kultur, Gesellschaft, Alltag* 11 (2003): 225–39.

Stoff, H. *Ewige Jugend: Konzepte der Verjüngung vom späten 19. Jahrhundert bis ins Dritte Reich*. Cologne, Germany: Böhlau, 2004.

Stoler, Ann Laura. *Carnal Knowledge and Imperial Power: Race and the Intimate in Colonial Rule*. Berkeley: University of California Press, 2002.

Stoler, Ann Laura. "Cultivating Bourgeois Bodies and Racial Selves." In *Cultures of Empire*. Edited by Catherine Hall. New York: Routledge, 2000.

Stoler, Ann Laura, ed. *Haunted By Empire: Geographies of Intimacy in North American History*. Durham, NC: Duke University Press, 2006.

Stoler, Ann Laura. *Race and the Education of Desire: Foucault's* History of Sexuality *and the Colonial Order of Things*. Durham, NC: Duke University Press, 1995.

Stoler, Ann Laura. "Sexual Affronts and Racial Frontiers." In *Tensions of Empire: Colonial Cultures in a Bourgeois World*. Edited by Frederick Cooper and Ann Laura Stoler. Berkeley: University of California Press, 1997.

Stoler, Ann Laura. "Tense and Tender Ties: The Politics of Comparison in North American History and (Post) Colonial Studies." In *Haunted by Empire: Geographies of Intimacy in North American History*. Edited by Ann Laura Stoler, 1–22. Durham, NC: Duke University Press, 2006.

Stoler, Ann Laura, and Frederick Cooper. "Between Metropole and Colony: Rethinking a Research Agenda." In *Tensions of Empire: Colonial Cultures in a Bourgeois World*. Edited by Frederick Cooper and Ann Laura Stoler, 1–56. Berkeley: University of California Press, 1997.

Stopes, M. *Married Love*. New York: Critic and Guide Company, 1918.

Storr, M. "Transformations: Subjects, Categories and Cures in Krafft-Ebing's Sexology." In *Sexology in Culture*. Edited by Lucy Bland and Laura Doan. Chicago: University of Chicago Press, 1998.

Stratz, C.H. *Die Rassenschönheit des Weibes*. 5th ed. Stuttgart, Germany: Ferdinand Enke, 1904.

Stratz, C.H. *Die Schönheit des weiblichen Körpers*. 2nd ed. Stuttgart, Germany: Ferdinand Enke, 1899.

Strother, Z.S. "Display of the Body Hottentot." In *Africans on Stage: Studies in Ethnological Show Business*. Edited by B. Lindfors, 1–55. Bloomington: Indiana University Press, 1999.

Sulloway, F. *Freud, Biologist of the Mind: Beyond the Psychoanalytic Legend*. New York: Basic Books, 1979.

Sünderhauf, E. *Griechensehnsucht und Kulturkritik: Die deutsche Rezeption von Winkelmanns Antikenideal*. Berlin: Akademie-Verlag, 2004.

Tandler, J., and Grosz, S. "Untersuchungen an Skopzen." *Wiener klinische Wochenschrift* 21 (1908): 277–82.

Taylor, Charles W. *Sources of the Self: The Making of Modern Identity*. Cambridge, MA: Harvard University Press, 1989.

Tennyson, Alfred. "In Memoriam." In *The Works of Alfred Tennyson*, Volume 10, 1875. London: Adamant Media, 2004.

Thomas, Nicholas. "Introduction." In *Tattoo: Bodies, Art, and Exchange in the Pacific and the West*. Edited by Nicholas Thomas, Anna Cole, and Bronwen Douglas, 7–29. London: Reaktion, 2005.

Thompson, E.P. *The Making of the English Working Class*. London: Victor Gollancz, 1963.

Thompson, E.P. *The Poverty of Theory and Other Essays*. London: Merlin Press, 1978.

Thompson, E. P. "Time, Work-Discipline and Industrial Capital." *Past & Present* 38 (12, 1967): 56–97.

Thompson, Sir Henry. *Cremation: The Treatment of the Body after Death*. London: Henry S. King & Co., 1874.

Thomson, Rosemarie Garland. *Extraordinary Bodies: Figuring Physical Disability in American Culture and Literature*. New York: Columbia University Press, 1997.

Thomson, Rosemarie Garland, ed. *Freakery: Cultural Spectacles of the Extraordinary Body*. New York: NYU Press, 1996.

Tichi, Cecelia. *Exposés and Excess. Muckraking in America, 1900–2000*. Philadelphia: University of Pennsylvania Press, 2004.

Tichi, Cecelia. *Shifting Gears: Technology, Literature, Culture in Modernist America*. Chapel Hill: University of North Carolina Press, 1987.

Tiefer, L. *Sex Is Not a Natural Act and Other Essays*. Boulder, CO: Westview Press, 1995.

Torchia, Robert W. "No Strangers to the Ravages of Death: John Neagle's Portrait of Dr. John Abraham Elkinton." *Nineteenth Century* 18 (Fall 1998): 12–19.

Trachtenberg, Alan. *Reading American Photographs: Images as History, Matthew Brady to Walker Evans*. New York: Hill & Wang, 1990.

Traub, Valerie. "The Psychomorphology of the Clitoris." *GLQ* 2 (1995): 81–113.

Trollope, Anthony. "The Turkish Bath." In *An Editor's Tales*. 1870. Repr. New York: Penguin, 1993.

Trumbach, R. *Sex and the Gender Revolution: Heterosexuality and the Third Gender in Enlightenment London*. Chicago: University Chicago Press, 1998.

Tucker, Jennifer. *Nature Exposed: Photography as Eyewitness in Victorian Science*. Baltimore: Johns Hopkins University Press, 2005.

Turner, J. *Reckoning With the Beast: Animals, Pain, and Humanity in the Victorian Mind*. Baltimore: Johns Hopkins University Press, 1980.

Tytler, Graeme. *Physiognomy in the European Novel: Faces and Fortunes*. Princeton, NJ: Princeton University Press, 1982.

Uhlmann, A. "'Der Sport ist der praktische Arzt am Krankenlager des deutschen Volkes': Wolfgang Kohlrausch und die Geschichte der deutschen Sportmedizin." Freiburg: Diss., 2004.

U.S. Census Bureau. "Historical Estimates of World Population." http://www.census.gov/ipc/www/worldhis.html.

Usborne, C. "The New Woman and Generation Conflict: Perceptions of Young Women's Sexual Mores in the Weimar Republic." In *Generations in Conflict: Youth Revolt and Generation Formation in Germany, 1770–1968*. Edited by Mark Roseman, 137–63. Cambridge: Cambridge University Press, 1995.

Vance, Norman. *The Sinews of the Spirit: The Ideal of Christian Manliness in Victorian Literature and Religious Thought*. Cambridge: Cambridge University Press, 1985.

Van Wyhe, J. *Phrenology and the Origins of Victorian Scientific Naturalism*. Aldershot, UK: Ashgate, 2004.

Vaughan, Megan. *Creating the Creole Island: Slavery in Eighteenth-Century Mauritius*. Durham, NC: Duke University Press, 2005

Vaughan, Megan. *Curing Their Ills: Colonial Power and African Illness*. Cambridge: Polity Press, 1991.

Vaughan, Megan. "Scarification in Africa: Re-reading Colonial Evidence." *Cultural and Social History* 4, no. 3 (2007): 385–400.

Vertinsky, P. "Embodying Normalcy: Anthropometry and the Long Arm of William H. Sheldon's Somatotyping Project." *Journal of Sports History* 29, no. 1 (2002): 95–133.

Virchow, R. "Der puerperale Zustand: Das Weib und die Zelle." In *Gesammelte Abhandlungen zur wissenschaftlichen Medizin*. Edited by R. Virchow, 735–79. Frankfurt am Main, Germany: Meidinger, 1856.

von Helmholtz, Hermann. "Über die Wechselwirkung der Naturkräfte und die darauf Bezüglichen neuesten Ermittelungen der Physik." In *Populäre Wissenschaftliche Vorträge. Drittes Heft*. Edited by Hermann von Helmholtz. Braunschweig, Germany: Druck und Verlag Friedrich Vieweg und Sohn, 1876.

von Schrenck-Notzing, A. *Therapeutic Suggestion Psychopathia Sexualis*. Translated by Charles Gilbert Chaddock. London: F. A. Davis, 1895.

Wahrman, D. *Imagining the Middle Class*. Cambridge: Cambridge University Press, 1995.

Walker, K., and J. Lumsden Cook. "Steinach's Rejuvenation Operation." *Lancet* 2 (February 1924): 223–26.

Walkowitz, J. *Prostitution and Victorian Society*. Cambridge: Cambridge University Press, 1980.

Ward, David C. *Charles Willson Peale: Art and Selfhood in the Early Republic*. Berkeley: University of California Press, 2004.

Warner, J. H. *Against the Spirit of System: The French Impulse in Nineteenth-century American Medicine*. Princeton, NJ: Princeton University Press, 1998.

Warner, J. H. *The Therapeutic Perspective: Medical Practice, Knowledge, and Identity in America, 1820–1885*. Princeton, NJ: Princeton University Press, 1984.

Warner, Michael. *The Letters of the Republic: Publication and the Public Sphere in Eighteenth-Century America*. Cambridge, MA: Harvard University Press, 1991.

Waters, C. "Sexology." In *Palgrave Advances in the Modern History of Sexuality*. Edited by H. G. Cocks and Matt Houlbrook, 41–63. Houndmills, UK: Palgrave Macmillan, 2006.

Wechsler, Judith. *A Human Comedy: Physiognomy and Caricature in 19th-Century Paris*. London: Thames and Hudson, 1982.

Wedemeyer-Kolwe, Bernd. *Der neue Mensch. Körperkultur im Kaiserreich und der Weimarer Republik*. Würzburg, Germany: Königshausen & Neumann, 2004.

Weeks, J. *Sexuality and Its Discontents*. London: Routledge and Kegan Paul, 1985.

Wege zu Kraft und Schönheit, director Wilhelm Prager (1925) in *Bundesarchiv-Filmarchiv Berlin*.

Weinbren, D. "Against *All* Cruelty: The Humanitarian League, 1891–1919." *History Workshop Journal* 38 (1994): 86–105.

Weindling, P. "Ernst Haeckel, *Darwinismus* and the Secularization of Nature." In *History, Humanity and Evolution: Essays for John C. Greene*. Edited by James R. Moore, 311–27. Cambridge: Cambridge University Press, 1989.

Weindling, P. *Health, Race, and German Politics Between National Unification and Nazism 1870–1945*. Cambridge: Cambridge University Press, 1989.

Weindling, Paul. "Theories of the Cell State in Imperial Germany." In *Biology, Medicine and Society 1840–1940*. Edited by C. Webster, 99–155. Cambridge: Cambridge University Press, 1981.

Weismann, A. *The Germ-Plasm: A Theory of Heredity*. Translated by W. Newton Parker and Harriet Rönnfeldt. New York: Scribner's, 1893.

Wellmer, Albrecht. *Critical Theory of Society*. New York: Seabury Press, 1974.

Welter, Barbara. *Dimity Convictions: The American Woman in the 19th Century*. Athens: Ohio University Press, 1977.

Wertz, Richard W., and Dorothy C. Wertz. *Lying-in: A History of Childbirth in America*. Exp. ed. New Haven, CT: Yale University Press, 1989.

White, Charles. *A Treatise on the Management of Pregnant and Lying-in Women*. London: W. Johnston, 1772.

White, Joanna. "Marks of Transgression: The Tattooing of Europeans in the Pacific Islands." In *Tattoo*. Edited by Nicholas Thomas, Anna Cole, and Bronwen Douglas, 72–89. London: Reaktion, 2005.

Whorton, J. C. *Nature Cures. The History of Alternative Medicine in America*. Oxford: Oxford University Press, 2002.

Wilkinson, J.J.G. *Smallpox and Vaccination*. London: F. Pitman, 1871.

Willard, E. *Sexology as the Philosophy of Life: Implying the Social Organization and Government*. Chicago, 1867.

Williams, Naomi. "The Implementation of Compulsory Infant Smallpox Vaccination in England and Wales, 1840–1890." *Journal of Historical Geography* 20, no. 4 (1994): 396–412.

Williams, Raymond. *Keywords: A Vocabulary of Culture and Society*. Rev. ed. London: Fontana, 1983.

Williams, Talcott. "Animal Locomotion in the Muybridge Photographs." *The Century Magazine* 34 (July 1887): 356–368.

Wood, A.D. "'The Fashionable Diseases': Women's Complaints and Their Treatment in Nineteenth-Century America." *Journal of Interdisciplinary History* 4 (1973): 25–52.

Wood, Amy Louise. "Lynching Photography and the Visual Reproduction of White Supremacy." *American Nineteenth Century History* 6 (September 2005): 373–399.

Worboys, M. "British Medicine and its Past at Queen Victoria's Jubilees and the 1900 Centennial. *Medical History* 45 (2001): 461–82.

Worboys, M. *Spreading Germs: Disease Theories and Medical Practice in Britain, 1865–1900*. Cambridge: Cambridge University Press, 2000.

Wordsworth, William. *Lyrical Ballads, with a Few Other Poems*. Bristol: T.N. Long man, 1798.

Wright, Thomas. *Clean and Decent: The Fascinating History of the Bathroom and the Water Closet*. New York: Viking Press, 1960.

Wyndham, D. "Versemaking and Lovemaking—W. B. Yeats' "Strange Second Puberty": Norman Haire and the Steinach Rejuvenation Operation." *Journal of the History of the Behavioral Sciences* 39 (2003): 25–50.

Yarrell, W. "On the Influence of the Sexual Organ in Modifying External Character." *Journal of the Proceedings of the Linnean Society (Zoology)* 1 (1857): 76–82.

Young, Robert. *Darwin's Metaphor: Nature's Place in Victorian Culture*. Cambridge: Cambridge University Press, 1985.

Zeising, A. *Neue Lehre von den Proportionen des menschlichen Körpers*. Leipzig, Germany: Rudolph Weigel, 1854.

Zelizer, Viviana A. Rotman. *Pricing the Priceless Child: The Changing Social Value of Children*. Princeton, NJ: Princeton University Press, 1994.

Zimmerman, A. "Adventures in the Skin Trade: German Anthropology and Colonial Corporeality." In *Worldly Provincialism: German Anthropology in the Age of Empire*. Edited by Glenn Penny and Matti Bunzl, 156–78. Ann Arbor: University of Michigan Press, 2003.

Zuboff, Shoshana. *In the Age of the Smart Machine: The Future of Work and Power*. New York: Basic Books, 1988.

CONTRIBUTORS

Clare Anderson is an associate professor in the Department of Sociology at the University of Warwick and the author of *The Indian Uprising of 1857–58: Prisons, Prisoners, and Rebellion* (2007) and *Legible Bodies: Race, Criminality and Colonialism* (2004). She is currently working on a book titled *Marginal Centres: Subaltern Biographies of the Indian Ocean World.*

Lisa Cody is an associate professor of history and the associate dean of the faculty at Claremont McKenna College. She is the author of *Birthing the Nation: Sex, Science, and the Conception of Britons* (2005), which won the Berkshire Conference of Women Historians Best First Book Prize.

Nadja Durbach is an associate professor of history at the University of Utah and the author of two books on aspects of the body in the nineteenth century: *Spectacle of Deformity: Freak Shows and Modern British Culture* (2010) and *Bodily Matters: The Anti-Vaccination Movement in England, 1853–1907* (2005).

Pamela K. Gilbert is the Albert Brick Professor of English at the University of Florida. Her books include *Cholera and Nation* (2008), *The Citizen's Body* (2007), *Mapping the Victorian Social Body* (2004), and *Disease, Desire and the Body in Victorian Women's Popular Novels* (1997).

Michael Hau is a senior lecturer in the School of Historical Studies at Monash University. He is the author of *The Cult of Health and Beauty in Germany: A Social History, 1890–1930* (2003) and is currently working on a book on the cultural history of performance enhancement in Weimar and Nazi Germany.

Thomas Laqueur is the Helen Fawcett Professor of History at the University of California, Berkeley. He has written numerous books and articles on the history of the human body from the ancient to the modern world, including *Solitary Sex: A Cultural History of Masturbation* (2003) and *Making Sex: Body and Gender from the Greeks to Freud* (1990).

Anson Rabinbach is a professor of history at Princeton University and a specialist in modern Europe. He is the author of *In the Shadow of Catastrophe: German Intellectuals between Apocalypse and Enlightenment* (2001) and *The Human Motor: Energy, Fatigue, and the Origins of Modernity* (1990).

Stephen P. Rice is a professor of American studies at Ramapo College of New Jersey and the author of *Minding the Machine: Languages of Class in Early Industrial America* (2004). He is working on a book on illustration and commercial wood engraving in nineteenth-century America.

Michael Sappol is a curator-historian at the National Library of Medicine and author of *Dream Anatomy* (2006) and *A Traffic of Dead Bodies: Anatomy and Embodied Social Identity in 19th-Century America* (2002). His current work focuses on the history of visual culture, museology, and medical film.

Chandak Sengoopta is a professor in the Department of History, Classics, and Archaeology at Birkbeck College, University of London. He is the author of several books on medical and scientific understanding in the nineteenth century, including *The Most Secret Quintessence of Life: Sex, Glands, and Hormones, 1850–1950* (2006) and *Imprint of the Raj: How Fingerprinting was Born in Colonial India* (2003).

Richard C. Sha is a professor of literature at American University and the author of *Perverse Romanticism: Aesthetics and Sexuality in Britain, 1750–1832* (2009) and *The Visual and Verbal Sketch in British Romanticism* (1998). He is working on a book on the imagination in art and science in British Romanticism.

Michael Worboys is the director of the Centre for the History of Science, Technology, and Medicine and Wellcome Unit for the History of Medicine at the University of Manchester. He is the author of *Spreading Germs: Disease Theories and Medical Practice in Britain, 1865–1900* (2000) and coauthor (with Neil Pemberton) of *Mad Dogs and Englishmen: Rabies in Britain, 1830–2000* (2007).

INDEX